DATE DUE			

Organic Chemistry of Coal

John W. Larsen, EDITOR
University of Tennessee

A symposium sponsored by the Division of Fuel Chemistry at the 174th Meeting of the American Chemical Society, Chicago, Illinois, August 29–September 1, 1977.

ACS SYMPOSIUM SERIES **71**

AMERICAN CHEMICAL SOCIETY
WASHINGTON, D. C. 1978

Library of Congress CIP Data

Main entry under title:
Organic chemistry of coal.

(ACS symposium series; 71 ISSN 0097-6156)
Includes bibliographies and index.

1. Coal—Analysis—Congresses.
I. Larsen, John W., 1940- . II. American Chemical Society. Division of Fuel Chemistry. III. Series: American Chemical Society. ACS symposium series; 71.

TP325.0685 662'.622 78-8114
ISBN 0-8412-0427-6 ASCMC8 71 1-327 1978

PRINTED IN THE UNITED STATES OF AMERICA

FOREWORD

The ACS SYMPOSIUM SERIES was founded in 1974 to provide a medium for publishing symposia quickly in book form. The format of the SERIES parallels that of the continuing ADVANCES IN CHEMISTRY SERIES except that in order to save time the papers are not typeset but are reproduced as they are submitted by the authors in camera-ready form. As a further means of saving time, the papers are not edited or reviewed except by the symposium chairman, who becomes editor of the book. Papers published in the ACS SYMPOSIUM SERIES are original contributions not published elsewhere in whole or major part and include reports of research as well as reviews since symposia may embrace both types of presentation.

CONTENTS

PREFACE

With the Arab oil embargo the American political structure became aware that oil and gas resources are finite and that ours would not suffice to fuel our economy. This has led to the current, awkward attempts to develop a rational, long-term energy policy for the United States. Since coal constitutes more than 90 percent of American fossil fuels which can be used with current technology, a natural result of governmental attention has been the flow of money into chemical research on coal. This area had been badly neglected since the last time the government was worried about oil supplies. Previous government commitments to coal research have not been sufficiently long lived to allow the development of satisfactory coal conversion technologies. There is hope that the current commitment to coal research will last long enough to allow both an increase in our knowledge of coal chemistry and the application of this knowledge to improve old processes and to develop new ones.

The new money for coal research has attracted some new people; various corporations have increased their activities in the area; and some individuals have discovered that the organic chemistry of coal is a fascinating research area. Thus, there has been a great increase in activity and much of it is now resulting in significant increases in our understanding. In each of the past several years there has been a meeting dedicated to the organic chemistry of coal. The papers in this volume were given at the 1977 meeting. Unfortunately, time limitations prevented the presentation of several excellent papers.

Coals are extraordinarily complex, insoluble organic mixtures. Complete elucidation of their structures has remained beyond the capabilities of the organic chemist and his instruments. However, the increased attention has begun to give results. Comparison of the proceedings of our annual coal chemistry meetings (University of Tennessee, 1975; Stanford Research Institute, 1976; this book, 1977) and conversations with workers active in the field will show that we are now just entering the rapid growth region of an exponential curve. We remain a long way from good structural models for coals and from a reasonable understanding of their chemistry, but the quality and quantity of work now being done is such that we will probably find answers to some of the fundamental questions in time to be of some aid to the development of conversion and cleaning processes.

But even without this practical goal, the problem of the structure and reactivity of this extraordinary complex material is worth solving simply because it is so complex and challenging.

University of Tennessee
Knoxville, Tennessee
February, 1978

John W. Larsen

1

A Primer on the Chemistry and Constitution of Coal

D. D. WHITEHURST

Mobil Research and Development Corp., P.O. Box 125, Princeton, NJ 08540

The purpose of this paper is to review what is known about the structure of coal and show how this information relates to the ultimate conversion of coal to conventional liquid fuels. Let us first consider some common beliefs about coal, as shown below:

- Coal is highly aromatic.
- Its structure contains predominantly condensed polycyclic aromatic rings.
- The high degree of condensation makes coal difficult to liquify.
- Extreme pressures and temperatures are required for coal conversion.
- Organic sulfur is much more difficult to remove than organic oxygen.
- Liquefaction requires high hydrogen consumption.

By the end of this paper I hope to have shown that all of these statements are wrong. To initiate this discussion, I propose to present three aspects of coal and coal product structure. These include, aromaticity, functionality, and molecular weight. I will then discuss reactivity of coal in non-catalytic hydrogenative processes and finally, how structure and reactivity interrelate.

Concerning the structure of coal, I would first like to say a few words about the origin of coal. It is generally agreed that coal originates primarily from plants. Through a series of evolutionary changes the primary products of the original decomposed plant matter becomes transformed through a series of steps in which the first product is humic acid. The humic acid

0-8412-0427-6/78/47-071-001$10.00/0

is then transformed sequentially into peat, lignite, subbituminous coal, bituminous coal, and finally to anthracite as shown in Figure 1. With these transformations, the carbon content increases and the oxygen content decreases. The result is that the calorific value of the coal increases with rank. Also shown in Figure 1, is the fact that coal as we know it today, can be identified as composed of a series of macerals, or fossilized plant fragments. These fossilized plant fragments are related to the original plant matter from which they are derived. The constituents of plants which could possibly give rise to coal and commonly associated structures are shown in Figure 2. The structures that we find in coal, or coal liquids, must be those related to the most stable of the structures from the original plant fragments. There are two schools of thought on the major constituent of coal. United States coals consist of primarily vitrinite, usually 80% or more. The composition of this vitrinite is believed to be the result of the coalification of either cellulose or lignin structures, which constitute the majority of the plant components (1). It has been shown by Given and others, however, that cellulose undergoes very rapid biodegradation in plants which are decomposing today (2). The same is true for protein. Plant constituents which are most resistant to bacterial attack are those of waxes, resins, tannins, lignins, flavonoids, and possibly alkaloids (3).

Although the previously discussed structures are present in plants today, and could be similar to those of plants of prehistoric times, it is not anticipated that the structures would survive intact over the long periods of time required for their transformation to coal. Some of the structural features however may possibly be recognizable even in today's coal. It has recently been shown by Given that certain components of coal can be related to structures evolved from lignins (4). It should also be remembered that the U.S. coals were layed down in two different geological ages; about 160 million years apart, and the structures associated with two geological ages may be substantially different.

Aromaticity of Coal

There is controversy on the proposed primary backbone structure of coal. Some workers contend that coal is primarily graphite-like, others argue that coal is of a diamond-like structure.

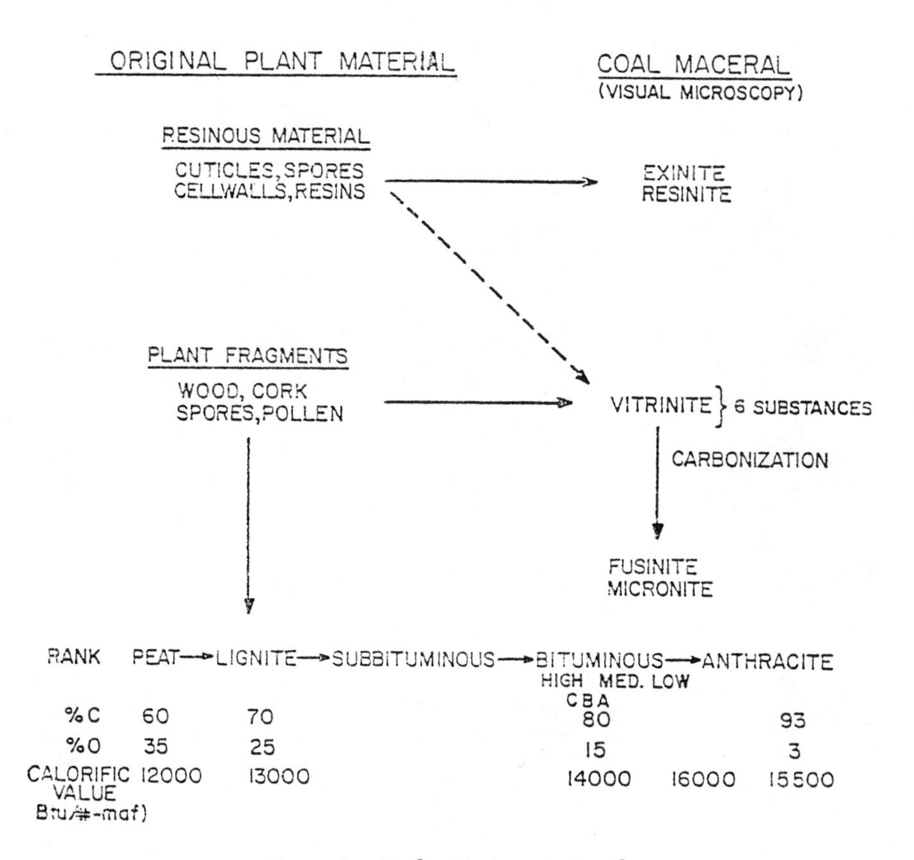

Figure 1. Mode of formation of coal

Cellulose

Protein

Waxes

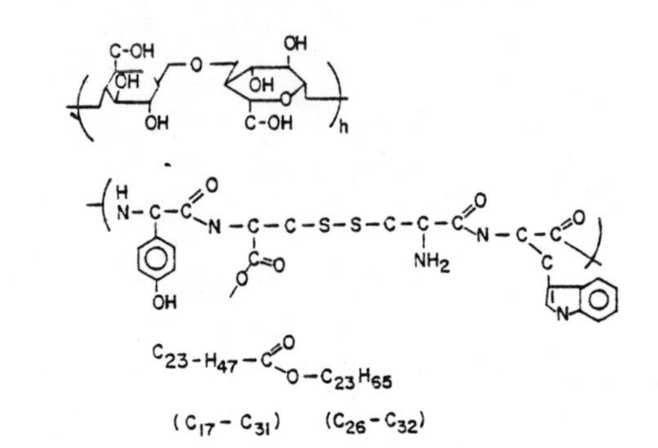

Resins

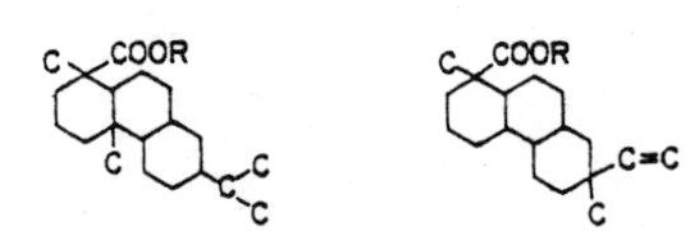

Terpenes

C-C-C=C-C-OH

COOH

HO

Sterols

Flavonoids

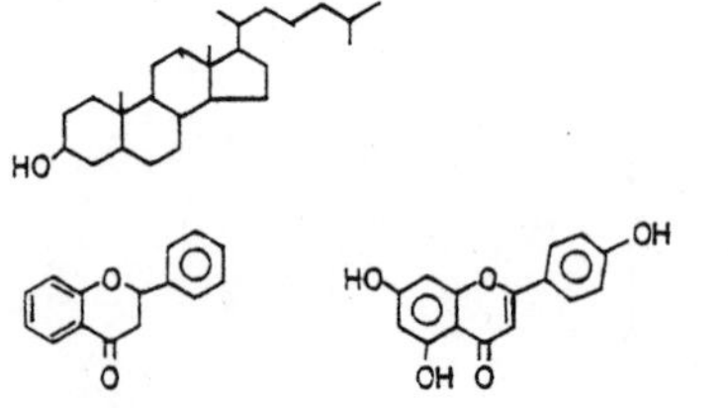

Tannins

Lignins

Alkaloids

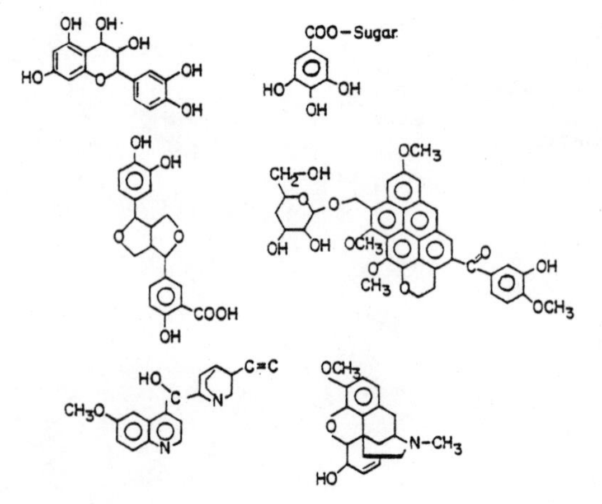

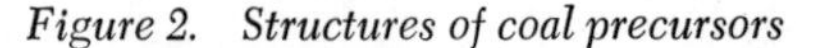

Figure 2. Structures of coal precursors

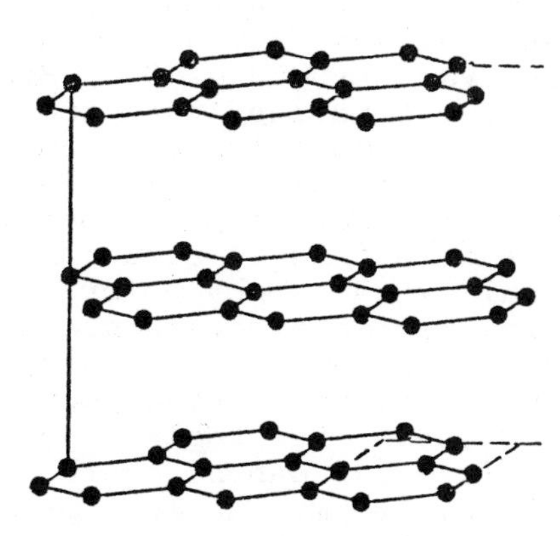

Graphite

Diamond

Both of these structures are low in H/C ratio which is consistant with the composition of coals.

To gain insight on the structure of coal, past workers have attempted to break coal down into recognizable units and then piece them back together as is done in natural product chemistry today. The most common technique presently pursued is that of oxidative degradation. With oxidants such as HNO_3, $K_2Cr_2O_7/HNO_3$, $KMNO_4/OH^-$, BuOOH/AIBN, or peracetic acid, workers have come to the conclusion that coal is predominately aromatic and contains many condensed rings (5,6). Other authors using $NaOCl/OH^-$ have come to a different conclusion, in that they believe coal contains large amounts of quarternary aliphatic carbon, or is diamon-like in structure and contains 50% aromatic carbon or less (7).

The preceeding methods of oxidation selectively oxidize only the aliphatic portion of coal. A new method pioneered by Dieno uses trifluoroacetic acids in combination with hydrogen peroxide. This method selectively oxidizes the aromatic rings. Combination of these two techniques could be a very powerful in structural characterization of coal (8).

Because of the difficulty of piecing back together the fragmented products of coal, a number of workers have attempted to do direct characterization of coal. Direct techniques suffer from the problem that coal is an opaque solid which is insoluble in its natural form and relatively few tools have been available up until the present time for such direct measurements. In the past, techniques such as X-ray scattering have been used and conflicting interpretations as to the predominant structure of coal have been reported. Hirsch first reported that coal was from 50-80% aromatic, with primarily 89% ordered structure (9). Ergun later, using X-ray scattering, concluded that coal is less aromatic and contains large quantities of amorphous regions (10). Friedel using ultraviolet techniques concluded that coal could not be polyaromatic and contained large amounts of aliphatic structure (11). Given, in characterizing coal extracts by polarographic reduction concluded that low rank coals were greater than 20% aromatic and high rank coals were greater than 50% aromatic (12). Polycyclic aromatic rings were believed to be prevalent. Recently, new tools have evolved and for the first time coal can be characterized directly in its natural form. The most promising of these tools is a solid state CP-C^{13} NMR developed by Pines (13).

Working in conjunction with Professor Pines, we have found that there is relatively little correlation between the hydrogen carbon mole ratio and the percent aromatic carbon found in coal or coal liquids, as shown in Figure 3. These data give some indication as to why there has been so much difficulty in the past correlating aromatic carbon content with the elemental composition of the coal. There is,however, some correlation between the rank of the coal and its aromaticity. This is shown in Figure 4. It can be seen that the aromatic carbon content increases from about 40-50% for subbituminous coal to over 90% for anthracite. It will be shown later that this aromaticity changes with conversion of the coal under liquefaction conditions. The CP-C^{13} technique is somewhat new and still evolving. It does show promise, however, in characterization of coal into aliphatic and aromatic components, but in addition, holds promise for further sub-division of the structural types. As shown in Figure 5, it is possible to distinguish in model compounds, aromatic, aliphatic, aliphatic ether, and condensed aromatic carbon. We hope eventually to use these same sub-divisions in the characterization of coal. As representative examples of the spectra that one can achieve, Figure 5 shows typical model compounds, the parent coal, SRC derived from a coal, unconverted residue and the spherical coke formed on extended thermal reaction (14).

Functionality of Coal

Figure 6 shows the major functional group types identified in coal. Oxygen occurs predominately as

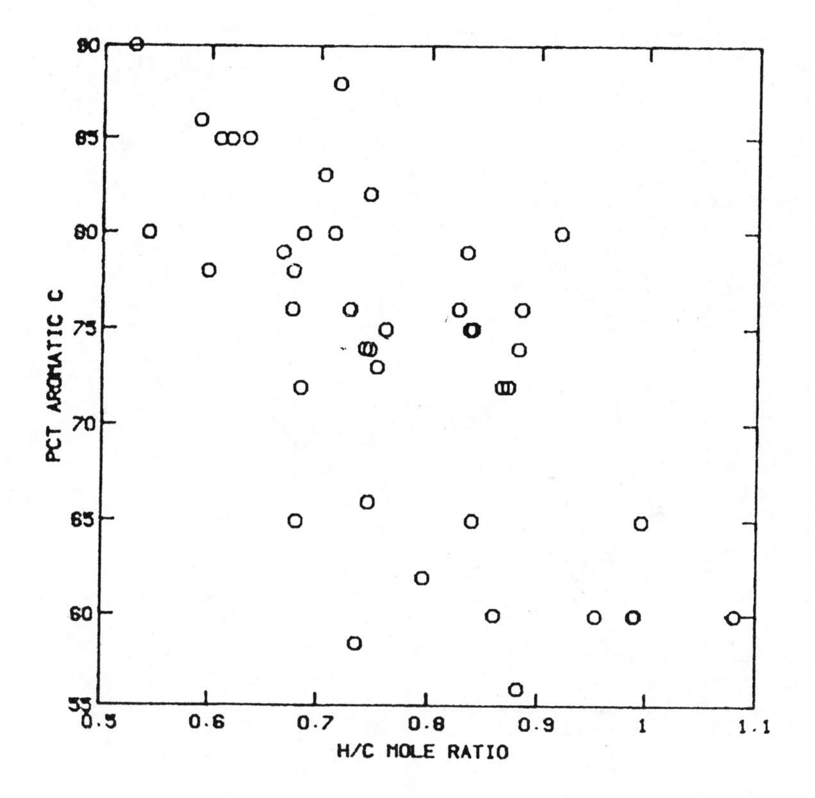

Figure 3. Aromatic carbon vs. H/C

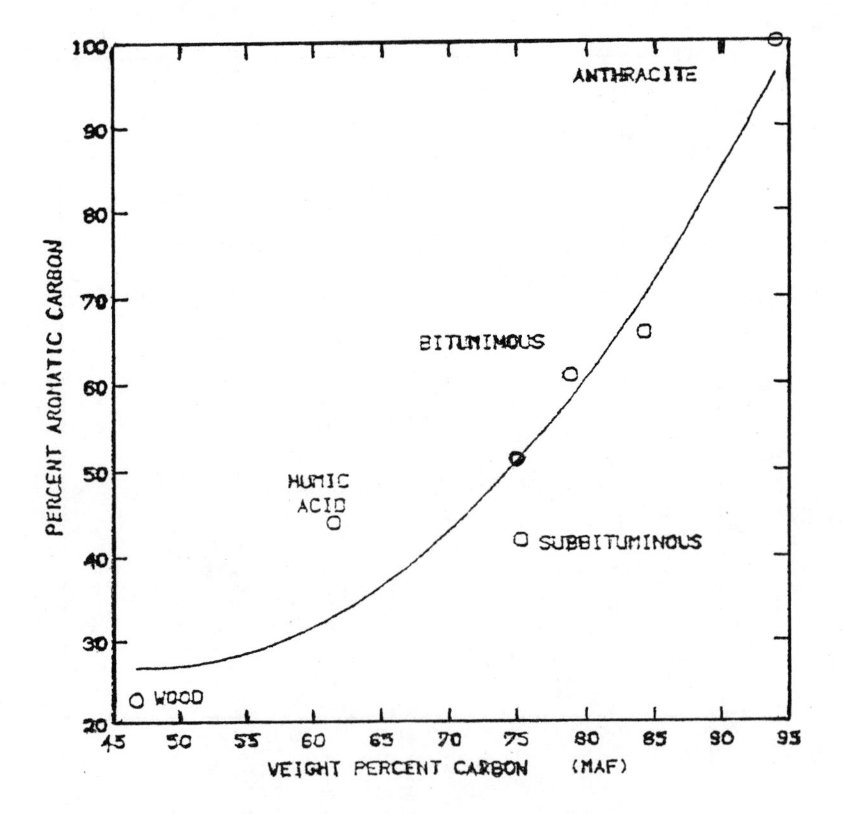

Figure 4. Aromaticity increase with rank

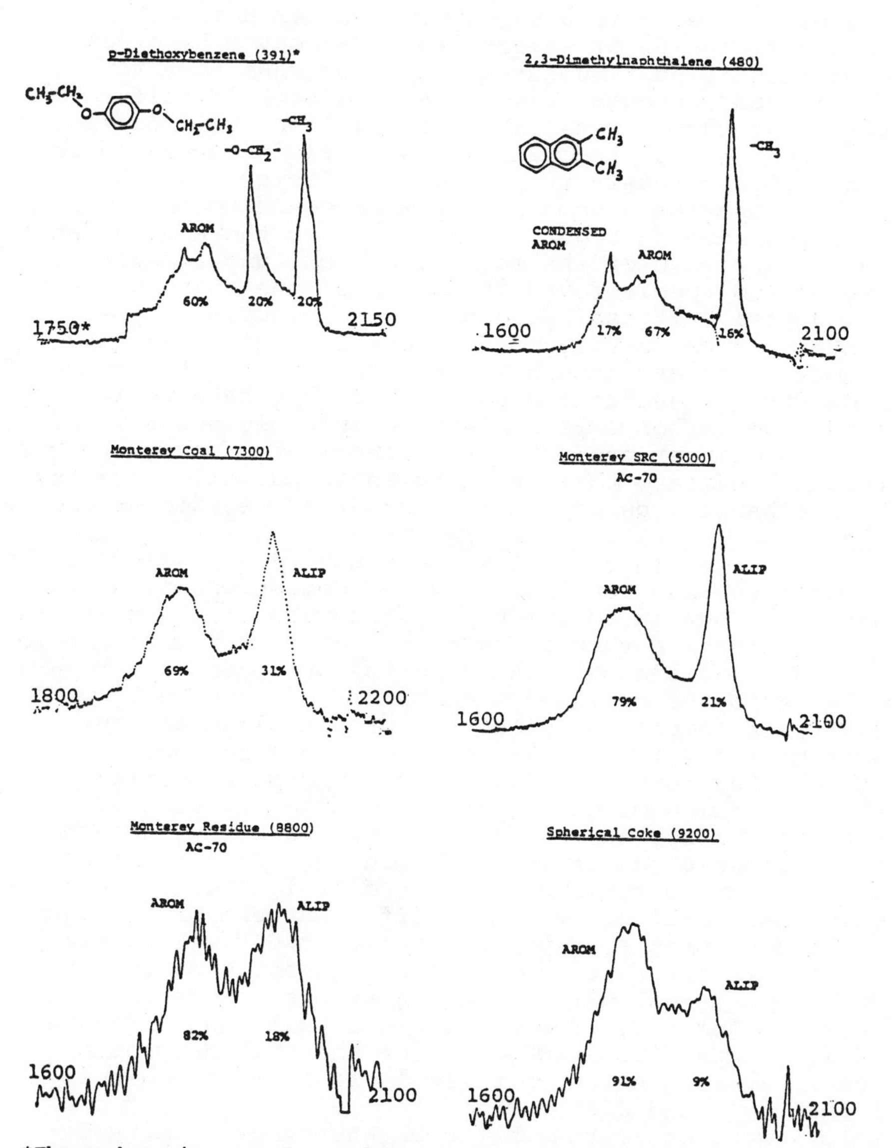

*The numbers in parentheses indicate the number of spectra accumulated to generate the spectrum presented. Note that the scales are not all the same.

Figure 5. CP–^{13}C NMR spectra of representative samples

phenolic or etheric groups with less amounts of carboxylic acids or esters; some carbonyls have also been identified. Sulfur has similar chemistry to oxygen and although sulfoxides have been identified in tar sands their presence in coal is less well defined. Nitrogen occurs predominately as pyridine or pyrrolic type rings. Metals are found as salts or associated with porphyrins. Some recent work conducted by Ruberto is summarized in Figure 7 (15). Shown there are quantitative analysis of the major types of functionality, oxygenated specie found in coals and one solvent refined coal. It can be seen that subbituminous coal contains considerably more carboxylic acid than bituminous coals and somewhat more carbonyl. After reaction under liquefaction conditions, the carboxylic acids and carbonyls are almost completely absent, and the predominent products are phenolic type oxygen. Our results indicate that in addition to phenolic type oxygen, etheric type oxygen is a major oxygenated specie (14).

As coal is converted in present day processes, the above described functionality, of course, changes with severity, but in addition the structure of the coal and its elemental carbon to hydrogen ratio must also change. Figure 8 shows a comparison of the hydrogen to carbon mole ratio for coal and a number of other natural products, in comparison with that of petroleum and the premium products that are desired from the coal. It can be seen that there is a very long path necessary in the conversion of coal to premium products such as gasoline, since coals contain about .8 hydrogen/carbon. The desired products contain about 2. This indicates that in any conversion process of coal, one of the primary goals will be extremely efficient use of hydrogen.

Most processes presently used today are initially thermal in nature since catalysts cannot contact the bulk of the coal matrix. But just where does the thermal chemistry of coal become significant? Figure 9 shows a superimposition of three thermal analyses of coal. These consist of thermal gravimetric, thermal mechanical, and differential thermal analysis of coal. This figure indicates that coal undergoes primary decomposition in the range of 400-450°, associated with this temperature range is the bulk of the swelling of the coal and significant changes in the thermal behavior of coal. It is not surprising, therefore, that most of the present processes being developed today, operate in the range of 400-450°C. But, at this temperature just how fast does coal react? We have shown that in the presence of hydrogen donors

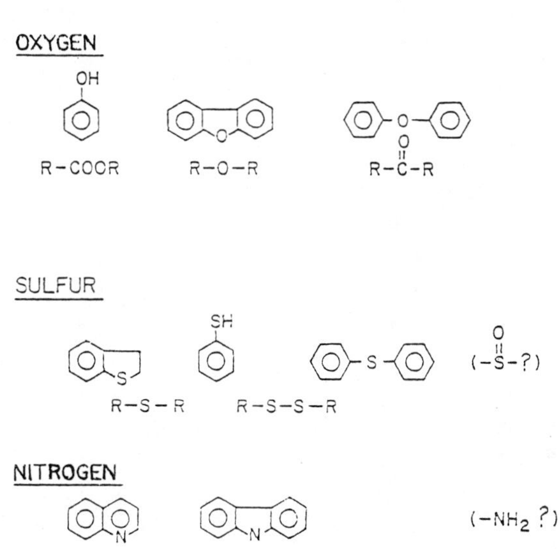

Figure 6. Functional groups found in coal

OXYGEN CONTENT AS:		COALS: BURNING STAR (BITUMINOUS)	COALS: BELLE AYR (SUB-BITUMINOUS)	SRC: BELLE AYR (400°C, 2000psig)
HYDROXYLIC	(−OH)	2.4	5.6	6.56
CARBOXYLIC	(−COOH)	0.7	4.4	.10
CARBONYLIC	(C=O)	0.4	1.0	.09
ETHERIC	(−O−)	2.8	0.9	.04
Δ H_2O				7.4
CO				.17
CO_2				3.71
	TOTAL	6.3	11.9	18.1
OXYGEN ANALYSIS				
ASH BASIS		5.9	16.2	
MINERAL MATTER BASIS		——	16.0	

Gulf Research and Development Company

Figure 7. Distribution of organic oxygen in coals and SRC (g O_2/100 g MAF coal)

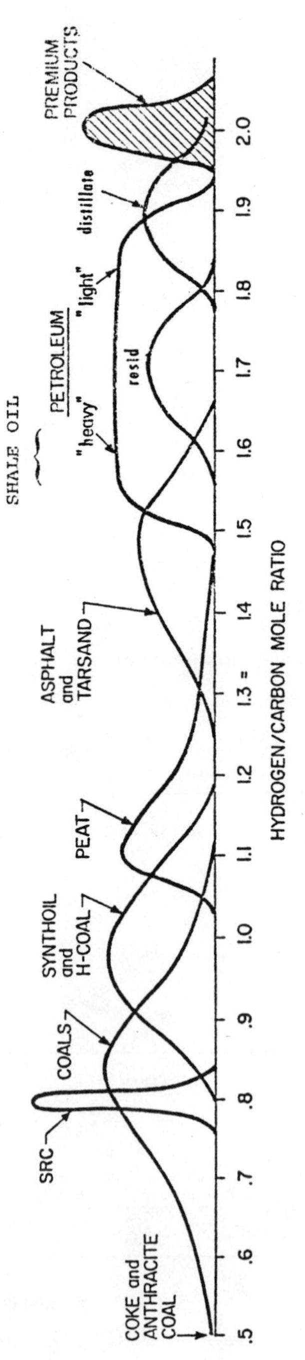

Coking 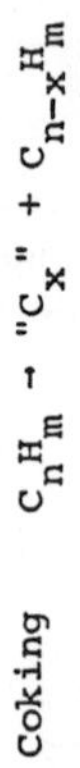 $C_nH_m \rightarrow "C_x" + C_{n-x}H_m$

(assume no H lost to coke)

Hydrogen Production $C_nH_m + .9n\ H_2O + .55n\ O_2 \rightarrow n\ CO_2 + (.9n + .5m)H_2$

(approximate balance based on energy requirements)

Figure 8. TGA, TMA, and DSC of West Kentucky coal

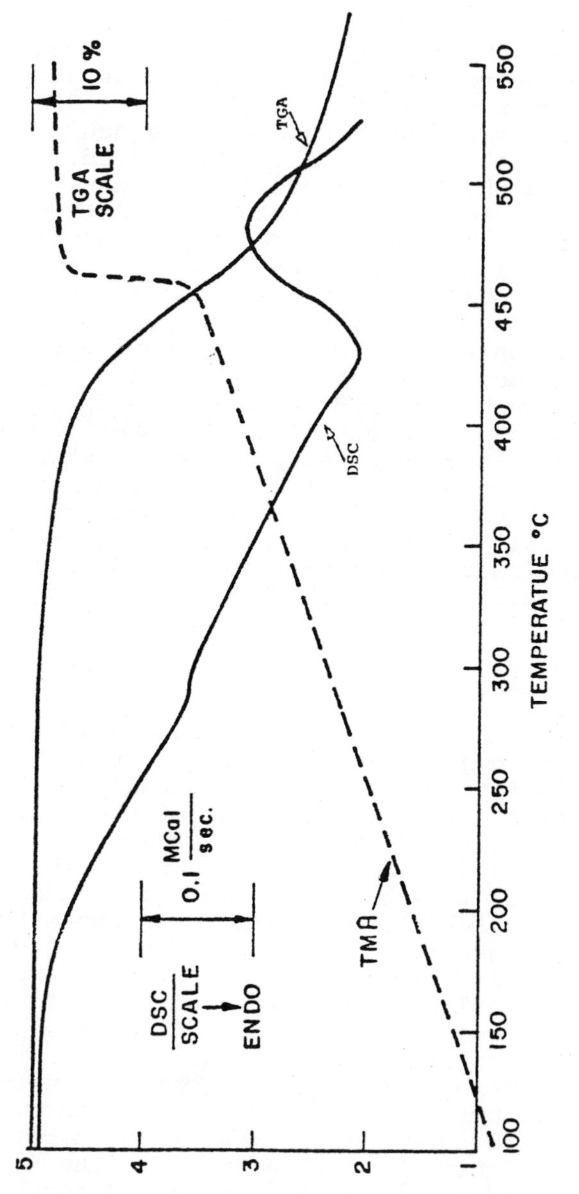

Figure 9. H/C ratios for various hydrocarbon sources

bituminous coals react very rapidly, 80% conversion can be achieved in one minute at 800°F (16,17). With subbituminous coals, however, at 800°F the reactivity is considerably slower, requiring in the order of 31 minutes to achieve about 70% conversion. In order to achieve higher conversions with subbituminous coals in shorter time, temperatures of in the order of 840° can be used, where about 80% conversion can be achieved in two to three minutes (14). It should be noted, however, that conversion in this case is defined as conversion of coal to material soluble in pyridine. Whether coal will actually dissolve or not, under the conditions of a liquefaction, will depend on the nature of the solvent. For example, solvents high in phenols and polyaromatic specie will truly dissolve the coal at these short times. Solvents which consist primarily of hydrocarbons will not dissolve the coal until it is converted into a much more hydrocarbon-like specie. To put these rapid conversion data into perspective, one must remember that in conversion processes such as the solvent refining of coal, a contact time of about 3 minutes occurs in the pre-heater of that process. This means that the coal will dissolve, at least bituminous coals, in the pre-heater, and what is called the dissolver is primarily a reactor in which the dissolved coal is upgraded.

Composition of Coal Products

The discussion thus far was concerned primarily with conversion of coal to soluble form, but just what does this soluble form consist of? Generally, it has been customary to classify the quality of the coal products in terms of solubility classes. These consist of hexane soluble materials (oils), benzene soluble materials, which are termed asphaltenes, and benzene insoluble materials, which have variety of names but we prefer the term asphaltols. We have recently described the systematic changes which occur on converting coal under typical solvent refining conditions to the various solubility classes as a function of time (18). A general description is shown in Figure 10 for subbituminous coal. Here it can be seen that the primary product of coal is pyridine soluble, benzene insoluble material. These are subsequently converted to both benzene soluble and hexane soluble specie. The classification of the products of coal in terms of solubility, is not very gratifying, however, from the standpoint of chemical understanding. Some of our characterization work has indicated that solubility is somewhat related to functionality but not particularly related to molecular weight or aromatic content. The products are still highly complex and many specie of similar chemical functionality can be found overlapping in the various solubility classes.

We have conducted a considerable amount of work on the characterization of the components of the various solubility classes (17-19). These are summarized as follows:

- Hexane soluble fractions (oils) have little or no functionality. They consist of hydrocarbons, ethers, thioethers, and non-basic nitrogen compounds (in decreasing order). Saturates are predominantly naphthenic. The molecular weight averages ~200-300.
- Benzene soluble fractions (asphaltenes) are predominantly mono-functional compounds. They consist of phenols or basic nitrogen compounds and may also contain ethers. The molecular weights range from 300-700.
- Benzene insoluble fractions (asphaltols) have multiple functionality. They consist of polyphenols (up to 5 OH/molecule) and multiple basic nitrogens or mixed functionality. Molecular weights range from 400->2000.

In addition to these high molecular weight products of coal, one produces hydrocarbon gases as well as distillate material as the severity of the reaction is increased. The composition of the distillate material which is generally used as a recycle solvent, has been found to be relatively simple as shown in Figure 11. Here it can be seen that the solvent is composed of primarily 6 compounds. These are naphthalene, methylnaphthalene, phenanthrene, paracresol and either biphenyl or diphenyl ether.

Looking once again at the non-distillable portion of coal products, we have found that the initial pyridine soluble benzene insoluble materials are higher molecular weight or highly associated specie. These are relatively rapidly converted to lower molecular weight materials in the order of 300-600 molecular weight. We have further characterized these non-distillable products of coal by liquid chromatography and separated them into discrete chemical classes; this is

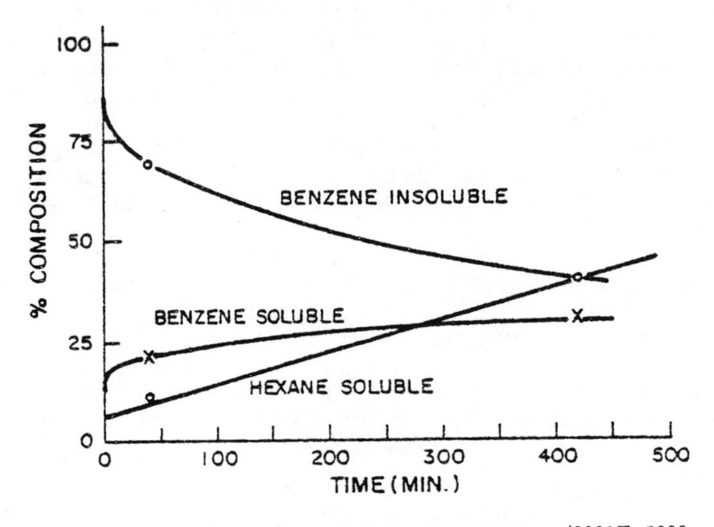

Figure 10. West Kentucky SRC composition vs. time (800°F, 1000–1300 psi H_2)

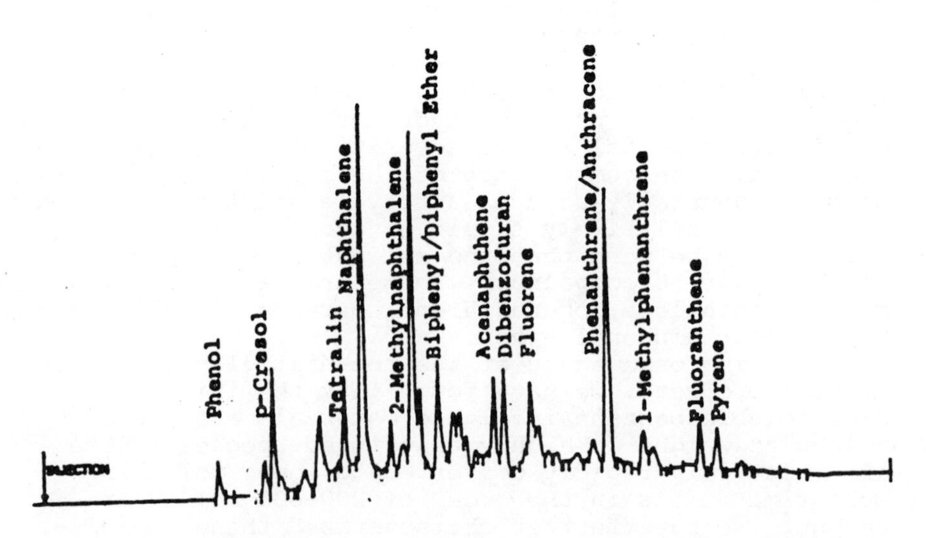

Figure 11. Gas chromatogram of solvent-range material of Illinois #6 coal

described elsewhere (14,17-19). A rather surprising result of these separations is that the molecular weights of common functionality classes of a given coal, were found to be the same independent of the degree of conversion, as shown in Figure 12. The predominant change which occurs to give lower molecular weights is the interconversion of the various functionality type classes. The significance of this result is that it indicates that coal structures are composed of rather stable nuclei which undergo functionality loss without major disruption. We do not wish to imply, however, that these nuclei are monodispersed, but rather are indicative of some common building blocks from which coal is composed. The molecular weight distribution of one of these functionality classes (in this case the heterocyclic compounds) from two different coals are shown in Figure 13.

I would like now to discuss the aromaticity of solubilized coal and how it changes with conversion under typical solvent refining conditions. As shown in Figure 14, the aromaticity of solvent refined coal can be considerably different from that of the parent coal, as is the case for subbituminous coal. For bituminous coal, the aromaticity is slightly higher but still quite similar to that of the parent coal.

As the coal is converted, first to soluble form and then to less functional material, the aromaticity systematically increases both for the carbon and for hydrogen, as shown in Figures 15 and 16, where the aromaticity of the coal is plotted versus the percent conversion to soluble form. The significance of these two figures is that the conversion to 80% soluble material occurs in as little as two to three minutes. Extrapolation of the aromatic hydrogen content to 0% conversion indicates that for this bituminous coal the aromatic hydrogen content was only 30%, which, in conjunction with the elemental analysis, indicates that the aromatics are highly substituted.

Aromatic content, although somewhat related to dealkylation of aromatic rings, is primarily due to the conversion of aliphatic carbon into aromatic carbon. The concentration of multiple aromatic rings were found by polarographic analyses to increase dramatically on extended reaction of the coal products (14,17). This increase in aromatic carbon content can occur by one of two processes, either hydrogen elimination from hydroaromatic rings or through rearrangement of polycyclic rings to more stable aromatic rings.

SESC FRACTION	LOW SEVERITY KENTUCKY SRC (RUN 10)	HIGH SEVERITY KENTUCKY SRC WILSONVILLE, ALA.	LOW SEVERITY MONTEREY SRC (AC-58)	HIGH SEVERITY MONTEREY SRC (AC-59)	HIGH SEVERITY WYODAK SRC HRI 177-114-2B
3	315[C]	309[C] and 313[T] (fractions 3 and 4)	440[T]	432[T]	660[C]
4	300[C]		470[T]	487[T]	630[C]
5	470[T]	476[T] and 481[C]	545[T]	492[T]	580[C]
7	610[T]	715[T] (fractions 7 and 8)			
8	960[T]		1290[T]	1142[T]	740[T]

C - Chloroform was used as the solvent for molecular weight determinations by vapor pressure osmometry.

T - Tetrahydrofuran was used as the solvent for molecular weight determinations by vapor pressure osmometry.

Figure 12. Average molecular weight determinations of SESC–SRC fractions

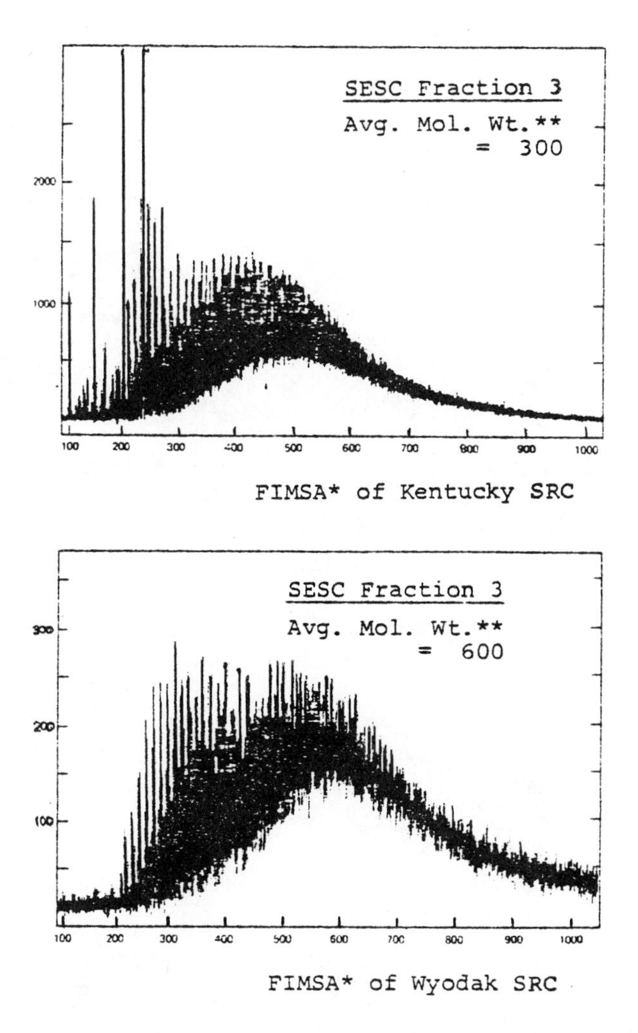

*Figure 13. Molecular weight distribution in a coal liquid fraction. *Field ionization mass spectroscopic analysis. **By vapor phase osmometry.*

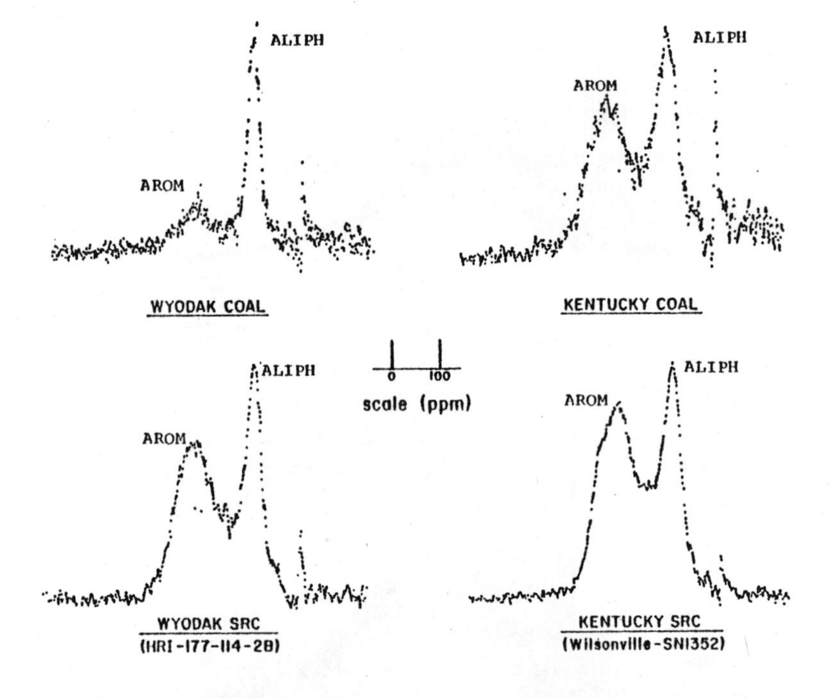

Figure 14. CP–^{13}C NMR spectra of coals and SRCs

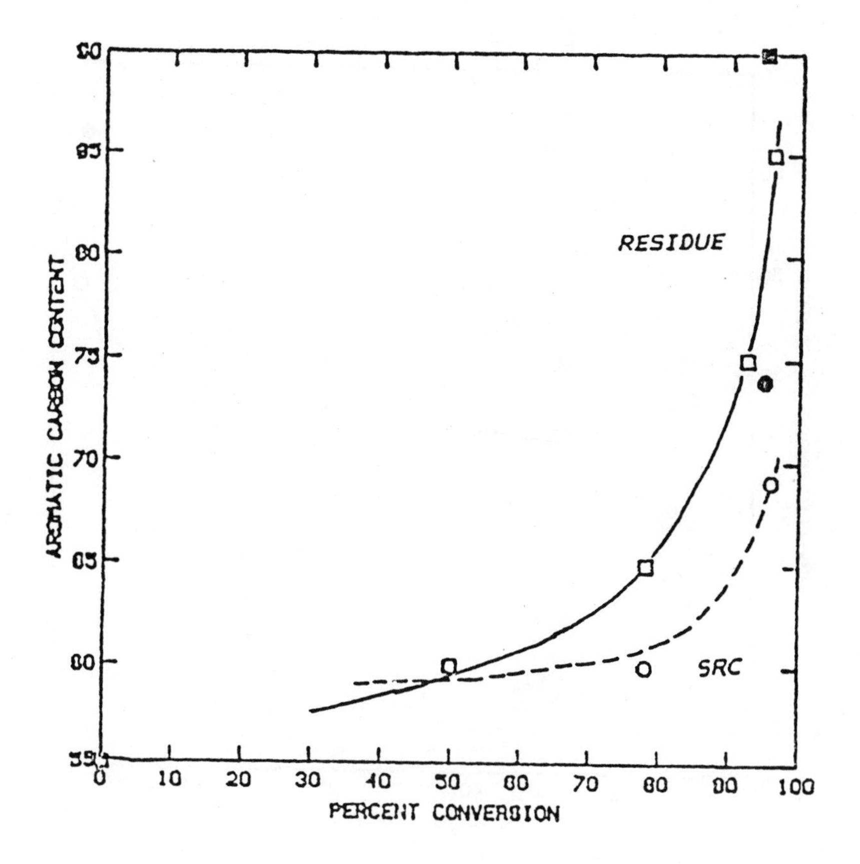

Figure 15. Increase in aromatic carbon with severity

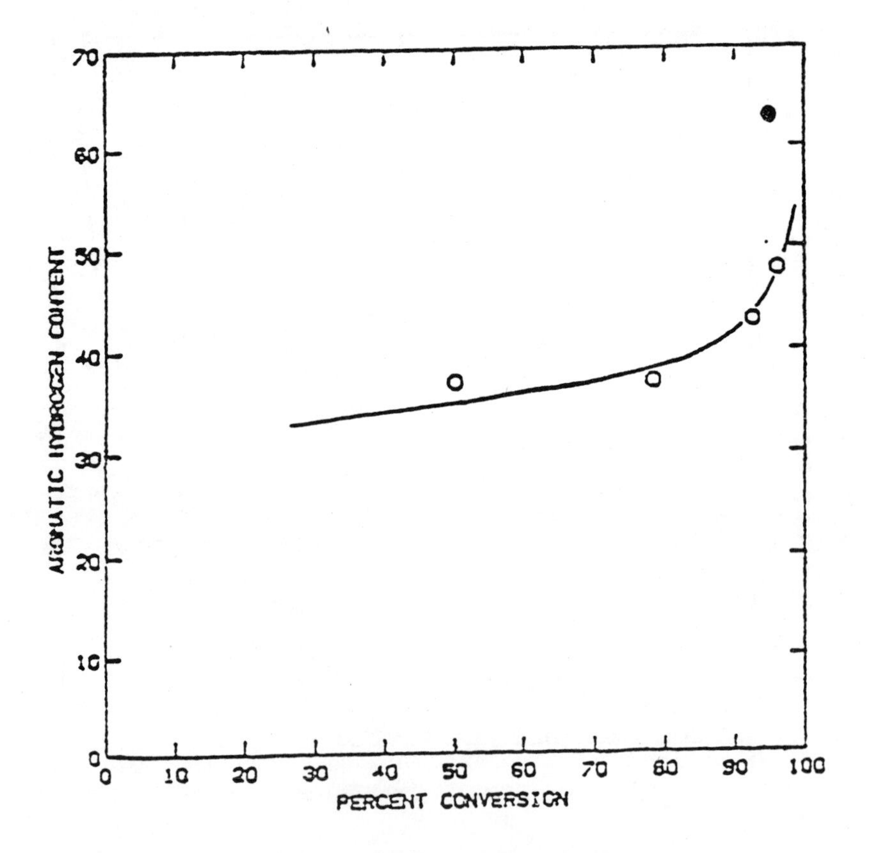

Figure 16. Increase in aromatic hydrogen with severity

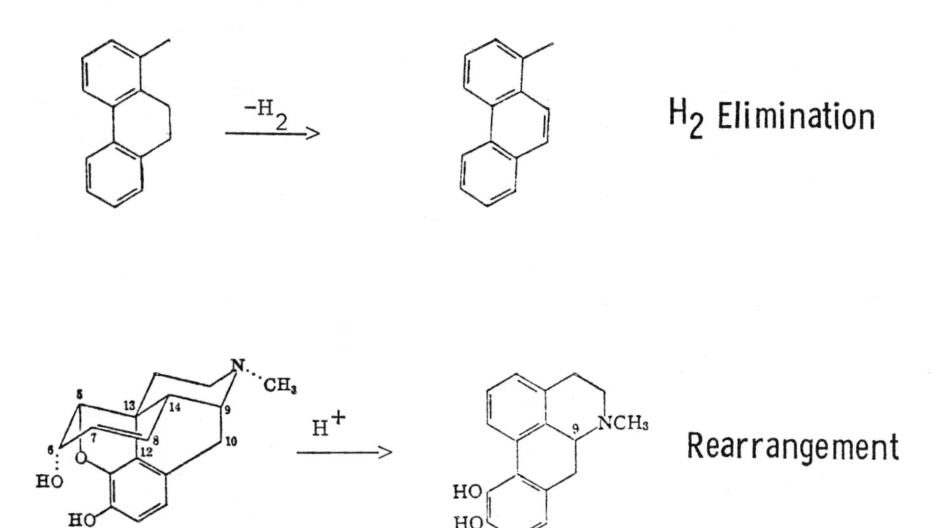

In conjunction with the loss of molecular weight and increase in aromaticity, the functionality of the initially dissolved coal undergoes major change. Oxygen is the primary element of concern as far as functionality and solubility class determination. If one examines the elemental composition of coal and products of coal at short and long times, a rather interesting result can be found. As shown below for every 100 carbon atoms in a coal, conversion at either short or long time causes essentially no change in the content of nitrogen. The hydrogen content is similar to the parent coal at short time, but becomes less at longer times. The oxygen content and sulfur contents both are reduced slightly at short time but are significantly reduced at longer times.

	General Formula	Number Heteroatoms/ 100 C
Monterey Coal (mml)	$C_{100}H_{88}N_{1.6}O_{13.2}S_{1.74}$	14.9
Monterey SRC, short Contact time (AC-59)	$C_{100}H_{88}N_{1.6}O_{9.1}S_{1.3}$	12.0
Monterey SRC, long Contact time (AC-58)	$C_{100}H_{84}N_{1.7}O_{5.1}S_{0.7}$	7.5

Oxygen is lost primarily as carbon dioxide and water, with smaller amounts of carbon monoxide. The rate of oxygen loss parallels the rapid initial dissolution of coal and rapid loss of high molecular weight material. About 40-50% of the oxygen is relatively easy to remove (17). The loss of sulfur is kinetically parallel to the loss of oxygen as shown in Figure 17. This might be anticipated in view of the origin of the organic sulfur of coal, which is believed to be the result of exchange of OH or carbonyl oxygen by sulfur, due to biological activity in the sediment (20,21). The significance of this is that 40-50% of the organic sulfur is also easily removed. The remaining sulfur is much more resistant to attack and is probably present in heterocyclic ring structures.

Hydrogen Consumption and Reactive Moieties

In conjunction with the loss of oxygen and sulfur, as well as molecular weight reduction of the soluble coal specie, there is hydrogen consumption required for the process. This hydrogen in the case of solvent refining is donated from the solvent to the coal or coal fragments. Initially, the loss of oxygen requires relatively little hydrogen consumption and is very close to stoichiometric requirements (16,17). This is shown in Figure 18, where it can be seen that only after about 30% of the oxygen is lost does the hydrogen consumption become greater than stoichiometry. This hydrogen consumption in excess of stoichiometry is due primarily to the formation of gaseous products or lower molecular weight distillates such as solvent and not due to the input of hydrogen into the higher molecular weight products of coal.

Another way to look at the conversion of oxygen is to compare the product composition, overall, with the percent oxygen removed from the total product. Figure 19 shows that in order to achieve maximum solubility of the coal about 60% of the oxygen must be lost. At the same time the SRC yield maximizes. The formation of solvent range material and light gases such as methane, then become major product constituents as the oxygen content is reduced further. What these results indicate is that high hydrogen consumption is not necessary in order to just dissolve the coal or to remove a major portion of the oxygen.

In order to gain an understanding of what kinds of reactions can be envisioned to explain the above results, we conducted a series of experiments using model compounds and reactions with typical solvents

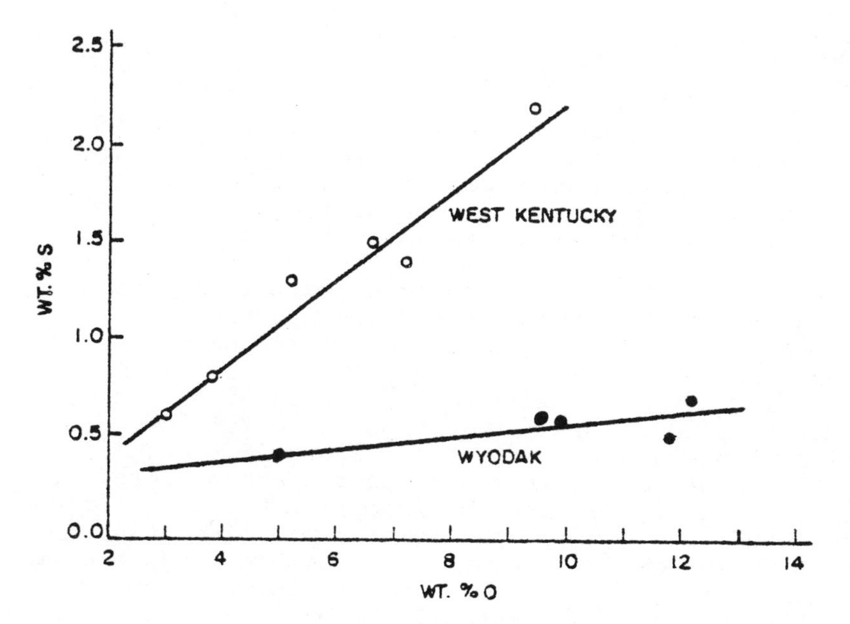

Figure 17. Percent S in SRC vs. percent O in SRC

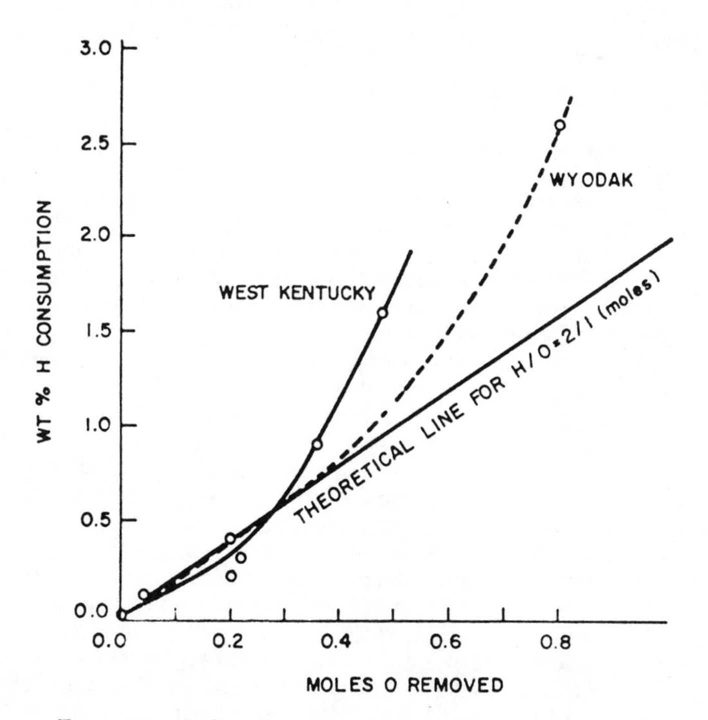

Figure 18. Hydrogen consumption vs. moles oxygen removed

for coal under conditions of coal liquefaction. In these studies we identified a number of chemical classes which could be converted at rates that were comparable to those of rapid dissolution of coal (14). These are summarized in Figure 20, where fast reactions are those in which 80% or higher conversion can be achieved in ten minutes at 800°F. It can be seen that benzylic ethers or benzylic thioethers certain esters and quinones react rapidly enough to account for the very short contact time coal dissolution. In addition, ring structures such as dihydrophenanthrenes will rapidly dehydrogenate under the same conditions. By contrast with these relatively fast reactions we have also identified a number of low reactivity specie which can be ruled out as being responsible for coal dissolution, these are summarized in Figure 21. Here it can be seen that structures such as aromatic ethers, or ring structured ethers, heterocyclic hydrogen compounds, polycondensed rings both aromatic and aliphatic, have relatively low reactivity. If aromatic rings are substituted onto a number of these structures, the reactivity increases dramatically. This increase in reactivity with higher aromatic substitution could possibly account at least in part for the higher reactivity of bituminous coals relative to subbituminous coals.

Speculations on Coal Structure

Up to this point, I have discussed primarily the chemistry of dissolved coal and how the chemical nature of the coal products change with the severity of conversion. It should be noted that at very short contact times the initial products of coal dissolution are very similar in both aromaticity and functionality to that of the parent coal. I will now discuss how this information can be used to help gain a better understanding of the original structure of parent coal. A number of workers have attempted to derive a representative structure of coal which is consistent in its observed chemistry. One of the first was that of Professor Given, shown in Figure 22. This structure was not intended to be the structure for coal but merely to represent what kinds of structures one should envision as constituting coal (22). The structure is consistent with highly substituted aromatics, which are not highly condensed, with functionalities which are known to be present in coal and with its elemental composition. A more recent, more sophisticated model, was presented by Professor Wiser (23) and is given in Figure 23. The significance of this figure is the location of a number of

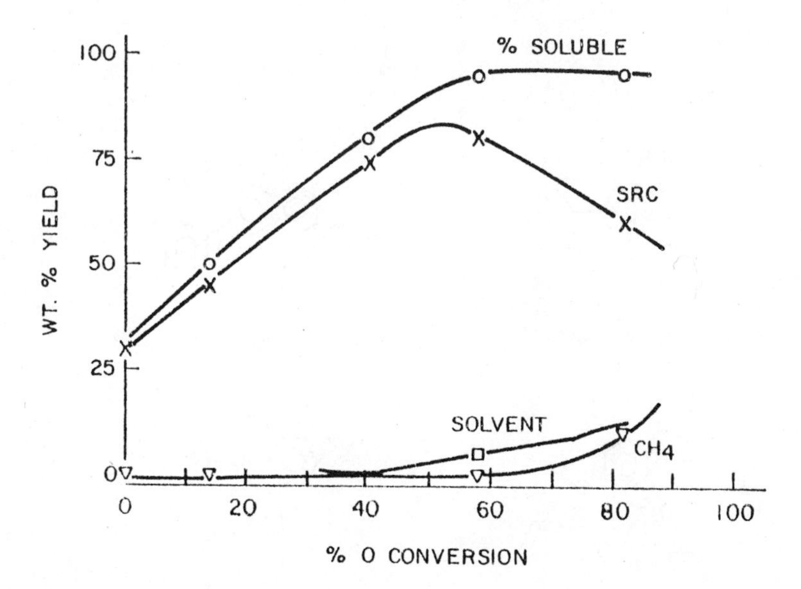

Figure 19. Product yields vs. percent oxygen conversion for West Kentucky coal

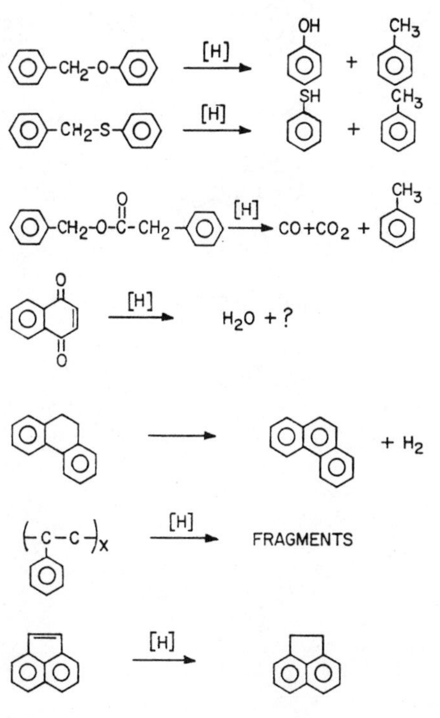

Figure 20. Fast reactions at 800°F

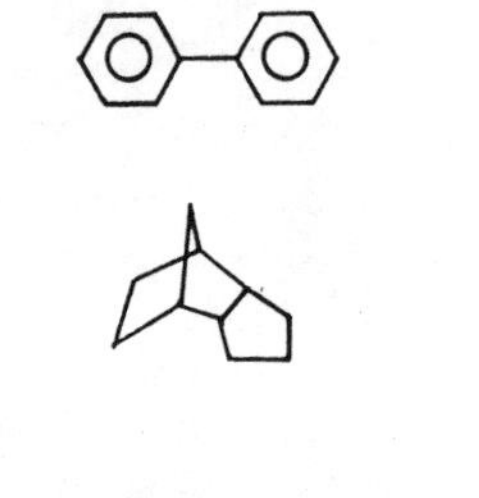

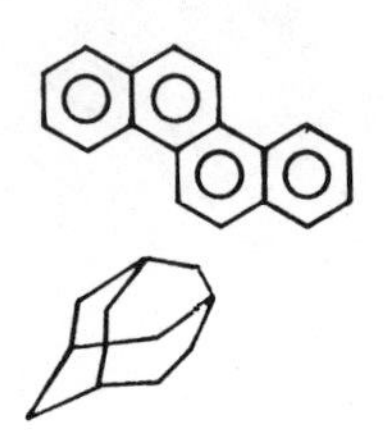

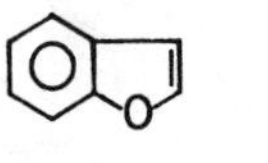

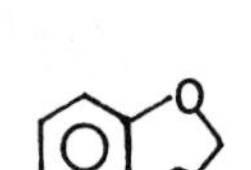

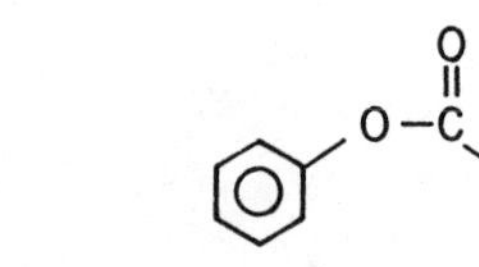

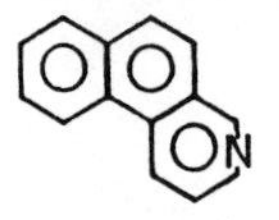

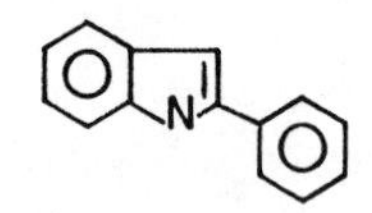

Figure 21. Low reactivity moieties at 800°F

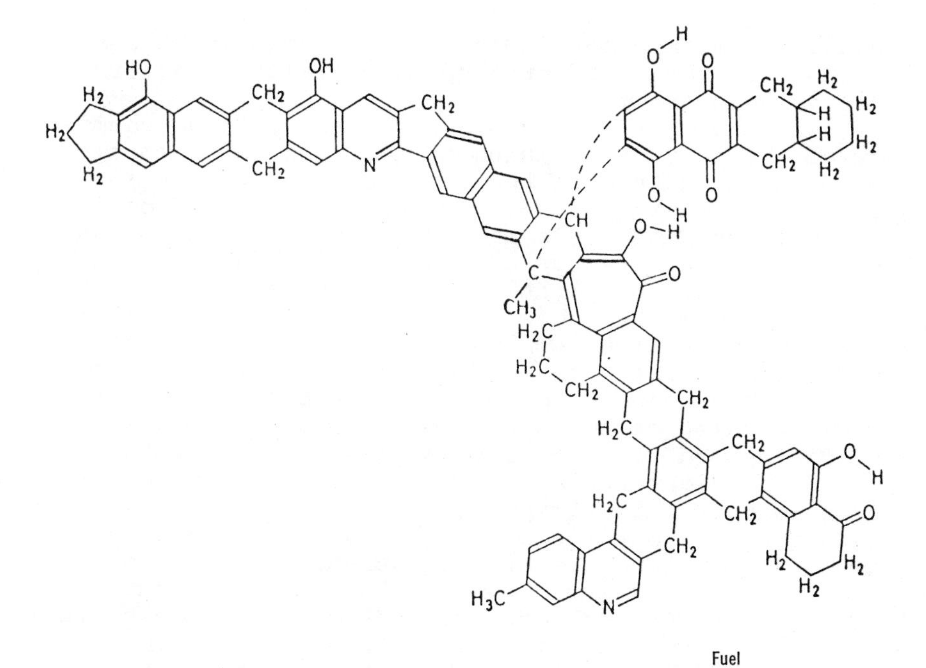

Figure 22. Proposed structural elements of coal by Given

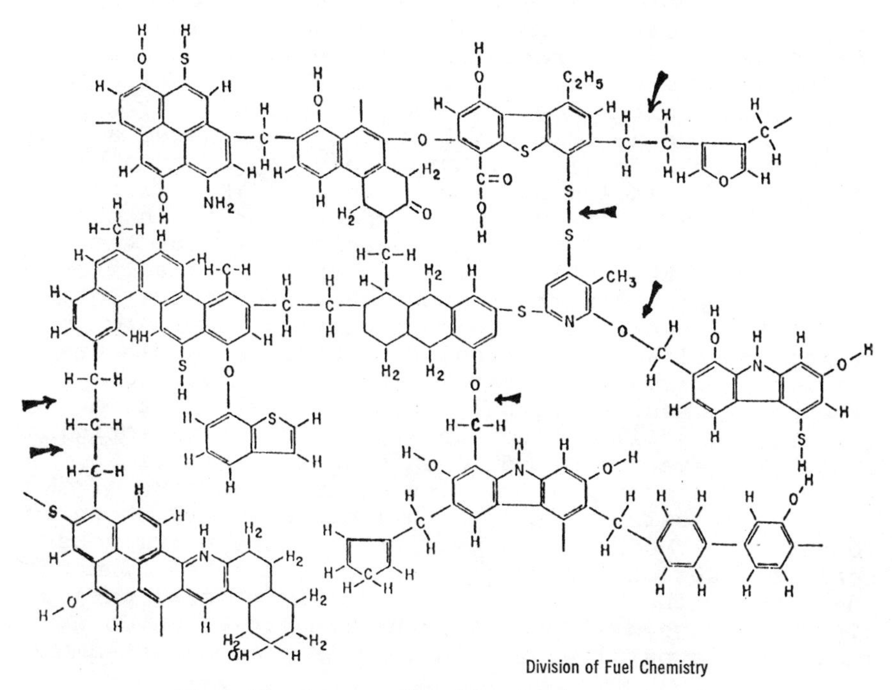

Figure 23. Representation of functional groups in coal

relatively weak bonds indicated by arrows which can account for the rapid breakup of coal into smaller more soluble fragments.

In view of the previous discussion of the chemistry of the solvent refining of coal and the composition and chemical nature of the soluble products, it is possible to envision the original structure of coal in a more gross sense as a highly crosslinked amorphous polymer, which consists of a number of stable aggragates connected by relatively weak crosslinks. As shown in Figure 24, this highly crosslinked structure on reaction at high temperatures thermally fragments into radicals which in the presence of hydrogen donor solvents are capped and appear as stable specie. These specie consist of aggragates or groups of aggragates and correspond to molecular weights in the 300 to 1000 molecular weight range. In the absence of hydrogen donor solvents, the original radicals or the smaller soluble specie may recondense to form char or coke. We have attempted in some of our characterization work to derive representative structures of such aggragates (14,17-19). Our work consisted of deriving the structures of the very initial products of coal dissolution which are pyridine soluble, benzene insoluble materials which we term asphaltols, and represent molecules which are high in molecular weight, low in both aromatic hydrogen and carbon, and high in functionality, in particular phenols. A representative example of such structure is given in Figure 25, for the asphaltol derived from a bituminous West Kentucky coal (18). A structure such as this should not be considered as the structure for asphaltols, but merely represent a kind of structure, including ring structures as well as functionality, that one would expect to find in these initial products of coal and therefore reflect the structure of the coal itself. As these initial products of coal are further converted, they increase in aromatic content and lose functionality, becoming more hydrocarbon like. The ring structures as we discussed previously, remain relatively intact and therefore the less polar less functional materials isolated from coal liquids can still reflect the structure of the original coal. We have attempted to show this in Figure 26, where the heterocyclic ring structure fractions of two bituminous coals and a subbituminous coal are shown for comparison. Here it can be seen that the bituminous coal consists of more planar ring structures, whereas the subbituminous coal contains as an integral part a highly condensed aliphatic structure (14). We do not know if this particular structure is predominant among

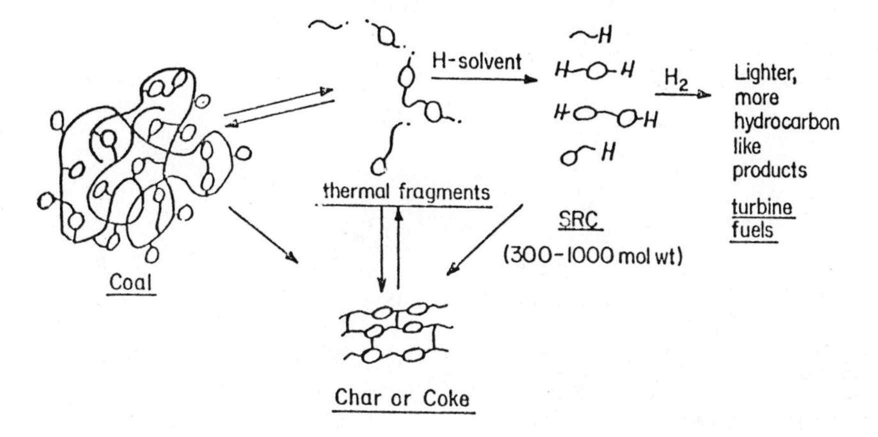

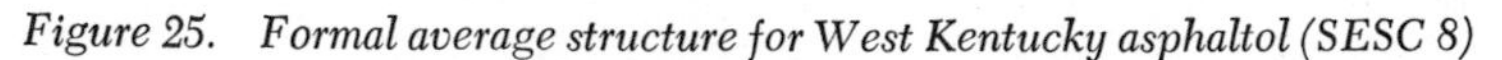
Figure 25. Formal average structure for West Kentucky asphaltol (SESC 8)

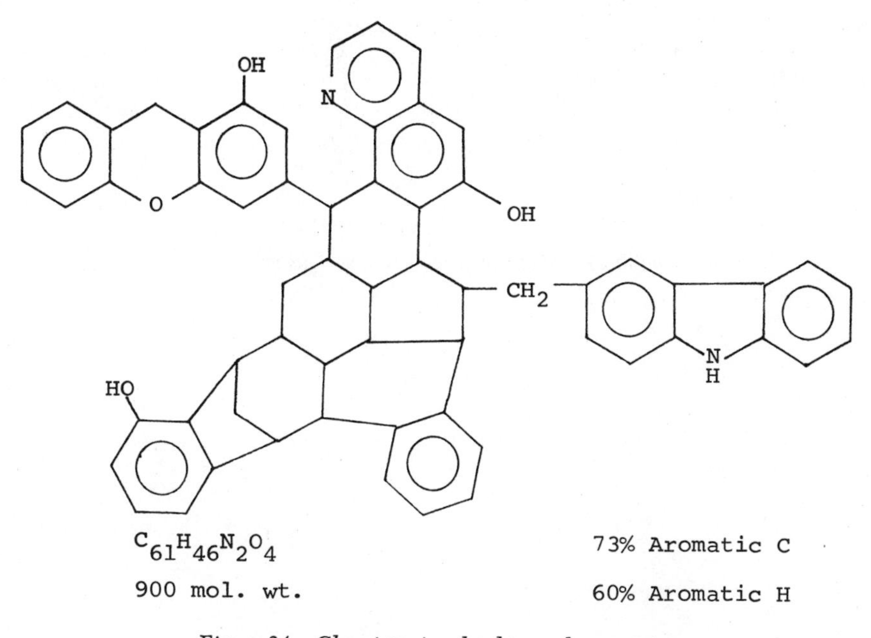

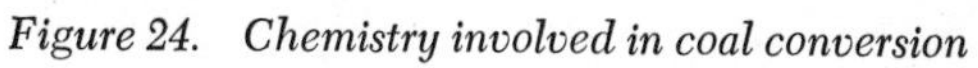
Figure 24. Chemistry involved in coal conversion

	COAL		
	Kentucky	Monterey	Wyodak
Composition	$C_{100}H_{82}N_{1.7}O_{9.4}$	$C_{100}H_{88}N_{1.6}O_{13.2}S_{1.74}$	$C_{100}H_{86}NO_{18}$
No. polycondensed saturated rings/100 c	av. 4	av. 6	av. 10
in Fr. 3 SRC short contact time	2	4	5
H consumption for 100% conversion short contact time g/100 g coal	0.43	0.63	0.89
long contact time	0.96	2.03	2.80
Structure Fraction #3 short contact time	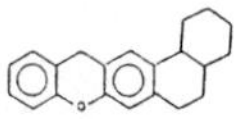	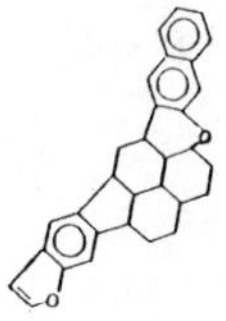	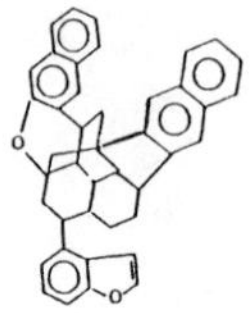

Figure 26. Skeletal structures in various coals

subbituminous coals or is specific to the subbituminous coal that we have examined. However, it does show major ring structural differences between the various coals that we have investigated.

In conclusion I would like now to modify slightly my initial statements as to what is believed about the chemistry and constitution of coal. These conclusions are summarized below.

- The aromaticity of coal varies with rank and can be as low as ~40% for subbituminous coals.
- There is little evidence for highly condensed aromatic rings in coal or the initial liquefaction products. Mono-diaromatic rings predominate.
- Subbituminous coal contains significant amounts of polycyclic aliphatic rings.
- High aromaticity in coal products is the result of the processes being used to convert coal and not an intrinsic property of the starting material.
- Coal is extremely reactive in the presence of hydrogen donors and "liquifies" easily.
- Although hydrogen pressure gives slight increases in yield at short times, it is not a necessity for coal conversion to soluble form. Temperatures above 750°F are necessary for rapid conversion.
- Organic sulfur and oxygen functionalities have similar chemical behavior. About 40-50% of each element can be easily removed. Higher conversions occur much more slowly.
- Coal liquefaction (conversion to soluble form) requires very little hydrogen consumption (~.3-.5% H).

Acknowledgments

Much of the work presented here was conducted under a program jointly sponsored by EPRI and Mobil Research and Development Corporation. Co-workers in this project are J.J. Dickert, M. Farcasiu, and T.O. Mitchell. Technical assistance was provided by B.O. Heady and G. Odoerfer.

Literature Cited

1. Francis, W., "Coal, It's Formation and Composition", Edward Arnold (Publishers) Ltd., London, 1961.
2. Given, P. H., Personal Communication.
3. Chen, A. S., "Flavanoid Pigments in the Red Mangrove, Rhizophora mangle L. of the Florida Everglades and in the Peat Derived From it", PhD Thesis, The Pennsylvania State University, September, 1971.
4. Given, P. H., Bimer, J., Raj, S., Oxidative Study of the Structure of Vitrinites, presented at the Fuel Division Chicago ACS Meeting, August 1977.
5. Reference 1, pp. 717-755.
6. Winans, R. E., Hayatsu, R., Scott, R. G., Moore, L. P. and Studier, M. H., "Examination and Comparison of Structure, Lignite, Bituminous, and Anthracite Coal", presented at the SRI Coal Chemistry Workshop, Menlo Park, California, August 1976.
7. Chakrabartty, S. K. and Berkowitz, N., Fuel (1974) 53, 240.
8. Deno, N. C., Fuel, in press.
9. Hirsch, P. B., Proc. Inst. Fuel Conf., "Science in the Use of Coal", Sheffield Eng. 1958, p. A-29.
10. Ergun, S. and Tiensuu, V., Nature (1959), 183, 1668. Acta Cryst. (1959), 12, 1050.
11. Friedel, R. A. and Queiser, J. A., Fuel (1959), 38, 369.
12. Given, P. H. and Peover, M. E., Fuel (1960), 39, 463.
13. Pines, A., Gibby, M. G. and Waugh, J. S., J. Chem. Phys. (1973), 59, 569.
14. Whitehurst, D. D., Farcasiu, M., Mitchell, T. O. and Dickert, J. J., "The Nature and Origin of Asphaltenes in Processed Coals", EPRI AF-480, Project 410-1, Annual Report July 1977.
15. Ruberto, R. G., "Oxygen and Oxygen Functional Groups in Coal and Coal Liquids" presented at the EPRI Contractor's Conference, Palo Alto, California, May 1977.
16. Whitehurst, D. D. and Mitchell, T. O., "Short Contact Time Coal Liquefaction"; "Liquid Fuels From Coal", R. T. Ellington, Ed., Academic Press, New York, 1977, p. 153.
17. Whitehurst, D. D., Farcasiu, M. and Mitchell, T.O., "The Nature and Origin of Asphaltenes in Processed Coals", EPRI AF-252, Project 410-1, Annual Report, February 1976.

18. Farcasiu, M., Mitchell, T. O. and Whitehurst, D.D., Chem Tech (1977), 7 680.
19. Farcasiu, M., Fuel (1977), 56, 9.
20. Orr, W., "Sulfur", Handbook of Geochemistry, Vol. II/4, Springer-Verlag Berlin·Heidelburg· New York.
21. Neavel, R., "Sulfur in Coal; It's Distribution in the Seam and in Mine Products", PhD Thesis, The Pennsylvania State University, 1966.
22. Given, P. H., Fuel (1960), 31, 147.
23. Wiser, W., preprints Fuel Division ACS Meeting, 20(2), 122 (1975).

RECEIVED March 6, 1978

2

Polymer Structure of Bituminous Coals

JOHN W. LARSEN and JEFFREY KOVAC

Department of Chemistry, University of Tennessee, Knoxville, TN 37916

Many structures have been proposed for bituminous coals.[1-3] The only point of agreement is that these coals contain varying amounts of small molecules (MW < 1000) which are extractable and that they also contain a mixture of larger molecules. While agreement is not universal, most coal chemists would accept van Krevelen's[4] statement that "coal has a polymeric character", that it consists of macromolecules. There is no agreement about the size distribution of the macromolecules and their degree of cross linking. Consideration of the bulk, plastic properties of coals leads not only to the conclusion that coal is a cross linked macromolecular network, but also provides estimates of the number average molecular weight per cross link (M_c).

It is convenient to consider coal structure on several levels. Following the protein chemists (at quite a distance) we will define three structural levels. This division has been done on the basis of experimental convenience but also provides a very useful conceptual framework allowing the complete structure to be treated as the sum of three nearly independent levels of description (substructures). This approach is much more useful than trying to solve the whole structure all at once. Different techniques and approaches are usually used to gain information about each of the structural levels. We find this division aids greatly in sharpening our thinking about coal structure.

First order structure is the size distribution of the macromolecules and molecules in coal and the degree of cross linking. The topology of the macromolecular network also falls in this category as does the amount and identity of the smaller volatile or extractable molecules present in the coal. While we know a good bit about the smaller, extractable molecules in coal it has not been recognized that a significant body of information leading to a coherent picture of the first order structure of bituminous coals exists. In this paper we attempt

0-8412-0427-6/78/47-071-036$05.00/0

to sketch this first order structure.

Second order structure is the chemical identity of the cross links and the structures of the carbon skeleton of coal. For example, consider the view that the macromolecules comprising bituminous coals are composed of polycyclic aromatic and hydroaromatic units linked together by methylene bridges and ether linkages. The secondary structure would then be the identity and quantity of the various links between the aromatic and hydroaromatic units as well as the average and extreme structures (carbon skeleton only) of those hydroaromatic units. Perhaps it is worth taking a little time to deplore the current emphasis on "average" structures. This emphasis is understandable; average structures are so easy to derive that even a computer can do it and they provide a convenient crutch when one wants to draw a coal structure. However, for most purposes, a knowledge of the extremes of the range of structures present will be much more useful in predicting the chemical behavior of coals.

Finally, the third order structure of coal is the nature and distribution of the functional groups. This needs little comment or explanation. Our knowledge of this area is more complete than the other two, but large holes remain.

Superimposed on all of this is the physical structure of coal, the Neavel fruitcake.[5] Each maceral will have first, second, and third order structure. Each coal will be a different mixture of macerals. Factors such as pore structure will have a strong effect on observed reactivity.

The Nature of the Macromolecular Network. In this paper we are concerned only with first order structure. We begin by considering the time dependent response of bituminous coals to a constant stress and show that this behavior is consistent only with bituminous coal being a cross linked, three dimensional macromolecule.

If a cross linked polymer network is subjected to a stress, it will deform until it reaches the limit set by the cross links. Chains will be stretched and straightened as the material deforms, but the presence of covalent cross links places a limit on this deformation. Thus, a plot of strain (eg. length of a piece of coal under tension) *vs* time will increase then level out. Furthermore, when the stress is removed, the polymer will return to its original dimensions. If the macromolecules are entangled or held together by weak forces, there will be no internal limit on the flow and the deformation will continue to increase with time. Thus with a polymer which is not cross linked a constant finite value will not be reached and the strain will not be recoverable.[6,7] Very high entangled materials may be exceptional, but given the highly planar structural units in coal this seems most unlikely.

There is some data in the literature describing the time

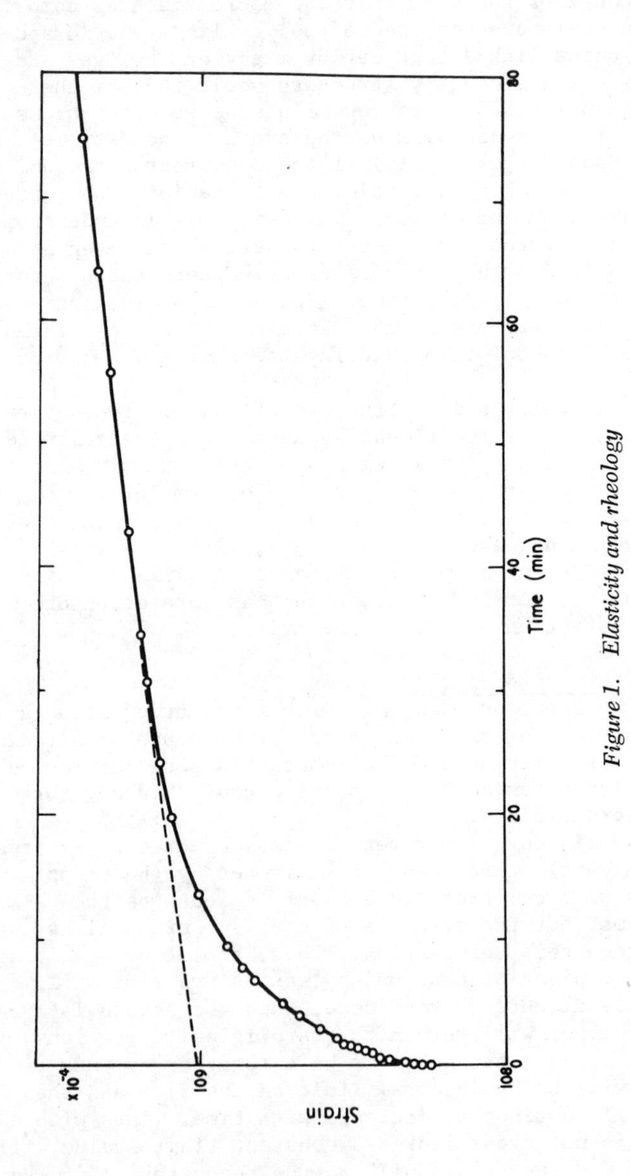

Figure 1. Elasticity and rheology

dependence of stress on strain for coals. Much of it is old work done to study the behavior of coal in various processing operations, such as grinding.[8] The results often have not been completely described and the work often was done on large samples and so some of the behavior might be due to physical flaws such as cracks and not be representative of the intrinsic molecular structure of the coal. The best study is that of Morgans and Terry and the strain-time curve for Barnsley Hards bituminous coal is shown in Fig. 1.[9] Only 0.5% of the total strain was irrecoverable. A Clarain showed 1% unrecoverable strain.[10] These data are quite consistent with bituminous coal being covalently cross-linked macromolecules in which hydrogen bonding and van der Waals forces contribute in only a small way to macromolecular association. This structure is also inconsistent with pure entanglement model. However, if there is a mixture of cross-links and entanglements, measurements such as these cannot distinguish this from a cross linked structure. If there are enough chemically bonded cross links to complete the polymer network, the modulus will be increased by entanglements and both structures in Fig. 2 will behave the same in a linear compression or extension experiments.

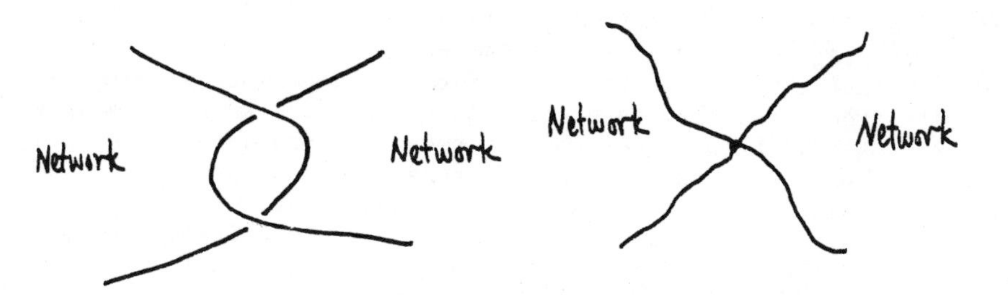

Figure 2. Entanglement and cross-linked networks

The principle point here is that bituminous coals are covalently cross linked macromolecules. Weak associative forces such as hydrogen bonds and van der Waals interactions cannot explain the results of these experiments.

Average Molecular Weight Per Cross Link.
Having established that coal is a cross linked macromolecule, the next step is to determine the frequency of the cross links. There are two independent approaches which can be used. The first is to utilize data from solvent swelling of coal and the

second is to use the moduli derived from the behavior of the polymer under stress to calculate the number average molecular weight per cross link (M_c).

If the macromolecules in coal are cross linked to form a network, it is possible to use a quantitative model of the elastic properties of a polymer network to obtain an estimate of the average frequency of the cross links. For an isotropic network only one parameter, the elastic modulus, controls the physical properties of the network. This parameter can be measured by any one of a variety of experiments. Data are available for solvent swelling and mechanical measurements of Young's modulus of bituminous coals so these experiments will be used as independent methods of determining the elastic modulus. It should be emphasized that the two experiments are measuring essentially the same quantity. In this paper the simplest statistical theory of polymer network elasticity[11,12] will be used to calculate the number average molecular weight per cross link (M_c) from the two different kinds of measurements. Before presenting the experimental data we will describe the statistical theory of polymer network emphasizing the physical model on which it is based and presenting the equations necessary to calculate M_c. The interested reader will find a lucid exposition of the subject of polymer network elasticity in _The Physics of Rubber Elasticity_ by L.R.G. Treloar.[11]

The statistical theory of polymer network elasticity assumes that the elastic restoring force in a polymer network results from stretching each chain in the network from its most probable (equilibrium) conformation to a less probable (stretched) conformation. Only the entropic factors are considered, the intermolecular forces are ignored. For the relatively long chains found in many polymers (eg. natural rubber) it is safe to assume that the probability of any chain conformation is given by a Gaussian distribution.[9] The entropy is then given by Boltzmann's formula

$$S = -k \ln P \tag{1}$$

where P is the probability, and the force (f) is obtained by differentiation with respect to length (l),

$$f = -\frac{\partial S}{\partial l} \tag{2}$$

It is also assumed that each chain in the network contributes independently to the force. Physically this means that entanglements between chains are not important and that in the model chains are allowed to pass through each other as the network is extended. This is sometimes called the phantom chain assumption.

With these assumptions one can derive stress-strain relationships for various experiments. For example, for a sample extended in one dimension by a factor

$$\lambda = \frac{\text{strained length}}{\text{unstrained length}}$$

$$f = G\,(\lambda - \lambda^{-2}) \tag{3}$$

where G is the elastic (shear) modulus

$$G = \frac{\rho RT}{M_c} \tag{4}$$

In eq. (4) ρ is the density, R the gas constant, T the absolute temperature and M_c the average molecular weight per cross link. G appears as the single elastic constant in all stress-strain measurements. Since ρ, R, and T are all experimental quantities any measurement of G allows a calculation of M_c.

A convenient method of determining M_c is by equilibrium solvent swelling. If a polymer network is brought into contact with a good solvent, the solvent will be absorbed by the network until the swelling pressure is exactly balanced by the elastic force of the network. Using the statistical theory of polymer network elasticity and the Flory-Huggins theory of polymer solutions, the following equation can be derived.[11]

$$M_c = \frac{1}{3}\rho_2 V V_2 \,/\, [-\ln(1-V_2) - V_2 - \chi V_2^2] \tag{5}$$

where ρ_2 is the original density of the polymer, V the molar volume of the solvent, V_2 the volume fraction of the polymer at equilibrium, and χ the Flory interaction parameter related to the heat of transfer of solvent from pure solvent to the pure polymer in units of kT.[13] χ must be determined experimentally but given χ, then M_c is easy to obtain from equilibrium swelling.

One measurement of solvent swelling was made by Sanada and Honda[14] who studied the swelling of a number of Japanese coals in pyridine after extracting them with pyridine. The choice of solvent is not a good one since some of it is incorporated into the coal.[15] Unfortunately a numerical error was made in calculating M_c so the values shown in the paper are incorrect. The correct values for the molecular weights are given in Table 1.

The results are numerically reasonable and behave as expected. The sudden drop with high rank coals (Yatake) is expected, due to the approach of a more graphite like, more highly cross linked structure. The general increase in M_c from low to high rank (excluding anthracites) can be ascribed to the presence of larger molecules assembled together to form the macromolecules or to fewer cross links in the higher rank coals. The first explanation is more in tune with current thought.

A rather different application of the same principle is contained in a paper by Kirov et al.[16] They studied the equilibrium swelling or three coals by 17 solvents. In this way

Table I. Average Molecular Weight per Cross Link in Some Japanese Coals

Coal	% C	% H	M_c
Odaira	65.1	5.0	1190
Nakago	74.3	5.3	684
Takamatsu	79.0	5.0	832
Bibai	80.9	5.9	825
Ashcbetsu	81.1	5.5	1120
Yūbari - I	84.9	6.2	1640
Yūbari - III	85.2	6.3	1500
Hoshima	86.6	5.6	1840
Yatake	88.7	4.4	49.5[a]
Hongei	93.0	3.3	

a) assuming $\chi = 0.8$

they were able to calculate the best values of χ and V_c to fit their data (V_c is the molecular volume per cross link). Their values of V_c were converted to M_c by multiplying by an estimated density for these coals of 1.3.

Table II. Average Molecular Weight per Cross Link for Some Australian Coals

Coal	% C	% H	M_c
Hebe	75.9	4.5	1530
Greta	82.4	6.2	1200
Bulli	88.2	5.1	522

The agreement between the two sets of results is surprisingly good. The drop off at very high rank is reproduced and the order of magnitude of the M_c values is the same. This strongly supports the notion that these values are reasonable.

Values of M_c can also be calculated from measurements of the mechanical properties of coal. From studies carried out by van Krevelen[17,18] it is clear that the shear modulus G is constant at about 1.6×10^{10} dyne/cm^2 for coals having between 82 and 92% carbon. After this it rises sharply, as expected due to the rapidly increasing cross linking in the anthracites and near-anthracites. Using this value for G, and assuming $p = 1.3$g. cm^{-3}, $M_c = 2$. The G values used here were derived from sound velocity measurements. Bangham and Maggs[19] have reported values for Young's Modulus for three coals measured by compression. For a cross linked polymer which undergoes no volume change on compression, the shear modulus is 1/3 Young's modulus.[6,7] Table III contains their data and the calculated values for M_c. The results are also very small. Other values for Young's modulus for bituminous coals can be found in the literature. All are similar and give very low values for M_c.

Table III. Values for M_c Calculated from the Young's Modulus of Bangham and Maggs[7]

Coal	Young's Modulus (dyne/cm^2)	Shear Modulus (dyne/cm^2)	M_c (g/mole)
Welsh Anthracite	40×10^9	13×10^9	2.5
Welsh Steam Coal	7×10^9	2.3×10^9	14
Northumberland House Coal	20×10^9	6.7×10^9	4.8

The $10^2 - 10^3$ discrepancy in M_c between the solvent swelling and the stress-strain measurements must be investigated. Both solvent swelling and the stress-strain measurements give information about the same property of the network, the elastic modulus. This modulus is then used to calculate M_c, using the same model in both cases. This descrepancy must be due to experimental differences and not to some failure of the model. There must be different factors operating in the two types of experiments. In some polymer systems, stress induced crystallization will produce a large modulus and therefore a small value of M_c. Perhaps an increase in intermolecular interactions due to the increased order in unidirectional stress is worth investigating. One sure source of some of the descrepancy is that extracted coals were used for the solvent swelling and unextracted coals were used in the stress-strain measurements. The presence of small molecules within the polymer network will have several effects; the density of the system will change, and small molecules will occupy volume in the network thus hindering the internal motions of the chains. The hindering of the motion will tend to increase the modulus of the network and hence lower the apparent molecular weight. We propose two qualitative arguments to explain part of the discrepancy.

1) Swollen network theory

Consider the unextracted coal to be a swollen polymer network. For concreteness assume a typical value of 30% extractable material and 70% network material. According to the simple theory of rubber elasticity the modulus of a swollen rubber is given by

$$G' = \frac{\rho' RT}{M'_c} V_2^{1/3} \tag{6}$$

where V_2 is the volume fraction of the network and ρ' is the density of the unswollen rubber.[11] If one did not consider the effect of swelling than the modulus would be given by

$$G = \frac{\rho RT}{M_c} \tag{7}$$

G (or G') is an experimental quantity and we are interested in the values of M_c obtained from different interpretations of G. Hence, we equate the right hand sides of (1) and (2)

$$\frac{\rho RT}{M_c} = \frac{\rho' RT}{M'_c} V_2^{1/3}$$

$$\frac{M'_c}{M_c} = \frac{\rho' V_2^{1/3}}{\rho} \tag{8}$$

The ratio M'_c/M_c represents the ratio of the true M_c to the apparent M_c. $V^{1/3}$ is always less than unit but for $V_2 = 0.7$, $V_2^{1/3} = 0.9$ so $V_2^{1/3}$ can be assumed to be near one. The question then is the change in density with swelling. If upon extracting the coal there is negligible volume change, then the swollen density ρ' will be greater than the unswollen density ρ. To get an estimate, assume unit density of the swollen network, equal densities of network and extractable material and no change of volume upon extracting. Then for 70% network we obtain

$$\frac{M'_c}{M_c} = \frac{1\ (0.7)^{1/3}}{0.7} = 1.27$$

Hence the true molecular weight will be larger than the apparent M_c.

(2) Intermolecular Obstructions.

The theory here was developed by J. L. Jackson and co-workers.[21,22] It is based on the notion that the polymer network is excluded from part of the volume available to it. The "intermolecular obstructions" are taken into account by means of a site fraction occupied by the polymer on a lattice. A smaller site fraction represents a larger number of obstructions. In the improved theory Jackson obtains the following stress-strain relationship for linear extension.

$$F = \frac{RT\rho}{M_c} (\lambda - {}^1/\lambda^2)\ [1 - \frac{1}{3M} (\ln(1-f)^{-1} - f)\ (2\lambda^2 + \lambda^{-1})] \quad (9)$$

This is eq. (18) in Jackson's paper.[21] f is the site fraction and M the number of segments between cross links. It is harder to compare the moduli here because of the form of the equation. We will define the modulus at a particular strain as

$$\frac{G}{\rho RT} = [\frac{1}{M_c} - \frac{1}{M\,M_c} (\ln(1-f)^{-1} - f) \quad (2\lambda^2 - \lambda^{-1})] \quad (10)$$

To obtain an estimate of the effects of the site fraction we will solve eq. (5) for various values of f for the special case

$$\frac{G}{\rho RT} = 10$$

$$M = 100$$

$$\lambda = 1.1$$

Jackson calculates $f \simeq 0.95 - 0.98$ for typical rubbers so we will assume that in the obstructed network $f = 0.95$ and will calculate the change in M_c for various degrees of extraction.

Table 4. Values of M_c for Various Site Fractions

Site Fraction (f)	M_c
0.95	5.61
0.85	4.03
0.75	3.15
0.65	2.51

The table shows clearly that for a fixed modulus the apparent chain length must increase as the number of obstructions increases. If we consider the extracted coal to have a site fraction of 0.7 and the unextracted coal to have a site fraction of 0.95 then there will be a factor of two difference in the apparent M_c.

Both the arguments developed here are qualitative but they both show that the molecular weight derived from the simple theory for the unextracted coal is probably smaller than the real M_c. Using the Jackson theory one obtains at least a factor of two difference. The explanations given here are based on simple models and are meant only to be suggestive. The very large modulus observed in linear extension experiments could also be due to intermolecular forces not accounted for in the statistical theory. For example, Su and Mark and co-workers have recently determined that the large increase in modulus at high extension of rubbers is due to stress-induced crystallization.[26] A similar phenomenon could occur in the unextracted coals. A proper resolution of the discrepancy must come from a systematic study of the molecular weight of extracted coals by both methods.

Use of the Statistical Model.

In this paper the statistical theory of polymer network elasticity has been used to calculate the average molecular weight per cross link in various coals from experimental data on solvent swelling and mechanical behavior. Although the numerical values for M_c obtained from the solvent swelling data are quite reasonable, they should only be regarded as order of magnitude estimates because the assumptions underlying the statistical theory are probably not fulfilled in the coal structure. Some of the assumptions will be discussed in the following paragraphs along with the possibility of correcting them.

The primary assumption of the statistical theory is that the chains obey Gaussian statistics, that is, that the probability distribution of the end-to-end distance (R) has the form

$$P(R) = \text{constant} \times e^{-AR^2}$$

It is easy to show that any chain will obey Gaussian statistics if it is both long and flexible.[13] The chains in the coal network are unlikely to be either long or flexible. The molecular weights obtained are of the order of 1000 which correspond to 10-15 aromatic rings per chain. A chain composed largely of fused aromatic rings is better characterized as short and stiff. A simple modified Gaussian theory is currently being developed to try to more accurately represent the coal network.[22] Preliminary calculations indicate that M_c will be larger by approximately a factor of two using the modified Gaussian model.

A further and more difficult problem is the assumption that the chains act independently. Entanglements and intermolecular forces which tend to increase the modulus could well be important as well as "loose-end" corrections which tend to decrease the modulus. The phenomenon of stress induced crystallization might also be found. Developing a model to account for cooperative network effects will be difficult. In spite of these drawbacks, the results of applying this model to coal are reasonable, and at least a good beginning in our attempts to determine the first order structure of coal.

If one accepts the M_c values for bituminous coals are of the order of 1500-1800, estimates of the number of subunits per cross link can be made. Here we assume that bituminous coals are composed of aromatic and hydroaromatic units linked together. The Heredy-Neuworth depolymerization is thought to cleave the alkyl chains linking the aromatic units.[23] Most of the bituminous coals which have been depolymerized give products having number average molecular weights between 300 and 500,[24] although a few are larger. Accepting these results at face value (molecular weight distributions have not been done), one concludes that the average cross link chain contains 3-6 aromatic units. In a very insightful paper, van Krevelen treated coal as a cross linked polymer gel and the extractable material as the unreacted "monomer".[25] This treatment leads to an average molecular weight value of 400 for the structural unit. This is in excellent agreement with the depolymerization results.

This highly cross linked structure for coal suggests that donor solvent liquefaction involves major degradation of the coal. It also has implications for the use of solubility differences to follow the progress of coal conversion (i.e. asphaltenes, preasphaltenes, etc.)

It must be emphasized that this is the first systematic use of bulk solid properties of coal to a determination of its structure. The conclusion that coal is a three dimensionally cross linked macromolecular network rests on a qualitative treatment of strain-time curves and seems sound. The use of two different techniques, stress-strain measurements and solvent swelling, to measure the elastic modulus of the network gives conflicting results which are probably due to differences in the sample pretreatment and differences which are artifacts of the experimental procedure. One technique was applied to a swollen or obstructed network (whole coal - the extractable material is dissolved in and swells or obstructs the network) and the other to an extracted, unswollen network. To examine this structural hypothesis further we are attempting to resolve the discrepancy between the stress strain and solvent swelling measurements and to derive a statistical mechanical theory of polymer network elasticity which fits bituminous coals. This work is in progress.

Acknowledgement:

We are grateful to the Department of Energy (JWL), NSF (JWL), and the Faculty Research Fellowship Fund of the University of Tennessee (JK) for support of this work.

Abstract

The physical properties of bituminous coals can be used to determine the nature of their macromolecular structure. Bituminous coals are cross-linked macromolecular networks and a preliminary estimate of the number average molecular weight per cross link is 1500-1800.

Literature Cited

1. Vahrman, M., Chem. in Britain, 8, 16 (1972); Fuel, 49, 5 (1970).
2. Sternberg, H., Storch Award Address, Fuel Div. Preprints, ACS National Meeting, Sept. 1976.
3. van Krevelen, D. W., Elements of Coal Chemistry Rotterdam, 1948, p. 170 Dryden, I.G.C., Chem. and Ind., 502 (1952); Fuel, 30, 39 (1952).
4. van Krevelen, D. W., Coal Elsevier Publishing Co., New York, N.Y., 1961, p. 440.
5. Neavel, R., (Exxon-Baytown) numerous lectures and seminars.
6. Ferry, John D., Viscoelastic Properties of Polymers, John Wiley and Sons, Inc., New York, 1961.
7. van Krevelen, D. W., and Hoftyzer, P. J., Properties of Polymers, Elsevier Publishing Co., New York, 1972.
8. The most recent summary is Brown, R. L. and Hiorno, F. J., in Chemistry of Coal Utilization, Suppl. Vol., H. H. Lowry Ed., John Wiley and Sons, Inc., New York, 1963.
9. Morgons, W. T. A., and Terry, N. B., Fuel, 37, 201 (1958).
10. Macrae, J. C. and Mitchell, A. R., Fuel, 36, 423 (1957).
11. Treloar, L. R. G., The Physics of Rubber Elasticity, Clarendon Press, Oxford, 1975.
12. Flory, P. J., Principles of Polymer Chemistry, Cornell University Press, Ithaca, N.Y., 1953.

13. Flory, P. J., Statistical Mechanics of Chem. Molecules, John Wiley and Sons, New York, 1969.
14. Sanada, Y., and Honda, H., Fuel, 45, 295 (1966).
15. Collins, C. J., Hagaman, E. W., and Raan, V. F., Coal Technology Quarterly Report, Oak Ridge National Laboratory, ORNL 5252, Dec. 31, 1976, p. 143.
16. Kirov, N.Y., O'Shea, J. M., and Sergeant, G. D., Fuel, 47, 415 (1968).
17. Schuyer, J., Dijkstra, H., and van Krevelen, D. W., Fuel, 33, 409 (1954).
18. van Krevelen, D. W., Chermin, H. A. G., and Schuyer, J., Fuel, 38, 438 (1959).
19. Bangham, D. H., and Maggs, F. A. P., Proceedings of a Conference on the Ultra-fine Structure of Coals and Cokes, BCURA, 1944, Lewis, H. K. and co., Ltd., London, p. 118.
20. Jackson, J. L., Shen, M. C., and McQuarrie, J. Chem. Phys., 44, 2388 (1966).
21. Jackson, J. L., J. Chem. Phys., 57, 5124 (1972).
22. Kovac, J., "Modified Gaussian Model of Rubber Elasticity," Macromolecules, in press.
23. Heredy, L. A., and Neuworth, M. B., Fuel, 41, 221 (1962).
24. Larsen, J. W. and Kuemmerle, E. W., Fuel, 55, 162 (1976).
25. van Krevelen, D. W., Fuel, 47, 229 (1966).
26. Su, T. K. and Mark, J. E., Macromolecules, 10, 120 (1977).

RECEIVED April 5, 1978

3

Oxygen and Oxygen Functionalities in Coal and Coal Liquids

RAFFAELE G. RUBERTO and DONALD C. CRONAUER

Gulf Research and Development Co., Pittsburgh, PA 15219

The work described in this paper is part of an on-going project sponsored jointly by Gulf and the Electric Power Research Institute of Palo Alto, California.

The purpose of this project is to develop an understanding of the reactions involving oxygen functional groups in coal during hydrogenative liquefaction. To achieve this goal, it is necessary to first liquefy coal under selected and controlled conditions. Subsequently, the concentrations of oxygen and individual oxygen-containing groups are determined. The overall material balance and conversion are then calculated.

This paper is concerned with the analytical aspects of this work. It outlines the various analytical methods that have been used and the results that have been obtained with selected experiments.

Analytical Methods

Analysis of Oxygen. Coal contains both organic and inorganic oxygen, and various methods are used in an attempt to quantify it (Table I).

The organic oxygen is usually calculated by subtracting from 100 the values for C, H, N, S, and ash. This method has two disadvantages: 1) all the errors incurred in the other determinations are combined in the oxygen value; and 2) the ash does not represent the mineral matter as it was originally present in coal. This, however, is still the procedure recommended as ASTM D3176-4.(1)

Chemical and nuclear methods to directly determine oxygen are also available. These methods are applicable to many types of samples. Due to their extensive use, they need no description here. However, it is important to note that the nuclear methods measure total oxygen, and the chemical methods typically measure organic oxygen and at least part of the inorganic oxygen. This is not a problem when inorganic oxygen is absent.

0-8412-0427-6/78/47-071-050$10.00/0

TABLE I

DETERMINATION OF OXYGEN

1. BY DIFFERENCE (ASTM D3176-74)

   O =100 - (C + H + N + S + ASH)

2. CHEMICAL

   A. OXIDATIVE METHODS: VARIOUS METHODS USING SOLIDS, LIQUIDS, OR OXYGEN AS OXIDANTS. VERY CUMBERSOME AND NOT MUCH USED.

   B. REDUCTIVE METHODS

      - HYDROCRACK SAMPLE IN H_2 TO CONVERT ALL THE OXYGEN TO H_2O
      - DECOMPOSE SAMPLE IN N_2 OVER C TO CONVERT ALL OF THE OXYGEN TO CO
      - DETERMINE H_2O OR CO

3. NEUTRON ACTIVATION

$$O^{16} + N \longrightarrow N^{16} + P - 9.62\ \text{MEV}$$

N^{16} DECAYS WITH $t_{1/2}$ = 7.38 SEC. $\rightarrow \beta + \gamma$

MEASURE ACTIVITY OF β OR γ RAYS.

In this work, oxygen is determined by a modified Unterzaucher method (2) by Micro-Analysis, Inc., Wilmington, DE. An in-house neutron activation method (3) and the ASTM difference method are also used for selected samples.

Optimization of Coal Particle Size. To obtain a meaningful analysis of coal, it is necessary to pulverize a sample to insure that the oxygen functional groups are exposed and can be reached by various reagents. The finer the coal sample is, the more sites will be exposed and analyzable. On the other hand, there is a possibility of obtaining excessively fine particles which retain a small static charge and are difficult to wet.

To determine an optimum particle size for analysis, samples of three particle sizes of each coal, namely, total pulverized coal, on-100 mesh, and through-325 mesh, were examined. For each coal sample, the total hydroxyl groups were determined by the acetylation method, and the carboxyl groups were determined by the ion exchange method. The data are reported in Table II.

Examination of the data indicates that the determined value of the O_{OH} was always lower for the on-100 mesh samples than for the through-325 mesh samples. The difference between the lowest and the highest values was small, but apparently significant, due to the consistency of the results.

The difference of the COOH concentrations between samples of different sizes was even smaller, and for the subbituminous and bituminous coals the results actually reversed. In performing these experiments, it was noted that the coal dust wetted poorly and tended to move to the walls of the reaction flask. It is possible that with the smaller particle size samples this effect was pronounced, and the low values were due to poor mixing. (Mixing was not a problem with the acetylation method for the determination of the OH groups, since extensive refluxing was necessary.) On the basis of the above observations, it was decided to use coal of particle size of 100% through 200 mesh.

An additional question remained, namely, were the differences in these analytical results due to site availability or to segregation of coal sub-components by crushing and sieving? It has been demonstrated that some concentration of macerals can be achieved by crushing and sizing. (4) Since macerals do contain different amounts of oxygen, (5) it is probable that the O-functionalities vary. To avoid any possibility of segregating these functionalities according to particle size, it was decided to grind the coals until the total sample passed through a 200-mesh sieve.

The grinding, sieving, transferring, and all the other operations were carried out in a nitrogen atmosphere. The results, therefore, are not due to possible alterations of the oxygen functional groups during the handling steps.

TABLE II

EFFECT OF PARTICLE SIZE ON DETERMINATION OF O-FUNCTIONALITIES (HCl-TREATED)

	O_{OH} WT % COAL	O_{COOH} WT % COAL
LIGNITE		
TOTAL	3.6	3.5
PLUS 100 MESH	2.0	3.1
MINUS 325 MESH	4.1	3.6
SUBBITUMINOUS COAL		
TOTAL	1.2	2.4
PLUS 100 MESH	1.1	2.3
MINUS 325 MESH	1.6	1.8
BITUMINOUS COAL		
TOTAL	1.7	.7
PLUS 100 MESH	1.5	.7
MINUS 325 MESH	2.1	.6

Effect of HCl Pretreatment of Coal. HCl pretreatment of coal is often recommended to insure that no salts are present and that all the oxygen functional groups are in a free state. As a test of the necessity of such pretreatment, a sample of on-100 mesh lignite was repeatedly analyzed for COOH groups before and after treatment with HCl. The results are shown in Table III. Since the content of the COOH groups was considerably higher for the HCl-treated samples, HCl treatment is deemed to be necessary. The procedure for the analysis of COOH groups is that of Blom et al. (6)

Table III

EFFECT OF HCl PRETREATMENT ON DETERMINATION OF -COOH IN LIGNITE (PLUS -100 MESH SAMPLE)

Run No.	Wt % O_{COOH}
1	1.0
2	0.8
3	0.7
4	0.8
5	0.9
6	1.1
After HCl Treatment	3.1

Determination of Total OH Groups. The concentration of -OH in coal is determined by the acetylation method of Blom et al. (6) This method consists of first acetylating the -OH groups with acetic anhydride in the presence of pyridine. The acetylated products are then hydrolyzed; the liberated acetic acid is distilled and quantified by titration with standard NaOH solution.

The overall reactions are as follows:

$$R - OH + (CH_3CO)_2O \xrightarrow[90°C]{Py} CH_3 - COOH + R - OOCCH_3 \quad (1)$$

$$R - OOCCH_3 + 1/2\ Ba(OH)_2 \xrightarrow[100°C]{} R - OH + 1/2\ Ba\ (OOCCH_3)_2 \quad (2)$$

$$1/2\ Ba(OOCCH_3)_2 + 1/3\ H_3PO_4 \longrightarrow CH_3\ COOH + 1/6\ Ba_3\ (PO_4)_2 \quad (3)$$

The -OH content of coal cannot be quantified from the amount of acetic acid liberated in reaction (1) because the results are not reproducible. (6)

This approach, however, should be applicable to coal liquids. To check this point and also to determine the accuracy of the method, a synthetic mixture was prepared (Table IV). A more complex or concentrated mixture could not be prepared due to limited solubilities of many compounds in tetralin. Tetralin was used as an inert solvent to insure homogeneity in sampling.

The synthetic mixture was analyzed repetitively, obtaining the following results:

$$\text{Wt \% } O_{OH} = 1.02,\ 1.01,\ 1.06,\ 0.99,\ 0.75,\ 0.82,\ 1.05$$

With the exception of two values, these results were reproducible but high, since the expected concentration of hydroxylic oxygen was 0.82%. An explanation was not obvious, and it was decided to check the acetylation of individual oxygen-containing compounds.

The list of compounds examined and the analytical results are shown in Table V. The following is a discussion of results obtained with these compounds.

NAPHTHOL, 2,3,5,-TRIMETHYLPHENOL and 2,6,-DIMETHYLPHENOL were acetylated quantitatively, even though 2,6-dimethylphenol is hindered. Phenol itself was not checked, but in view of the above results, it should not present difficulties.

DECAHYDRO-2-NAPHTHOL could not be completely acetylated in the first set of experiments, but it was quantitatively acetylated by doubling the concentration of the acetylating agent (second set of data, Table V).

TRIPHENYLMETHANOL could not be completely acetylated with any of the conditions attempted. Increasing the concentration of the acetylating agent and tripling the reaction times (with respect to the other determinations) did not prove to be helpful.

PYROGALLOL appeared to be sensitive to the concentration of the acetylating agent, probably because it contains three OH groups. The first two acetylations were incomplete as indicated by the data in Table V. The products of these reactions were soluble in water but could not be titrated with an indicator. In fact, pyrogallol is very soluble in water and gives it a yellowish color that masks the color of the indicator. The acetate derivative is, instead, a white, water-insoluble material, and this makes the colorimetric titration very easy. Correct values were determined by increasing reactant concentration.

BENZYL ETHER could not be acetylated, as expected. However, the HEXADECYL ETHER was partially acetylated, indicating that the aliphatic ether bond is cleaved, thereby forming an alcohol. This alcohol can then be acetylated.

BENZOIC ACID cannot be acetylated. Its concentration was determined by direct titration in aqueous solution.

Carbonyl compounds such as BENZOPHENONE (I) could not be acetylated, while ANTHRONE (II) was quantitatively acetylated. This behavior is believed to be due to the fact that I is a

TABLE IV

TEST MIXTURE OF OXYGEN COMPOUNDS

COMPOUND	OXYGEN FUNCTION	WT % COMPOUND IN MIXTURE	WT % OXYGEN IN MIXTURE
ANTHRONE	CARBONYL	0.66	0.05
PHENOL	HYDROXYL	1.89	0.32
DECAHYDRO-2-NAPHTHOL	HYDROXYL	1.29	0.14
TRIPHENYLMETHANOL	HYDROXYL	0.87	0.05
NAPHTHOL	HYDROXYL	1.24	0.14
2,6-DIMETHYLPHENOL	HYDROXYL	1.23	0.17
BENZYL ETHER	ETHER	2.24	0.18
N-HEXADECYL ETHER	ETHER	0.76	0.03
BENZOIC ACID	CARBOXYLIC	0.54	0.14
TETRALIN	NONE	89.27	0.00
		100.00	1.22

TOTAL CARBONYLIC OXYGEN, WT %	0.05
TOTAL CARBOXYLIC OXYGEN, WT %	0.14
TOTAL ETHER OXYGEN, WT %	0.21
TOTAL HYDROXYL OXYGEN, WT %	0.82
	1.22

TABLE V

TESTING OF MODEL COMPOUNDS BY ACETYLATION FOR -OH FUNCTIONALITY

COMPOUND	OXYGEN FUNCTION	OXYGEN CONTENT	WT % O DETERMINED AS -OH OXYGEN
2,6-DIMETHYLPHENOL	-OH	13.1	14.2, 13.7, 12.7, 12.9
NAPHTHOL	-OH	11.1	10.7, 10.5
DECAHYDRO-2-NAPHTHOL	-OH	10.4	7.8, 5.7, 6.9, 10.4*, 10.4*, 9.8*
TRIPHENYLMETHANOL	-OH	6.2	2.1, 4.9, 3.7, 2.5, 3.9
2,3,5-TRIMETHYLPHENOL	-OH	11.7	11.9, 11.5
PYROGALLOL	-OH	38.1	3.9, 10.3, 38.5*, 36.1*
BENZYL ETHER	-O-	8.1	0.9, 0.5, 0.4
N-HEXADECYL ETHER	-O-	3.4	2.7, 1.3
BENZOIC ACID	-COOH	26.2	26.2, 24.2 (DIRECT TITRATION)
ANTHRONE	=C=O	8.2	8.3, 9.6, 9.2, 9.0
BENZOPHENONE	=C=O	8.8	1.7, 0.4, 0.6
COAL LIQUID F11-16	-	1.98	1.6, 1.5, 1.6

*THE ANALYTICAL METHODS WERE MODIFIED FOR THESE DETERMINATIONS, SEE TEXT.

stable ketone, while II is tautomerized to the anthrol which can be acetylated.

From the above results, it is evident that most -OH compounds believed to be present in coal liquids can be acetylated and this type of oxygen-containing group determined. However, some of the OH functionalities can be "hindered" with a resulting low value detected. In contrast, some compounds such as anthrone (containing a carbonyl oxygen) have the potential to show up as containing -OH functionality. In summary, several problems exist in the exact determination of -OH functionality. However, the results should fall in a reasonable range considering the complexity of coal liquids.

Determination of Carboxylic Groups. The carboxylic groups in coal and coal liquids are determined by an ion exchange method using calcium acetate. (6) The method consists of reacting the sample with excess calcium acetate at room temperature for 18 to 20 hours, and then determining the acetic acid liberated by titration with standard NaOH solution.

$$RCOOH + 1/2\ Ca(COOCCH_3)_2 \longrightarrow RCOO(1/2\ Ca) + CH_3COOH$$

Determination of Carbonylic Groups. This method takes advantage of the reaction of carbonyl groups with hydroxylamine to form the corresponding oxime. (6)

$$R_2\text{-}C{=}O + OHH_2N \cdot HCl \longrightarrow R_2\text{-}C{=}NOH + H_2O + HCl$$

The products are recovered and an aliquot is used to determine the nitrogen content. Another aliquot is reacted with acetone and sulfuric acid to decompose the oxime, and the nitrogen content of the new products is determined. The carbonyl content of the coal can be calculated from the difference in nitrogen content between the oxime derivatives and the decomposed oximes. Blom (6) found that the determination of the carbonyl content from the HCl released in the above reaction was not reproducible with coal.

The carbonyl content of coal liquids, however, can be determined from the HCl released in the above reaction. To test the method, a solution of two carbonyl compounds (anthrone and benzophenone) was prepared and analyzed. Duplicate oxygen determinations were 0.63 and 0.65 wt %, while the actual content was 0.64 wt %.

Determination of Ether Groups. Ether groups in coal are determined by reaction of coal with HI to convert the ethers into the corresponding hydroxyls and halides. (7) The hydroxyls are determined by the acetylation method, and the difference in hydroxyl content between the HI-treated and the untreated sample represents the amount of ether groups. This method is commonly referred to as the Zeisel method, and it is generalized by the following reactions:

$$Ar - O - CH_3 \xrightarrow{HI} Ar - OH + CH_3I$$

$$AR - O - AR \xrightarrow{HI} AR - OH + AR - I \xrightarrow{HI} AR - OH + AR - H$$

$$AR - O - CH_2 - AR \xrightarrow{HI} AR - OH + Ar - CH_2 - I$$

A direct method for determining ether oxygen in coal liquid is not available. A minimum amount of ether oxygen can be determined by subjecting the oils to a SARA separation. (8) This separation concentrates all of the neutral ethers in the neutral oil fraction. The oxygen content of the neutral oils is a measure of the ether oxygen present. Neutral compounds containing carbonyl groups may also appear in this fraction, and a correction may be necessary. The ether oxygen obtained in this way is a minimum because compounds which contain both an ether linkage and a polar group will be isolated with the resins fraction and not with the neutral oils. We have made no effort to determine the ether oxygen in coal liquids.

Production of the Coal Liquids

Various experiments have been carried out to produce coal liquids at different reaction conditions.

The details of this part of the experimentation are beyond the scope of this paper and can be found in Reference 9. Only a generalized flow diagram is given below to indicate what are the feeds and the products.

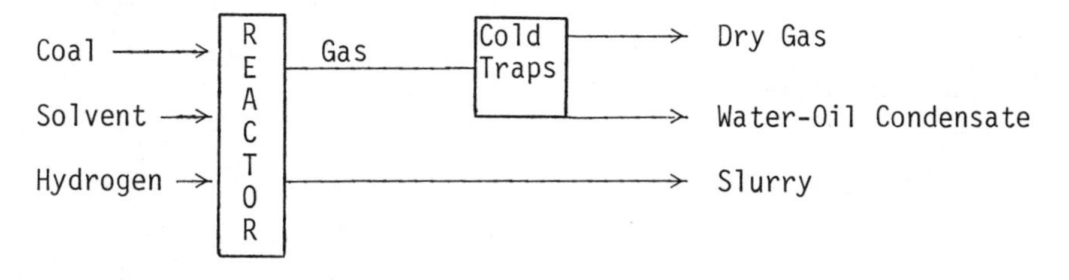

Analysis of Feed Streams

Coal. The ultimate and proximate analyses of Belle Ayr subbituminous coal, obtained by standard ASTM methods, (1) are shown in Table VI.

The analysis of the oxygen functional groups, obtained by the methods outlined above, is shown in Table VII.

These two tables also contain the analyses of the Burning Star (Illinois No. 6) bituminous coal which will be used in the next phase of the project.

TABLE VI

ULTIMATE AND PROXIMATE ANALYSES OF COALS

COAL RANK	BELLE AYR SUBBITUMINOUS	BURNING STAR BITUMINOUS
ULTIMATE ANALYSIS (WT % DRY BASIS)		
CARBON	69.28	71.55
HYDROGEN	4.34	4.46
NITROGEN	1.01	1.23
OXYGEN, DETERMINED	19.93	12.57
DIFFERENCE	14.52	5.85
SULFUR	0.51	3.64
ASH, TOTAL	10.34	13.27
METALS	4.93	6.55
PROXIMATE ANALYSIS (AS IS)		
% MOISTURE	30.25	13.45
% ASH	7.01	11.19
% VOLATILE	28.26	31.68
% FIXED CARBON	34.48	43.68
HEAT OF COMBUSTION (BTU/LB)		
AS IS	8,141	10,711
DRY	11,671	12,375

TABLE VII

ANALYSES OF OXYGEN FUNCTIONALITY IN COALS

(WT % MAF COAL BASIS)

COAL	BURNING STAR (BITUMINOUS)	BELLE AYR (SUBBITUMINOUS)
OXYGEN CONTENT AS:		
HYDROXYLIC (-OH)	2.4	5.6
CARBOXYLIC (-COOH)	0.7	4.4
CARBONYLIC (=C=O)	0.4	1.0
ETHERIC (-O-)	2.8	0.9
TOTAL	6.3	11.9
OXYGEN BY DIFFERENCE:		
ASH BASIS	5.9	16.2
MINERAL MATTER BASIS	--	16.0

Solvent. To avoid the problem of distinguishing the oxygen functionalities of the solvent from those of the coal-derived liquids, the experiments were carried out using hydrophenanthrenes as solvent. The solvent's analysis is shown in Table VIII. This solvent is essentially free of oxygen and is an effective coal liquefaction agent. (10)

Analysis of Product Streams

Analysis of Cold Trap Materials. The materials collected in the wet ice and dry ice traps were quantitatively combined, an organic solvent was added to take up the water-insoluble organic materials, and the two layers were separated by filtration through a phase separating filter paper. The hydrocarbons in the organic phase were quantified by GLC. The phenols in the aqueous phase were initially quantified by GLC techniques; (10) however, the quantities were always very small, and they are no longer being determined.

Analysis of Gases. The gas stream was analyzed by an on-line GLC unit to quantify the content of the hydrocarbon gases and carbon oxides.

Analysis of the Slurry. The slurry stream containing the liquid products, the unreacted coal, and the coal mineral matter was divided into "filtrate" and "wet cake" by filtration. These two fractions were subjected to the solvent extraction separation scheme illustrated in Figure 1. From this, the various components were calculated:

oils = samples - pentane insolubles
asphaltenes = pentane insolubles - benzene insolubles
preasphaltenes = benzene insolubles - pyridine insolubles
pyridine insolubles = as determined.

A sequential extraction technique was not used because experimentation has shown that the distribution of these components varies depending on whether the pentane to pyridine or the reverse sequence is used. It is our opinion that the scheme illustrated in Figure 1 minimizes the solvent-solute interactions and gives the best estimate of these components.

The total "filtrate" and "wet cake" and selected solvent extraction fractions were analyzed for oxygen functional groups using the methods described above.

The normalized material balances of three selected runs are shown in Table IX, together with pertinent run conditions.

The oxygen and oxygen functional groups distributions are shown in Table X. This table reports the total oxygen and the oxygen in the hydroxylic, carboxylic, carbonylic, and etheric functionalities, as weight percent of the total products. It is

TABLE VIII

ANALYSIS OF HYDROPHENANTHRENE SOLVENT

ELEMENTAL ANALYSIS	WT %
CARBON	91.08
HYDROGEN	8.83
NITROGEN	0.006
OXYGEN	-
SULFUR	0.08
GAS CHROMATOGRAPHIC ANALYSIS	
HYDROPHENANTHRENES	
PER-	9.8
OCTA-	24.5
HEXA-	2.8
TETRA-	18.3
DI-	6.2
PHENANTHRENE	14.9
UNKNOWNS	23.5

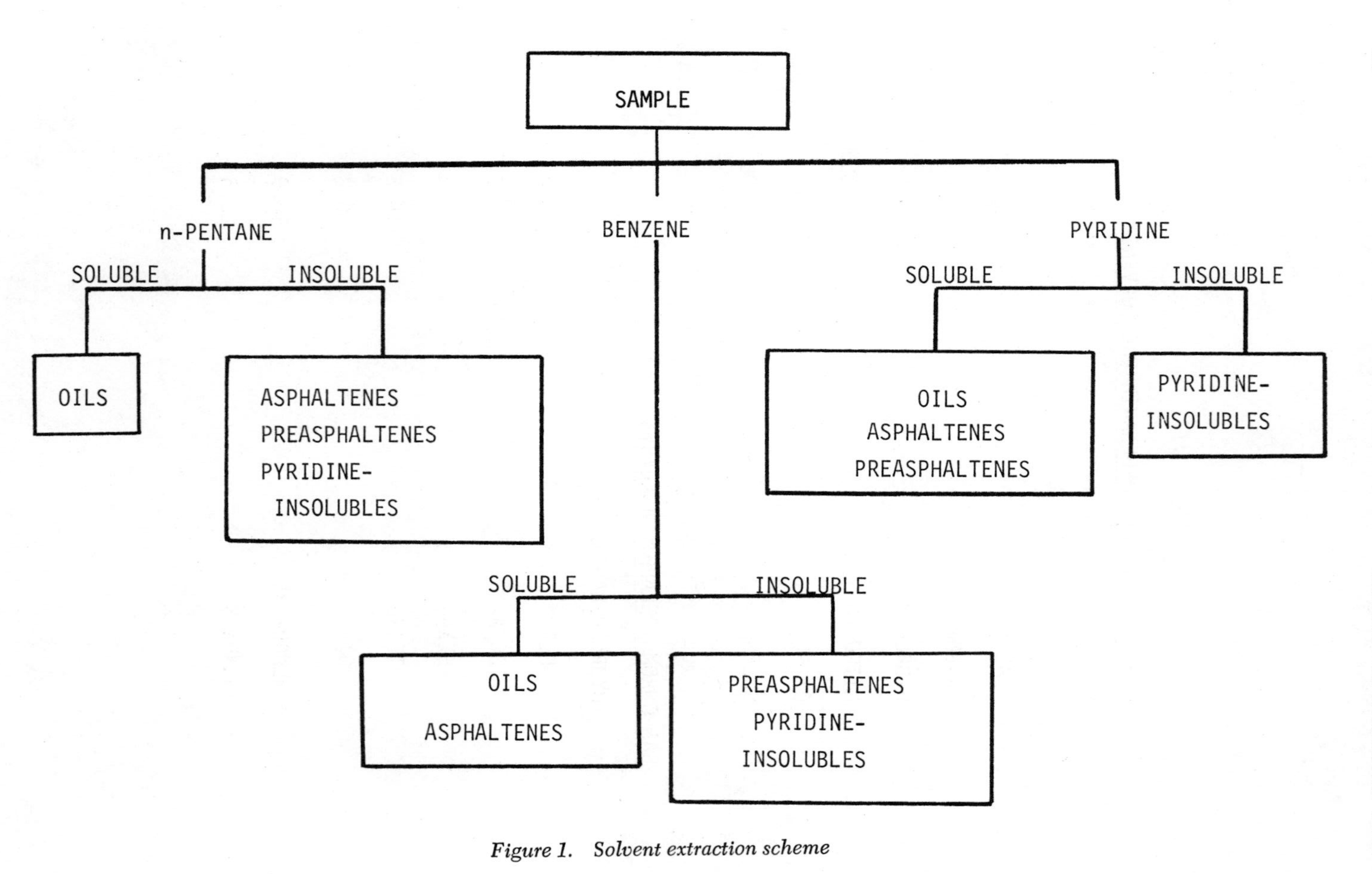

Figure 1. Solvent extraction scheme

TABLE IX

NORMALIZED MATERIAL BALANCES

RUN NO.	F11-32	EP-2	EP-18
CONDITIONS			
TEMPERATURE, °C	400.0	450.0	400.0
PRESSURE, MPa	13.79	13.79	13.79
SPACE TIME, MIN.	12.49	12.64	23.85
FEED			
COAL, MAF (BELLE AYR)	26.43	26.37	26.34
ASH	2.83	2.79	2.82
MOISTURE	10.82	10.79	10.79
SOLVENT (HPh)	60.10	59.97	59.93
HYDROGEN CONS.	-0.18	0.05	0.12
TOTAL	100.00	100.00	100.00
PRODUCTS			
GAS	1.87	4.07	2.08
OILS	62.63	70.85	68.64
ASPHALTENES	3.23	4.24	3.48
PREASPHALTENES	5.71	2.93	2.85
PYRIDINE INSOL.	14.11	5.48	10.34
WATER	12.45	12.43	12.60
TOTAL	100.00	100.00	100.00

TABLE X

FRACTION AND OXYGEN DISTRIBUTIONS
(WT % OF PRODUCTS)

	F11-32	EP-2	EP-18
FILTRATE			
TOTAL	28.15	54.46	55.06
HYDROXYLIC OXYGEN	0.25	0.48	0.34
CARBOXYLIC OXYGEN	0.02	0.01	0.0
CARBONYLIC OXYGEN	0.01	0.05	0.03
ETHERIC OXYGEN	0.03	0.14	0.32
TOTAL OXYGEN	0.31	0.69	0.69
ASPHALTENES	3.23	1.85	1.54
HYDROXYLIC OXYGEN		0.12	
CARBOXYLIC OXYGEN		0.0	
CARBONYLIC OXYGEN		< 0.01	
TOTAL OXYGEN		0.12	
WET CAKE			
TOTAL	61.61	31.88	33.58
HYDROXYLIC OXYGEN	2.65	1.05	2.89
CARBOXYLIC OXYGEN	0.01	0.0	0.0
CARBONYLIC OXYGEN	0.0	0.0	0.06
TOTAL OXYGEN	5.02	3.65	4.56
n-C_5 INSOL.	39.57	10.20	15.14
HYDROXYLIC OXYGEN	6.17	1.07	1.97
CARBOXYLIC OXYGEN	0.01	0.0	0.0
CARBONYLIC OXYGEN	0.13	0.0	0.05
TOTAL OXYGEN	5.56	1.14	2.06

TABLE X (cont'd)

PY. INSOL.	14.11	5.42	10.34
HYDROXYLIC OXYGEN	1.57	0.66	1.03
CARBOXYLIC OXYGEN	0.0	0.0	0.0
CARBONYLIC OXYGEN	0.01	0.0	0.04
ETHERIC OXYGEN	0.08	0.0	0.09
TOTAL OXYGEN	2.20	0.68	1.80
WATER	4.08	2.84	3.32
OXYGEN	3.63	2.52	2.95
GAS	1.87	4.07	2.08
OXYGEN	0.65	1.41	0.99
WATER (TRAPS)	8.37	9.59	9.28
OXYGEN	7.44	8.52	8.25

TABLE XI

OXYGEN BALANCE

(WT %)

	Feed	F11-32	DCEP-2	DCEP-18
HYDROXYLIC OXYGEN				
COAL	1.48			
SOLVENT	-			
TOTAL	1.48	2.10	1.53	3.23
CARBOXYLIC OXYGEN				
COAL	1.16			
SOLVENT	-			
TOTAL	1.16	0.03	0.01	0.0
CARBONYLIC OXYGEN				
COAL	0.26			
SOLVENT	-			
TOTAL	0.26	0.01	0.05	0.9
ETHERIC OXYGEN				
COAL	0.24			
SOLVENT	-			
TOTAL	0.24	0.11	0.14	0.41
WATER OXYGEN	9.59	11.07	11.04	11.20
ASH OXYGEN	1.48	-	-	-
GAS OXYGEN	-	0.65	1.41	0.99
UNKNOWN OXYGEN	1.08	2.36	2.60	1.69
TOTAL OXYGEN	15.29	17.41	17.33	17.61
% FEED	-	113.9	113.3	115.2

noted that the etheric oxygen in the total filtrate was obtained by subtracting the hydroxylic + carboxylic + carbonylic oxygen from the total oxygen, while that in the "pyridine insolubles" was determined by the same method used for coal.

Table XI summarizes the various forms of oxygen according to the fractions where they appear and compares the results with the feed. These results indicate that carboxylic, carbonylic, and etheric oxygen functionalities of coal are reduced during the liquefaction reaction resulting in an increase in hydroxylic oxygen functionality and in the production of carbon oxides and water.

Conclusions

Analytical techniques for determining the content of various oxygen functional groups in coals have been confirmed and have been modified to quantify the same groups in coal liquids produced from Belle Ayr subbituminous coal. It has been shown that carboxylic, carbonylic, and etheric oxygen functional groups of coal are reduced during liquefaction, resulting in carbon oxides, water, and an increase of hydroxylic groups.

Abstract

This on-going work was undertaken to develop an understanding of the reactions involving oxygen and oxygen functional groups during the production of synthetic fuels from coal by hydrogenative liquefaction. Analytical procedures have been adopted to determine the major oxygen-containing groups in coal and coal liquids, namely, hydroxylic, carboxylic, carbonylic, and etheric. The changes of concentration of these groups have been observed during the liquefaction of Belle Ayr subbituminous coal in hydrophenanthrene solvent. It has been found that carboxylic, carbonylic, and etheric oxygens are rapidly removed, resulting in the production of carbon oxides, water, and hydroxylic groups.

Acknowledgment

The authors wish to express their gratitude to the Electric Power Research Institute for partial support of this work; to K. A. Kueser for preparing the samples; and to W. C. Rovesti, D. M. Jewell, R. G. Goldthwait, and A. Bruce King for helpful discussions.

Literature Cited

1. 1975 Annual Book of ASTM Standards, Part 26.
2. Aluise, V. A., Hall, R. T., Staats, F. C., and Becker, W. W., Anal. Chem. (1947), 19, 347.

3. Stallwood, R. A., Mott, W. E., and Fanale, D. T., Anal. Chem. (1963), 35, 6.
4. Given, P. H., Cronauer, D. C., Spackman, W., Lovell, H. L., Davis, A., and Biswas, B., Fuel, London (1975), 54, 40.
5. Van Krevelen, D. W., "Coal," Elsevier Publishing Company, New York, 1961.
6. Blom, L., Edelhausen, L., and Van Krevelen, D. W., Fuel, London, (1957), 36, 135.
7. Bhaumik, J. N., Mukherjee, A. K., Mukherjee, P. N., and Lahiri, A., Fuel, London (1962), 41, 443.
8. Ruberto, R. G., and Jewell, D. M., Proceedings of the NSF Workshop, "Analytical Needs of the Future as Applied to Coal Liquefaction," Greenup, KY, 1974.
9. Cronauer, D. C., and Ruberto, R. G., Annual Report to EPRI on Research Project 713-1, January 1977.
10. Ruberto, R. G., Cronauer, D. C., Jewell, D. M., and Seshadri, K. S., Fuel, London (1977), 56, 25.
11. Brady, R. F., Jr., and Pettitt, B. C., J. of Chromat. (1974) 93, 375.

RECEIVED February 10, 1978

4

Asphaltenes and Preasphaltenes—Components of an Original hvb Bituminous Coal

ANNA MARZEC, DANUTA BODZEK, and TERESA KRZYZANOWSKA

Petroleum and Coal Chemistry Department, Polish Academy of Sciences, 44-100 Gliwice, 1-go Maja 62 St., Poland

Asphaltenes are considered to be the principle intermediates in the conversion of coal to oil products. Weller (1) stated that the catalytic conversion of coal involves two consecutive first order reactions:

$$\text{Coal} \xrightarrow{k_1} \text{Asphaltenes} \xrightarrow{k_2} \text{Oil} \qquad (1)$$

At 400°C, k_1 was reported to be 27 times higher than k_2 and at 400°C k_1 was 10 times higher than k_2.

Liebenberg and Potgleter (2) derived another mechanism which includes the following reactions:

$$\text{Coal} \xrightarrow{k_1} \text{Asphaltenes} \xrightarrow{k_2} \text{Oil}$$

$$\text{Coal} \xrightarrow{k_3} \text{Asphaltenes} \qquad (2)$$

$$\text{Coal} \xrightarrow{k_4} \text{Oil} \qquad (3)$$

They tentatively determined the sum $k_1 + k_2$ at 400°C and 440°C in the conversion of coal with tetralin and no added catalyst.

More recently Yoshida et al.(3) established that the mechanism for catalytic conversion of Hokkaido coal includes Reaction 1 and Reaction 3. It is worth emphasizing that according to their experiments, k_4 is considerably higher or lower than k_1 and k_2, depending upon the type of coal.

Sternberg et al. (4,5) have proposed that preasphaltenes are the intermediates between coal and asphaltenes. Contrary to this, Schwager and Yen (6) considered that "preasphaltenes may arise from reactive coal depolymerization moieties, which are... repolymerized into materials more difficult to degrade than the original coal substance".

Collins et al. (7) claim that "carbon-carbon scission must be considered as an important factor in asphaltene formation" and in

0-8412-0427-6/78/47-071-071$05.00/0

preasphaltene formation as well. The statement is based on results from the thermal treatment (over 300°C) of model compounds-arylalkanes, diphenylalkanes and aryl alkyl ethers. It leads to the conclusion that asphaltenes and preasphaltenes are not expected to be present in any product if it had not previously been treated at high temperature.

Schweighardt and Sharkey (8) presented in 1977 the following scheme for liquefaction:

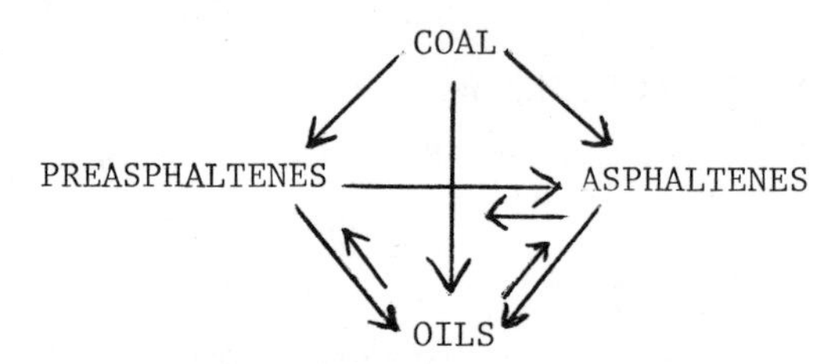

The scheme includes almost all previous concepts (1,2,3,4,5), introduces a new reaction (preasphaltenes → oils), and emphasizes the reversible character of k_2, k_6, and K_7 reactions.

We have found that asphaltenes and preasphaltenes are the components of low temperature coal extracts. Coals were extracted at ambient temperature and precautions in the analytical procedures were observed. Therefore, the components of the extracts are considered to be the components of an original coal.

EXPERIMENTAL

High volatile A bituminous coal J (vitrinite 60%, inertinite 33%, exinite 6%) and hvcb coal W (vitrinite 45%, inertinite 45%, exinite 10%) have been investigated. Proximate and ultimate analyses are presented (Table 1). Coal samples having a particle size below 1,4 mm were used.

Table 1. Proximate and ultimate analyses of the coals[a)], yields and ultimate analyses of the extracts.

	Coal J	BE extract from Coal J	DMF extract from Coal J	Coal W	BE extract from Coal W
M^{ad}	4,34	-	-	4,09	-
Ash^{ad}	4,02	-	-	9,00	-
V^{daf}	39,05	-	-	37,40	-
C	82,34	82,43	81,07	80,67	82,65
H	5,40	7,45	7,04	5,11	7,08
N	1,64	1,00	1,41	1,40	1,15
S	0,57	0,64	0,84	0,87	1,55
O[b/]	10,05	8,48	9,64	11,95	7,57
Yield[c/]	-	4,51	8,76	-	4,47

a) Ultimate analysis in wt % of daf coal,
b) by difference, c) yields in wt % of daf coal,

Extraction The samples were successively extracted by benzene-ethanol 7:3 vv mixture and by dimethylformamide.

Benzene-ethanol extraction was performed in a Soxhlet apparatus for 150 hours. Yields and ultimate analysis of the extracts are given in (Table 1). Although the yields are low, they are considerably higher than the separate benzene extract and ethanol extract yields (ca. 0.1 wt %).

Temperature curriculum of the extracted compounds: after extraction at ambient temperature the extract solution was exposed only to temperatures below 100°C (the bottom flask of the water bath heated Soxhlet) for 18 hours. After each 18-hour period, the solution was removed and fresh solvent used. The total extract solution was carefully filtered and the solvents were evaporated from it in a rotary apparatus at 50°C and reduced pressure.

Dimethylformide extraction of the residue (i.e. the benzene-ethanol insolubles) was carried out at ambient temperature by mechanical agitation, filtration, and addition of fresh DMF. This was repeated as many as 25 times. Yields and ultimate analysis of the DMF extract are given (Table 1).

Temperature curriculum of the extracted compounds: the total extract solution was freed from the solvent (DMF) in a rotary apparatus at 70°C and reduced pressure.

Fractionation of the extracts This was carried out according to a scheme (Figure 1) based on the procedure described by Schweighardt et al. (9,10) for analysis of hydrogenated coal liquids. The following group components have been isolated:
preasphaltenes i.e., benzene insolubles/pyridine sol.
asphaltenes i.e., hexane insolubles/benzene solubles
basic fraction of asphaltenes
acidic/neutral fraction of asphaltenes
benzene and hexane solubles.
Moreover, the preasphaltenes derived from BE extract have been separated into basic and acidic/neutral fractions by dry HCl treatment (Figure 1). An attempt to separate the preasphaltenes derived from DMF extract results in the separation of a significant amount of HCl adducts insoluble in DMF. However, the next steps of the procedure failed since

- free bases cannot be obtained as they are insoluble in toluene; on the other hand, toluene cannot be replaced by DMF which easily mixes with NaOH aqueous solution,
- some HCl adducts are soluble in DMF and cannot be separated from the acidic/neutral fraction.

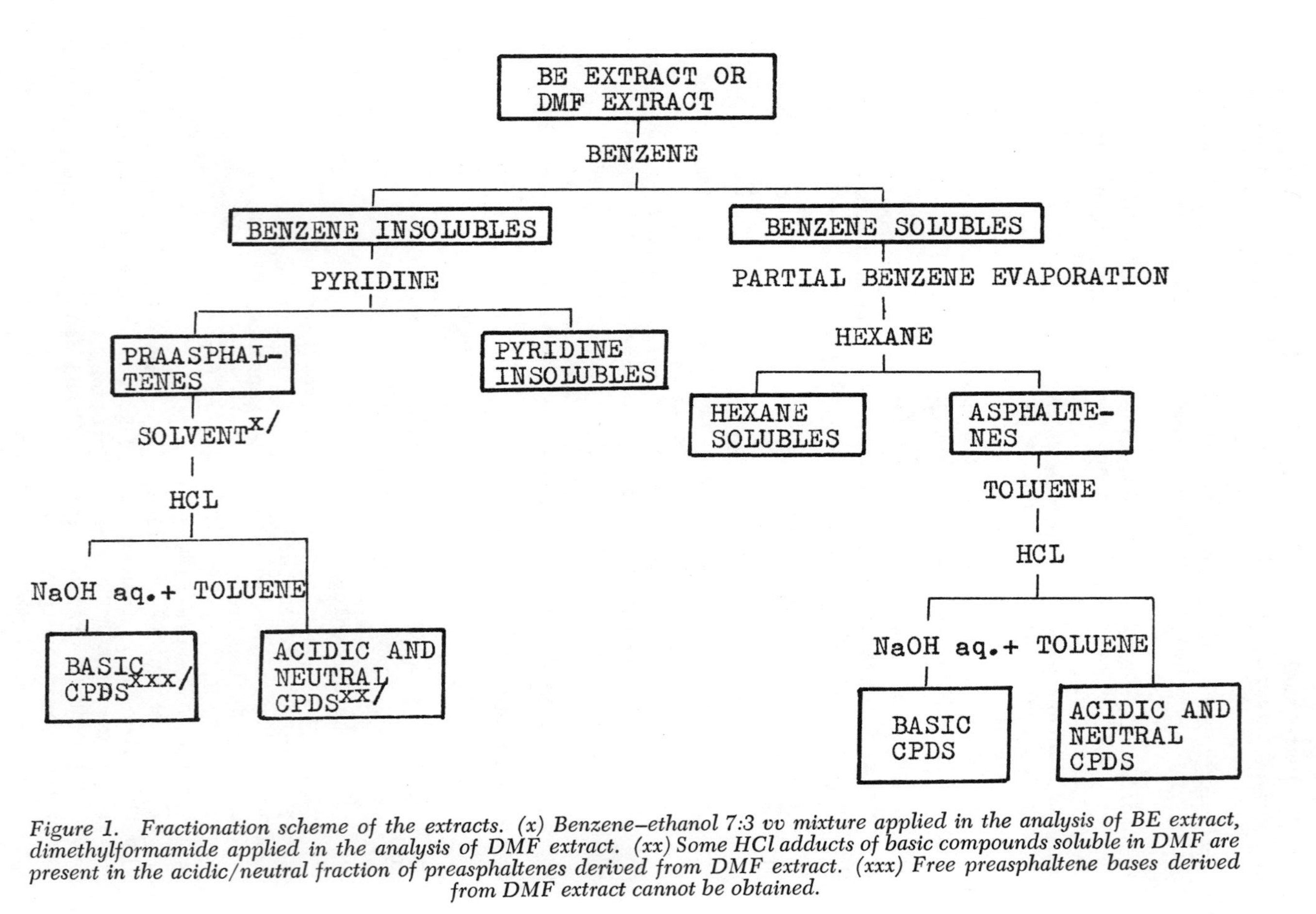

Figure 1. Fractionation scheme of the extracts. (x) Benzene–ethanol 7:3 vv mixture applied in the analysis of BE extract, dimethylformamide applied in the analysis of DMF extract. (xx) Some HCl adducts of basic compounds soluble in DMF are present in the acidic/neutral fraction of preasphaltenes derived from DMF extract. (xxx) Free preasphaltene bases derived from DMF extract cannot be obtained.

In all cases, the content of preasphaltenes was determined by the difference: benzene insolubles minus pyridine insolubles.

Temperature curriculum of the fractionated compounds: all group components were freed from the solvents in a rotary apparatus at 50°C or at 70°C (preasphaltenes) and reduced pressure except pyridine insolubles - they were dried at 110°C. Other analytical work was done at ambient temperature. Samples were stored in nitrogen.

Solubility test of isolated group components

Small samples of each group component isolated from BE extract and DMF extract were redissolved in the same solvent which had been applied in the coal extraction. All of them were easily dissolved at ambient temperature. Therefore, no indication was found of any polymerization (6) or other chemical change accompanying the fractionation procedure and causing any solubility decrease. Contrary to this, the pyridine insolubles was derived from both extracts and freed from pyridine at 110°C are almost insoluble in the BE mixture and DMF respectively.

The group composition of the extracts

The yields of the group components of the extracts are given (Figure 2, Figure 3, Figure 4). The successive extraction of the hvab coal J leads to the conclusion that at least 1 wt % of the asphaltenes and 6 wt % of the preasphaltenes on a daf basis are present in the original coal. However, the real content of asphaltenes in the coal is still not known since we do not know a solvent which is capable of extracting all of them.

TLC analysis of the extracts and their group components

The samples were developed on the TLC plates using the indicated solvent mixtures (Table 2). Neutral gel (type MN-Kieselgel HF) on 20x20 cm plates was used. Full development cycle was 16 cm high; 1/2 cycle was realized as follows: the plate was kept in the chamber until the solvent reached 1/2 height. It was then removed from the chamber, dried in nitrogen, and reinserted in the chamber.

The spray reagents used for the functional group detection are presented in Table 3. The preparation of the reagents is based on the E. Merck handbook - "Anfarbereagenzien fur Dunnschicht- und Papier-Chromatographie" (Darmstadt, 1970). The same reagents were used by us in analysis of the hydrogenated coal liquids (11).

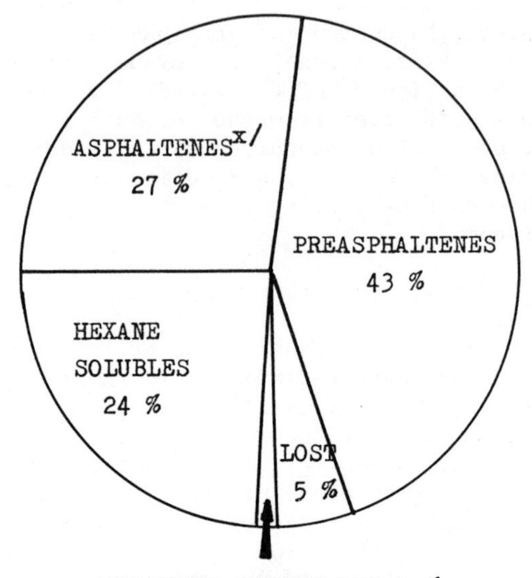

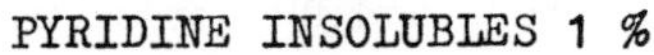

Figure 2. Benzene–ethanol extract from hvcb coal W 4.5% on daf coal. (x) The ratio of acidic/neutral cpds and bases is 3.1.

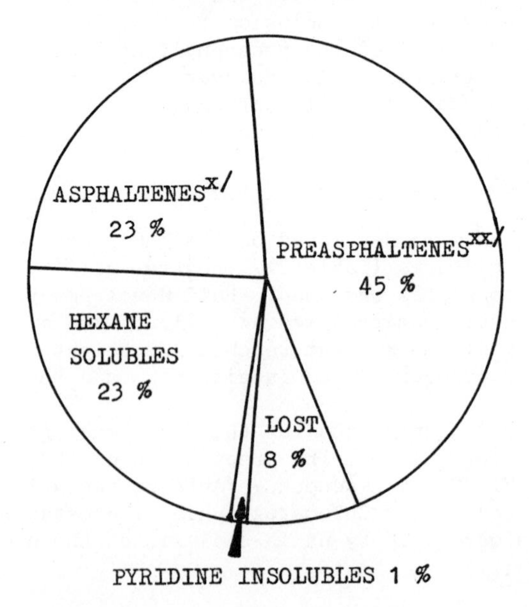

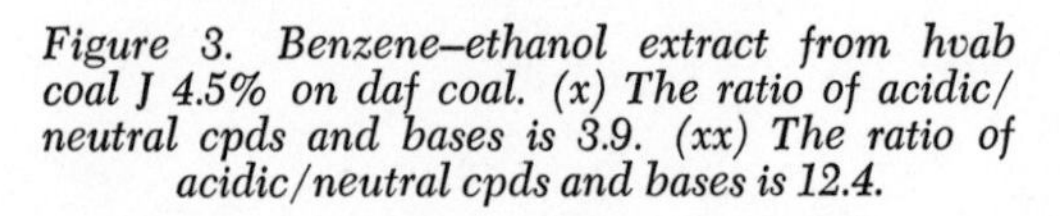

Figure 3. Benzene–ethanol extract from hvab coal J 4.5% on daf coal. (x) The ratio of acidic/neutral cpds and bases is 3.9. (xx) The ratio of acidic/neutral cpds and bases is 12.4.

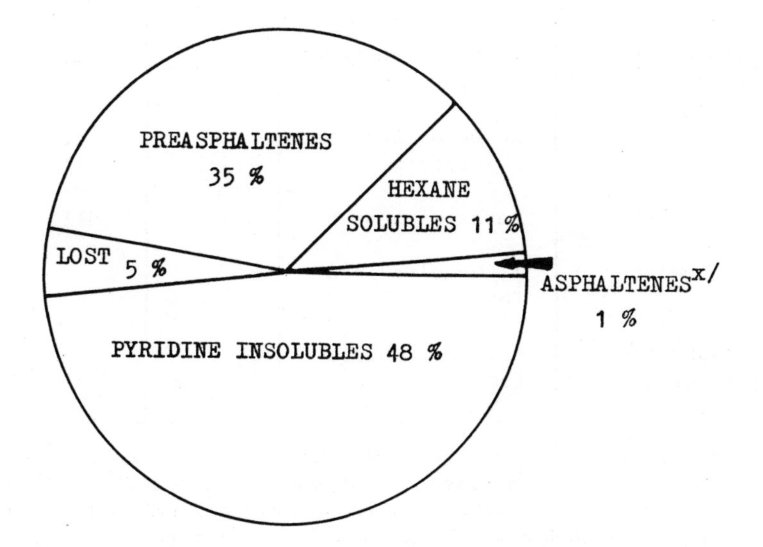

Figure 4. Dimethylformamide extract from hvab coal J 8.8% on daf coal. (x) The ratio of acidic/neutral cpds and bases is 3.4.

Table 2. Thin-Layer Chromatography Conditions

	Solvent system	Solvent ratio	Development cycle
BE,DMF extracts	methanol:benzene: chloroform	1:1:1	1/2, full
BE preasphaltenes	methanol:benzene	3:7	full
DMF preasphaltenes	methanol:benzene: chloroform	1:1:1	full
BE preasphaltene bases	methanol:benzene: chloroform	1:1:1	1/2, 3/4, full
BE preasphaltene acidic/neutral fr.	methanol:benzene: chloroform	1:1:1	1/2, 3/4, full
BE,DMF asphaltenes	methanol:benzene	3:7	full
BE,DMF asphaltene	methanol:chloroform	1:1	full
BE,DMF asphaltene acidic/neutral fr.	methanol:chloroform	1:1	1/2, full

Results of TLC analysis of the coal extract as well as its separation products are summarized in Table 4. The results indicate that asphaltenes and BE preasphaltenes derived from coal extracts contain heterocompounds which may be separated into basic and acidic/neutral fractions. Phenol and pyrrol derivatives are present in the acidic fraction, pyridine derivatives in the basic portion. Therefore, analogous results have been obtained in TLC analysis of asphaltenes extracted from coal at ambient temperature (this paper) and asphaltenes derived from the high temperature coal conversion products (5, 9, 10, 11).

Table 3. Thin-Layer Spray Reagents

	Functional group indication				Detection limit (ug)
	Ar-OH Phenol type	N-H ring Pyrrol type	=N-ring Pyridine type	$Ar-NH_2$ Amines	
Fast Blue Salt B	red-violet	brown-violet	-	yellow-orange	0,01
FeCl 3% in 0, 5n HCl	violet	dark green	-	blue	0,01
Wachmeister's Reagent	yellow-violet	purple-brown	-	brown-	0,1
Erlich's Reagent	-	violet[a]	light yellow	light yellow	0,05
Dragendorff's Reagent	-	-	orange	pink-red	0,1
Iodoplatinate	-	-	brown	beige	0,5

[a] after few hours

Table 4. Results of Spray Reagent Test

	Functional group			
	Ar-OH Phenol type	N-H ring Pyrrol type	=N-ring Pyridine type	$Ar-NH_2$[x] Amine
BE,DMF extracts	+	+	+	+
BE,DMF preasphaltenes	+	+	+[xx]	+
BE preasphaltene bases	-	-	+	+
BE preasphaltene acidic/neutral fr.	+	+	-	-
BE,DMF asphaltenes	+	+	+	+
BE,DMF asphaltene bases	-	-	+	+
BE,DMF asphaltene acidic/neutral fr.	+	+	-	-

[x] TLC-Spray Reagent indication of amine presence is not an irrefutable proof.

[xx] Meaningless test - preasphaltenes are contaminated by pyridine during separation procedure.

Mass spectrometry of some group components

Field Ionization MS. Molecular weight distributions of the samples listed (Table 5) were determined using FI MS technique. The technique offers a unique way to characterize coal derived products and to understand their nature in molecular terms (12). The spectra were obtained on a Varian MAT 711 and recorded at various probe temperatures (50°C, 100°C, 150°C, 200°C, 250°C, 300°C). The heights of each molecular peak were summed and the total intensity of all ions was calculated for each sample. The molecular weight distribution (Table 5) was calculated using FI sensitivities from Severin's paper (13).

Table 5. Molecular Weight Distribution

Sample	Molecular Range of Ions (% of total amount)							Maximum M observed
	78–186	187–300	301–400	401–500	501–600	601–700	701–800	
EB hexane solubles	4	25	35	22	12	2	-	630
DMF hexane solubles	8	29	23	23	9	6	2	790
EB asphaltene bases	16	41	29	14	-	-	-	500
DMF asphaltene bases	7	21	35	25	8	3	1	750
EB asphaltene acidic/neutral fr.	6	30	40	19	5	-	-	600
DMF asphaltene acidic/neutral fr.	10	34	26	20	9	1	-	690
EB preasphaltenes	7	27	39	22	5	-	-	590
DMF preasphaltenes	5	26	24	21	15	8	1	735

The instrument's capability for recording high molecular weight masses was checked by running the spectra of FPC calibration standards (800 and 1200 GPC polypropyleneglycols). Peaks up to 800 amu and 1200 amu respectively, have been detected.

All investigated samples were abundant in compounds within 187-500 amu range. They constituted over 70% of each sample. No molecular masses over 630 amu and over 790 amu were detected for BE and DMF derived group components respectively. Nearly all molecular peaks within the molecular ranges presented in the Table 5 were observed in the FI spectra of each sample. It means that at least a few hundred compounds are present in each sample.

It should be noticed that the total intensity of the preasphaltene and asphaltene base spectra (spectra of nitrogen concentrates) are considerably lower than the other group components. The reason is unknown, however lower sensitivities of the nitrogen compounds should be considered first.

High Resolution MS is the analytical technique that can furnish significant information on the hetero-compounds in coal liquids (14,15).

HR MS analyses of the asphaltenes derived from BE and DMF extracts were performed using Varian MAT 711 and Jeol JMS D-100 spectrometers. The spectra were recorded at a resolving power of approximately 1/8000, 70 eV ionizing voltage and at various temperatures from ambient temp. to 300°C. A direct insertation probe was used. Calculation of the precise masses and formulas were carried out on a Texas Instruments 980 B computer.

The identified formulas are listed in Tables 6 and 7. Only those formulas which had been detected by both spectrometers were considered. It is evident that the presented data do not contain all compounds occuring in the samples since more molecular masses were recorded using FI MS than are reported data (Tables 6 and 7). However, some interesting conclusions can be drawn from HR MS data. The most striking is that there are significant differences between the composition of asphaltenes derived from BE extract and from DMF extract. For instance $C_nH_{2n-23}N$ up to $C_nH_{2n-33}N$ and $C_nH_{2n-18}N_2$ up to $C_nH_{2n-26}N_2$ and $C_nH_{2n-6}O$ up to $C_nH_{2n-12}O$ formulas found in BE asphaltenes are entirely absent in DMF asphaltenes.

CONCLUSIONS

Asphaltenes and preasphaltenes are components of the original coal. Therefore some amount of these group components in coal liquefaction products is not the product of thermal or catalytic conversion.

The real content of asphaltenes and preasphaltenes in an original coal is still an unsolved problem. We have to know more about electron-donor and -acceptor properties of solvents as well as about the strength of electron donor-acceptor bonds occuring in coals in order tc select a solvent capable of extracting the total amount of these components from any original coal.

The functional groups detected in the preasphaltenes and asphaltenes (TLC and HR MS data) occuring in the original coal, have proton-donor and electron-acceptor characteristics (OH-phenol, NH-pyrrol type) and electron-donor characteristics (oxygen compounds, =N-pyridine type, possibly amines). Therefore, hydrogen bonds as well as electron donor-acceptor bonds should be considered to be responsible for binding together the extractable heterocompounds and the cross-linked macromolecules that form three-dimensional network of an original coal (16).

The molecular weights of the bulk of the extractable compounds are in the 187-500 amu range.

Table 6. HR MS Data on Nitrogen and Hydroxy-Nitrogen Compounds of Asphaltenes derived from BE and DMF Extracts.

	Carbon Number Distribution				Possible Structural Types
	Asphaltenes from BE extract		Asphaltenes from DMR extract		
	Bases	Acidic/ neutral	Bases	Acidic/ neutral	
1	2	3	4	5	6
$C_nH_{2n-z}N$					
3	-	-	7,11	8,9	Pyrroles
5	6,7,12-15	-	7	9	Anilines, Pyridines
7	8,9	8	7,9,14,15	9,12,14	Azaindanes
9	8-10, 13-15	8-12	10,13,14	10-14	Indoles, Dihydroquinolines
11	9,10, 12-14	-	14,15	12	Quinolines, Naphthylamines
13	12-15	-	14-16	9,11,15	Phenylpyridines, Tetrahydroacridines, N-Benzyl-methyl-anilines
15	11-14	-	12-15	12,13	Azafluorenes, Dihydroacridines
17	13-19	12-14	15	13	Acridines, Phsnylindoles
19	13-19	14	-	15	Azabenzo(ghi)fluorenes, Phenylquinolines
21	15-22	15,17	15	-	Azapyrenes, Benzocarbazoles
23	17-20	-	-	-	Benzacridines
25	18-23	-	-	-	Azabenzo(ghi)fluoranthenes
27	19-22	22	-	-	Azabenzopyrenes, Dibenzocarbazoles
29	21-23	19,21	-	-	Dibenzacridines
31	21-24	24	-	-	Azabenzoperylenes
33	23,24	-	-	-	Azadibenzopyrenes

Table 6, continued

1	2	3	4	5	6
			$C_nH_{2n-z}N_2$·		
8	9	9	10	-	Azaindoles, Benzimidazoles
12	-	-	13	-	Acenaphthydrines
14	-	-	11-14	12	Carbolines
16	12	11	11-12	11-14	Phenylazaindoles, Phenylbenzimidazoles
18	-	19	-	-	Phenylquinazolines
20	-	15-18	-	-	Diazapyrenes
22	19	-	-	-	Diazachrysenes
24	20	-	-	-	Diazabenzofluoranthenes
26	22	22	-	-	Diazahenzopyrenes
			$C_nH_{2n-z}ON$		
5	-	-	11	-	Aminocresoles
7	-	8	9,10	-	Nitrosobenzes, Acethylpyridines
9	-	9,11,13	-	-	Hydroxyindoles, Aminobenzofuranes
11	-	10-13	-	-	Hydroxyquinolines
15	-	13-14	12	-	Azahydroxyfluorenes
17	19	14-15	-	16,19	Azafluorenones
19	-	15-16	-	17	Pyridodibenzofuranes
21	-	17	-	-	Benzophenoxazines
23	22	-	-	-	
			$C_nH_{2n-z}ON_2$		
10	-	-	9	-	Hydroxy-Nitrogen Compounds
16	-	15	-	-	"
18	-	18	-	-	"

Table 7. HR MS Data on the Hydroxy- and Dihydroxy Compounds of Acidic/Neutral Parts of Asphaltenes derived from BE and DMF Extracts[x].

z	Carbon Number Distribution: Acidic/Neutral asphaltenes from BE extract	Carbon Number Distribution: Acidic/Neutral asphaltenes from DMF extract	Possible Structural Types
		$C_nH_{2n-z}O$	
6	8,9	-	Phenoles
8	17	-	Naphthenophenoles
10	13,14	-	Benzofuranes, Indenoles
12	12-14	-	Naphtoles, Furylbenzoles
16	12-16	18,21	Dibenzofuranes, Hydroxy-fluorenes
18	14-16	16,17	Hydroxyantracenes/hydroxy-phenantrenes
20	-	16,17	Diphenylfuranes, Naphthylphenoles
22	17-19	21	Hydroxypyrenes
24	19,23	-	Hydroxychrysenes
		$C_nH_{2n-z}O_2$	
6	-	9,10	Dihydroxybenzenes
18	15	-	Dihydroxyantracenes
22	16-20	-	Dihydroxypyrenes

[x] Hydroxy- and dihydroxy compounds have not been detected in the asphaltene bases derived from BE and DMF extracts.

ACKNOWLEDGEMENT

This research was supported by Polish Academy of Sciences's Program No. 03.10.

The authors gratefully acknowledge Prof. Dr. Wlodzimierz Kisielow who firmly encouraged us to begin research on coal organic chemistry by a solvent extraction and for his fruitful discussions.

We wish to thank also: Dr. J. Szafranek, Eng. M. Stobiecki and Eng. A. Plaziak for providing mass spectra; Mrs. J. Piecha, Miss S. Urbaniec and Mr. K. Kowalski for their technical assistance and analytical work; Dr. M. Boduszyński for providing GPC standards.

LITERATURE CITED

1. Weller S., Pelipetz M.S., Friedman S., Ind. Eng. Chem., (1951) 43, 1572 and 1975.
2. Liebenberg B.J., Potgleter H.C.J., Fuel (1973) 52, 130.
3. Yoshida R., Maekawa R., Ishii T., Takeya G., Fuel (1976) 55, 337.
4. Sternberg H.W., ACS Div. Fuel Chem., Preprints (1976) 21 (No.7) 1.
5. Sternberg H.W., Raymond R., Schweighardt F.K., ACS Div. Petroleum Chem., Preprints (1975) 20 (No. 4) 763.
6. Schwager I., Yen T.F., ACS Div. Fuel Chem., Preprints (1976) 21 (No.5) 199.
7. Collins C.J., Raaen V.F., Benjamin B.M., Kabalka G., Fuel (1977) 56, 107.
8. Schweighardt F.K., Sharkey A.G., "Coal Chemistry Workshop - Preprints", Paper No.6, Stanford Research Institute, Stanford, 1976.
9. Sternberg H.W., Raymond R., Schweighardt F.K., Science (1975) 188, 49.
10. Schweighardt F.K., Retcofsky H.L., Raymond R., ACS Div. Fuel Chem., Preprints (1976) 21 (No. 7) 27.
11. Bodzek D., Krzyzanowska T., Lotocka Z., Marzec A., Nafta (1977) 33, 96. (In Polish).
12. John G.A. St., Butrill S.E., Anbar M., ACS Div. Fuel Chem., Preprints (1977) 22 (No.5) 141.
13. Severin D., Oelert H.H., Bergman G., Erdol u. Kohle (1972) 25 (No.9) 514.
14. Sharkey A.G., Shultz J.L., Schmidt C.E., Friedel R.A. PERC/RI-77/7, Pittsburgh Energy Research Center, April 1977.
15. Aczel T., Lumpkin H.E., ACS Div. Petroleum Chem., Preprints (1977) 22 (No.3) 911.
16. Kovac J., Larsen J.W., ACS Div. Fuel Chem., Preprints (1977) 22 (no.5) 181.

RECEIVED February 10, 1978

5

Phenols as Chemical Fossils in Coals

J. BIMER
Instiute of Organic Chemistry, Polish Academy of Sciences, Warsaw, Poland

P. H. GIVEN and SWADESH RAJ
Fuel Science Section, Material Sciences Department, Pennsylvania State University, University Park, PA 16802

It is generally considered that vitrinite, the principal maceral in most coals, represents coalified, partly decayed wood. Hence lignin should be one of the important precursors to the vitrinites in coals. Accordingly, it would be interesting to know whether any chemical fossils related to lignin could be found in coals. The purpose of this paper is to report what we believe to be a successful search for such fossils. The experimental approach exploited a degradation reaction developed in a study of soil humic acids by Burges et al. (1)

This reaction involves a reductive degradation with sodium amalgam and hot water. Thin layer chromatography of the ether soluble part of the product (yield, about 20%) showed the presence of a number of phenols and phenolic acids, most of whose structures bore obvious relationships to known microbial and chemical degradation products of lignin (I) but some to the A ring of flavonoids (II).

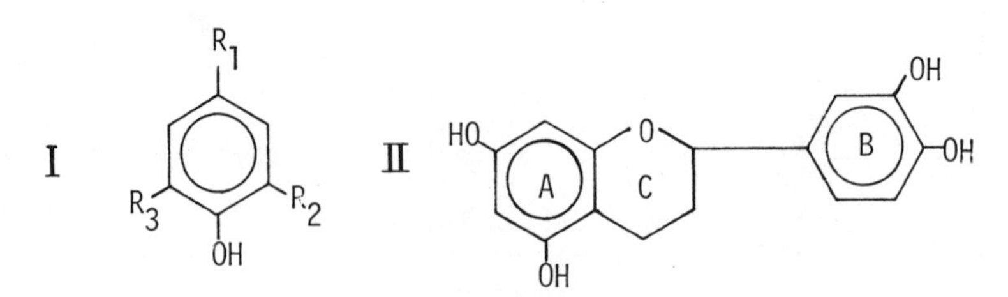

where R_1 = COOH, CHO, -CH=CH-COOH,-CH_2-CO-COOH, etc.; R_2 = H or OCH_3 or OH; R_3 = H or OCH_3 or OH. The R_1 group may represent a side chain of 1, 2 (rarely) or 3 carbon atoms, in various states of oxidation. Burges et al. therefore concluded that the humic acids they studied were condensates of phenolic compounds from the degradation of plant products with amino acids (see also Flaig) (2).

0-8412-0427-6/78/47-071-086$05.00/0

Humic acids can be extracted from peats and lignites but not from bituminous coals. However, oxidation of bituminous coals with aqueous performic acid generates in high yield (80-110% by weight) materials that closely resemble humic acids (3). In preliminary studies performed some years ago by one of us (J.B.) while on leave of absence from his Institute, the Burges reductive degradation was applied to humic acids extracted from some peats and lignites, and produced by oxidation of a number of bituminous coals. A number of identifications of products were made by gas chromatography with co-injection of standards, but at that time facilities were only rarely available to permit confirmation by mass spectrometry. Such confirmation has been more fully obtained recently by another of the authors S.R.), who also studied a wider range of coals (totalling 43 samples).

It is proposed in this preliminary publication to describe the experimental procedures and to give a sufficient selection of the data to show what was found by co-injection and later confirmed by mass spectrometry.

The major coal measures of the United States were laid down either in the Carboniferous era (ca. 300 m. years B.P.), or in the Cretaceous and Tertiary (130-60 m. years B.P.). Between these two eras a great deal of evolution occurred in the plant kingdom; associated with this were some changes in the nature of groups R_2 and R_3 in structure I above. However, having stated the fact, we need not pursue this matter here.

Experimental

Humic acids were extracted from peats and lignites with 0.5N sodium hydroxide following standard procedures.

The oxidation of coals with aqueous performic acid is highly exothermic. Five gm of coal was dispersed in 50 ml of anhydrous formic acid, and 50 ml of 30% hydrogen peroxide was added in 2 ml portions at such a rate that the temperature did not exceed 55°C (however, in the earlier phase, the coal/formic acid suspension was cooled in ice during the addition of H_2O_2). The mixture was then allowed to stand at room temperature with stirring for 24 hours. The washed and dried solid product was extracted with 1N NaOH under nitrogen and centrifuged. The extract was precipitated by acidification to pH 1. The washed and dried humic acids were redissolved in alkali and, following the procedure of Burges et al. (1), treated with 3% sodium amalgam while the solution was heated in an oil bath at 110-115° for 4-5 hours.

After removal of mercury, the resulting solution was acidified to pH 1 and centrifuged; the supernatant was carefully removed and the residue extracted twice by centrifugation with ether and twice with methylene chloride. Solvents were removed and the residues mixed. The residues were treated with Sylol HTP reagent (Supelco, Inc., Bellefonte, Pa.) under the conditions recommended for converting phenols to trimethylsilyl ethers and carboxylic acids to the corresponding esters.

Experiment showed OV 101 column packing (3% on 80/100 mesh Supelcoport) to be the most effective for gas chromatography of the six packings tested. GC analyses, with and without co-injection of standards, were performed with a Hewlett-Packard No. 5750 instrument, equipped with flame ionization detectors. GC/MS analyses in the later phase were made by Mr. David M. Hindenlang, using a Finnigan model 3000 instrument under the charge of Dr. Larry Hendry of the Chemistry Department of this University. OV 101 columns were again used. The GC/MS instrument was provided with a data system, and this was used to subtract the mass spectrum at the foot of each peak just before it began to elute, or just after it had done so, from the spectrum recorded as the maximum of the peak was eluted. Such a procedure is certainly arbitrary, but disc space could not be monopolized for continued storage of our data while other procedures were tested and interpretations worked out. Consequently, the raw MS data, massaged as described above, were reported for standards and unknowns as printouts tabulating m/e values and relative intensities, and comparisons were made by visual inspection of the printed data.

Results

On the dry basis the weight of crude oxidized coal was usually 85-105% of the weight of raw coal. The yields of daf humic acids were in the range 65-90% of dmmf coal. Within the range of rank studied (78-87%C dmmf), the yield tended to increase with increasing rank. Petrographic studies of the oxidized products of three coals (by Dr. Alan Davis and Mr. Harvey Zeiss, to whom we are indebted) showed that the vitrinites were greatly altered compared with their appearance in the raw coals, while sporinite and the inert macerals had changed little or not at all. These apparently unaltered macerals could still be recognized in the NaOH-insoluble materials. Thus the humic acids were derived very largely from the vitrinitic macerals. The yields of ether soluble products from the reductive degradation were in the range 10-40% of the weight of humic acids taken; yields when the reaction is applied to soil humic acids were about 20% (1). Yields from humic acids from the younger Western coals tended to be somewhat higher than those from Carboniferous coals of the eastern U. S.

Chromatograms obtained in the earlier phase of the work are shown in Figures 1 and 2, where the names of compounds identified, mostly by co-injection, are entered against the corresponding peaks (trivial names are used in the figures because they are usually

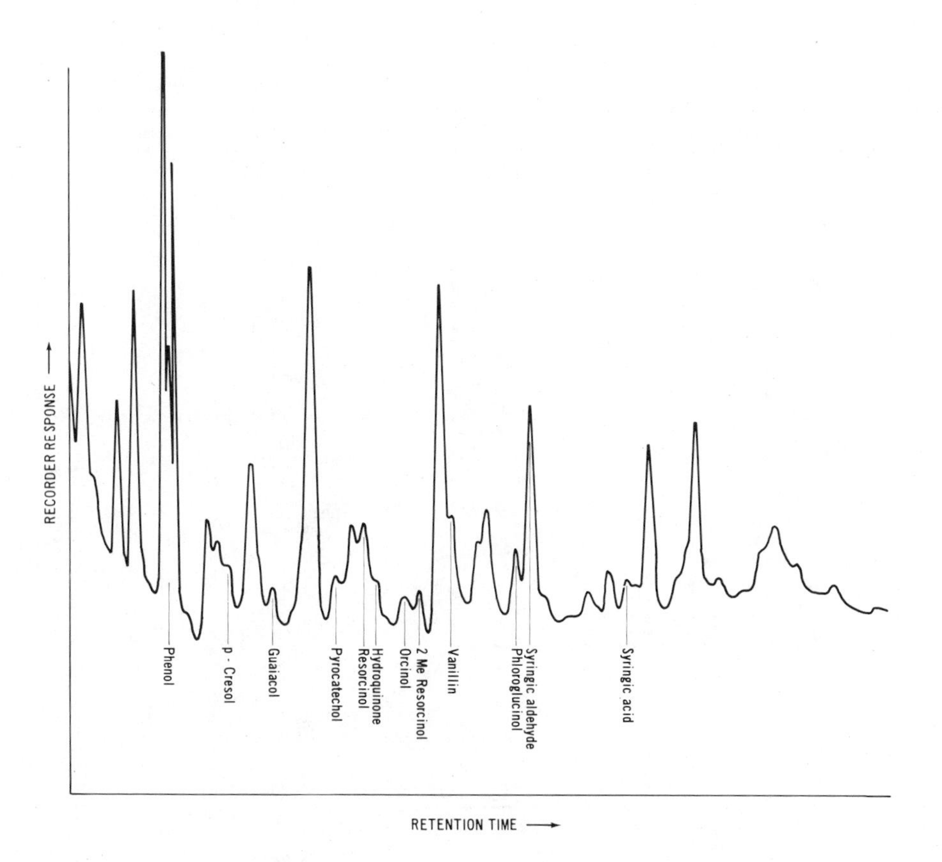

Figure 1. Chromatogram of degradation products (as TMS derivatives) from C Seam Coal, Benham, Kentucky (HVA, PSOC 13)

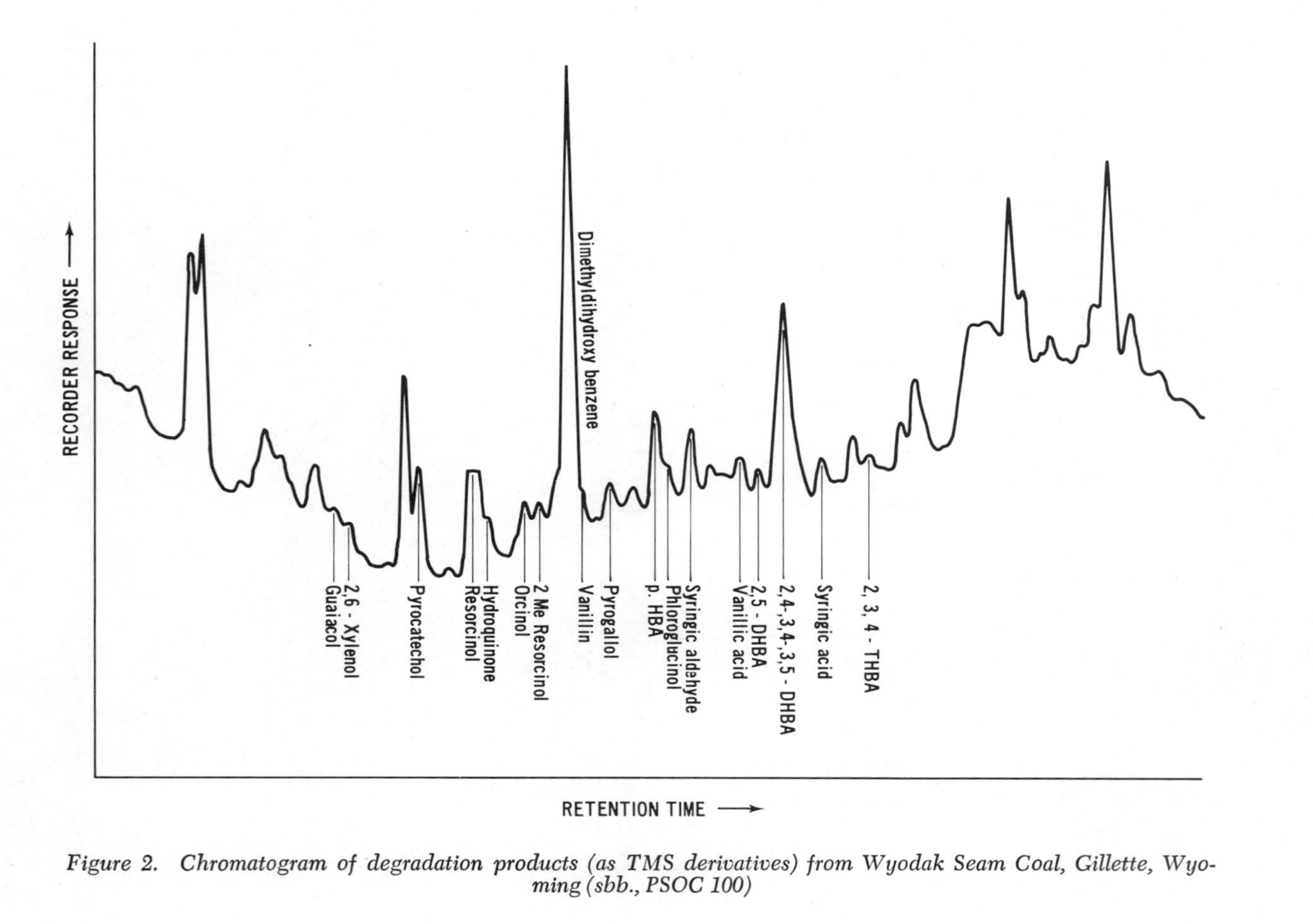

Figure 2. Chromatogram of degradation products (as TMS derivatives) from Wyodak Seam Coal, Gillette, Wyoming (sbb., PSOC 100)

shorter; a list of equivalent systematic names is given in the Appendix. Comparison of the structures with I and II above will show which may be biologically related). It will be seen that resolution is moderately good but that even so a number of major peaks are unidentified.

These curves are typical of what was found for products from the six coals studied. All 3 dihydroxy-benzenes were commonly found and 2 of the trihydroxy-benzenes. 2,6-Xylenol was frequently found in both the earlier and later work, and was the only one of six xylenols to be identified. Several of the compounds frequently encountered (vanillin and vanillic acid, syringic aldehyde and acid, p. hydroxy-benzoic acid) are well known as degradation products of lignin. A peak is seen in Figure 1 labelled 2,5-dihydroxy-benzoic acid (2,5-DHBA), and another labelled 2,4-, 3,4-, 3,5-dihydroxy-benzoic acid. Experiments with known compounds showed that the latter three isomers could not be resolved under the conditions used. Later work, using GC/MS, showed that of the three only the 3,4-isomer was in fact present, and this is lignin-related. However, the 2,5-isomer is not, though it could be derived from the A ring of flavonoids. 2,3,4-Trihydroxy-benzoic acid has no obvious biological associations, though the 3,4,5-isomer (gallic acid) occurs widely in the plant kingdom (both isomers were identified).

When, later, a wider range of samples was studied (with somewhat more severe conditions of oxidation), some very poorly resolved chromatograms were obtained; the products evidently often represent very complex mixtures of substances. Surprisingly, the products from nearly all of the younger western coals were resolved poorly or very poorly, while the resolution of those from the Interior and Eastern provinces ranged from mediocre to good. Representative examples of each type are shown in Figures 3 and 4. The resolution in Figure 3 is so bad that one might question whether any identifications are possible. In Table 1, we have assembled details of the mass spectra of four substances alleged to be identified from the GC/MS run shown in Figure 3, with details of the spectra of the standards. The agreement is surprisingly good, and the identifications are, taken with matching of retention times, reasonably secure. When resolution is better, as in Figure 4, one can surely be confident in the identifications; examples of the correspondence of mass spectral data are shown in Table 2.

It should be added that even when the GC/MS system is used, many prominent peaks remain unidentified. However, many of these showed fragments at m/e=73 in the corresponding mass spectra, so that they evidently contained the trimethylsilyl group and were therefore phenols or acids or both.

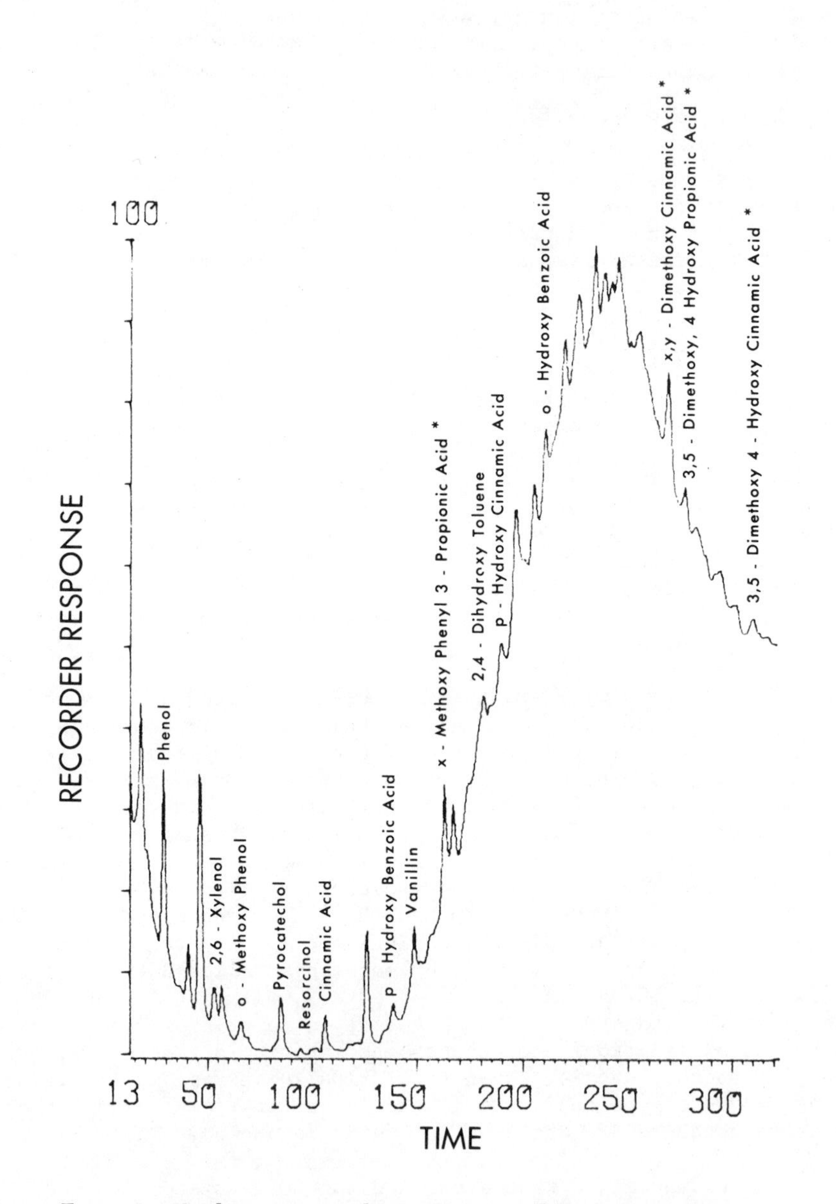

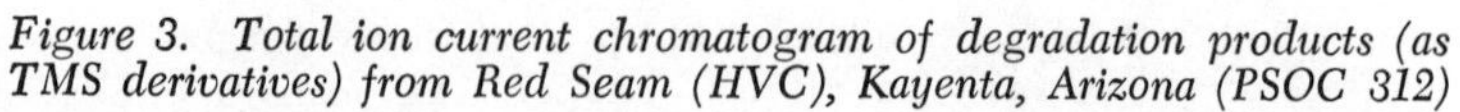

Figure 3. Total ion current chromatogram of degradation products (as TMS derivatives) from Red Seam (HVC), Kayenta, Arizona (PSOC 312)

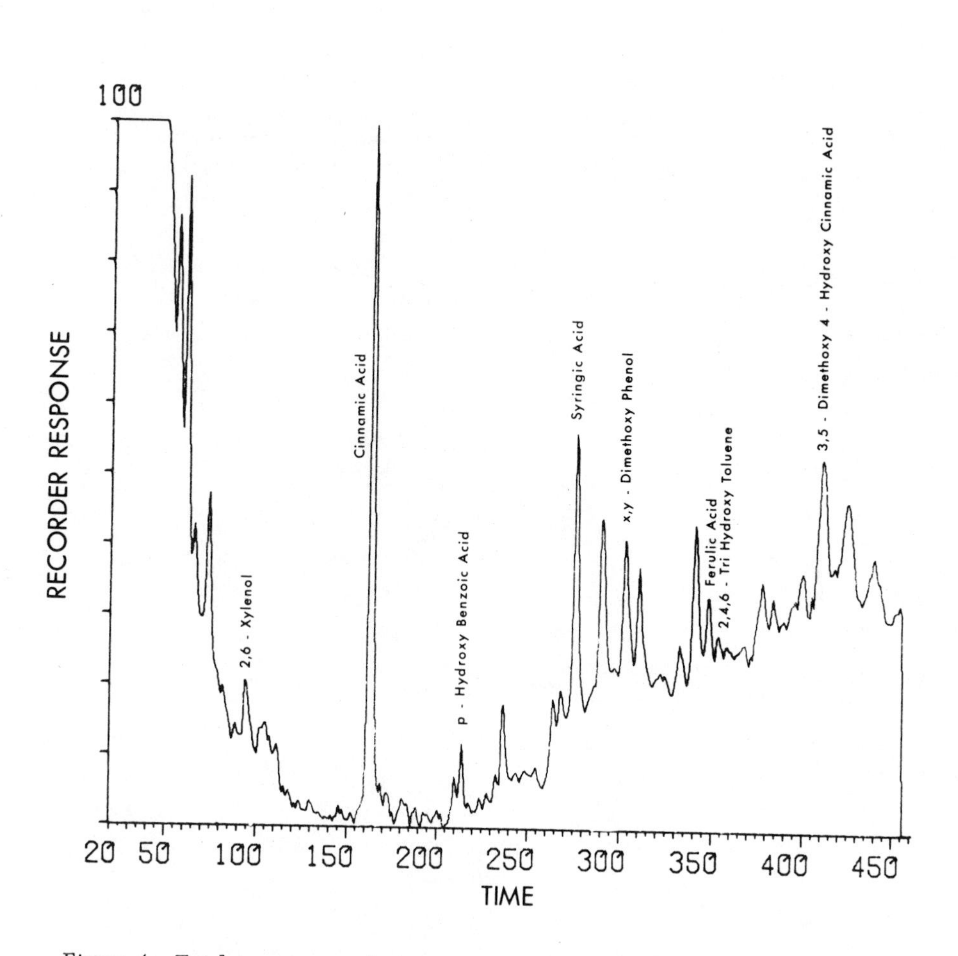

Figure 4. Total ion current chromatogram of degradation products (as TMS derivatives) from Ohio No. 1 Seam (HVC), Jackson, Ohio (PSOC 202)

Table I

Data for MS Identification of Substances in Degradation Products of Coal from Black Mesa, Arizona (HVC)(PSOC 312)

	Cinnamic Acid			p. Hydroxybenzoic Acid	
m/e	Standard	Unknown	m/e	Standard	Unknown
p/p+1	3.98	4.13	p/p+1	3.28	2.73
205	100	100	73	100	100
57	82	86	193	45	63
45	48	39	267	44	59
145	27	23	223	42	56
55	25	20	45	21	41
89	20	16	75	15	23
67	18	14	126	13	17

	Vanillin			p. Hydroxycinnamic Acid	
m/e	Standard	Unknown	m/e	Standard	Unknown
p/p+1	6.95	5.15	p/p+1	6.00	5.27
194	100	88	13	100	83
23	83	100	219	76	100
193	53	45	235	61	79
209	37	60	293	32	41
151	29	38	191	21	39
45	27	33	45	16	25
165	9	17	75	15	19

Table II

Data for MS Identification of Substances in Degradation Products of Coal from Ohio No. 1 Seam (HVC)(PSOC 202)

p. Hydroxybenzoic Acid		
m/e	Standard	Unknown
p/p+1	3.28	3.18
73	100	100
193	45	52
267	44	48
223	42	41
45	21	34
75	15	19
126	13	14

Syringic Acid		
m/e	Standard	Unknown
p/p+1	3.72	3.15
73	100	100
45	18	29
141	15	23
327	12	19
312	10	13
89	10	15
297	9	10

Ferulic Acid		
m/e	Standard	Unknown
p/p+1	2.71	2.93
73	100	100
75	48	49
117	38	35
225	24	32
181	22	30
129	22	25
297	20	31

In considering the origin of the substances shown as identified in Figures 1-4, one must assume with bituminous coals that carboxylic acid groups result largely or entirely from the performic acid oxidation. There is every reason to suppose that some of the hydroxyl groups were present as such in the raw coals, but one must necessarily enquire to what extent is new OH introduced by the oxidation (performic acid is a known reagent for converting alkenes to epoxides and cleaving them; this process occurs by a concerted electrophilic mechanism. However, the mechanism of attack on coals could be quite different and might, for example, follow a free radical mechanism).

The substances named in Figures 1-4 can be arbitrarily classified as follows: (1) "non-committal" substances - those like phenol, p. cresol and benzoic acid, whose structures are relatively simple and contain no obvious clues to origin, (2) those whose structures could obviously be related to biological precursors, such as vanillic, syringic, ferulic and other substituted cinnamic acids, (3) those whose structures are relatively complex but do not display any obvious association with biological precursors, such as the dimethyldihydroxybenzene shown in Figure 1 (isomers not identified). Of these classes, (1) requires no further comment here, while (3) presumably may include - or consist largely of - compounds whose OH groups are an artifact of the oxidation process. This leaves open the question, to what extent are the compounds of class (2) chemical fossils or artifacts?

Many of the substances identified have methoxyl groups in the 3-position or the 3,5-positions, as in lignins. These groups could hardly have beeen introduced by the performic acid oxidation, and therefore indicate a chemical fossil status for the substances. On the other hand, Blom et al. (4) state that methoxyl groups are eliminated by metamorphism in the subbituminous range. We must admit that the structures we identify as methyl phenyl ethers represent a small weight fraction of the whole coals: but can also point out that Blom et al. present no evidence that the Zeisel procedure, which they used in their analyses, was completely effective (the reagent HI may not have penetrated the pore structure fully), or even that it produced with coals the results found with simple compounds. Whatever the mechanism of the performic acid oxidation with coals, one would expect that if it introduces new OH groups, more than one isomer would usually be produced. Therefore, where a peak in the chromatograms was identified as due to a lignin-related substance like vanillin or caffeic acid, the mass spectra corresponding to neighboring chromatographic peaks were carefully examined to determine whether they could represent spectra of position isomers. Such isomers were rarely detected. On the other hand, some of the relatively complex substances not obviously related to biological precursors evidently did have isomers present.

We conclude that some of the OH groups in the products may well have been introduced by the performic acid oxidation, but that most of the substances of structure related to lignins were probably not artifacts.

The question, to what extent do the findings reported here relate to the composition of whole coals, must be left for discussion in a later paper when a more complete set of data can be presented. In concluding this paper, we should draw attention to some implications of the findings, if the provisional assumption can be made that the findings do indeed have relevance to the structure of vitrinite macerals in coals:

1. o-Dihydroxybenzenes, and still more trihydroxybenzenes, are notoriously easily oxidized. This, if such structures are indeed present in vitrinites, could explain why it is so difficult to nitrate or sulfonate coals, even with mild reagents, without accompanying oxidation (3), and also why coals so readily oxidize in weathering. Partially methylated polyphenols are less reactive, but still quite readily susceptible to oxidation.

2. *Ortho*-dihydroxybenzene derivatives are capable of chelating boron and other elements. Less information is available about o.hydroxy-methoxy derivatives, but there seems no reason why they should not be capable of chelation.

3. During catalytic hydrogenation of coals of any rank to liquid fuels, under conditions that give high conversion, a substantial fraction of the oxygen is removed. However, in interactions of coal with hydrogen donor solvent alone, or in applications of the SRC process, oxygen removal is less complete and the products may retain o.dihydroxybenzene structures, a point perhaps worthy of note by those concerned with the composition of coal liquids.

It is worth noting here that in a study of the products of laser pyrolysis of coals in the ionization chamber of a time-of-flight mass spectrometer, homologous series of what appeared to be dihydroxybenzenes were noted (5). The technique could not distinguish positional isomers, but the finding is suggestive when taken with the data in this paper.

Acknowledgements

The authors are indebted to Mr. David Hindenlang and Dr. Larry Hendry for assistance with obtaining GC/MS data. They are also grateful to Dr. William Spackman for fully characterized coal samples from the Penn State/DOE Coal Sample Base assembled under his direction.

Appendix. Trivial and Systematic Names of Relevant Phenolic Substances

Trivial name	Systematic name
resorcinol	1,3-dihydroxybenzene
pyrocatechol	1,2-dihydroxybenzene
phloroglucinol	1,3,5-trihydroxybenzene
pyrogallol	1,2,3-trihydroxybenzene
orcinol	3,5-dihydroxytoluene
2-methyl resorcinol	2,6-dihydroxytoluene
guaiacol	2-methoxyphenol
vanillin	3-methoxy-4-hydroxybenzaldehyde
vanillic acid	3-methoxy-4-hydroxybenzoic acid
syringic acid	3,5-dimethoxy-4-hydroxybenzoic acid
cinnamic acid	phenyl-3-propenoic acid
ferulic acid	3-methoxy-4-hydroxyphenyl-3-propenoic acid
caffeic acid	3,4-dihydroxyphenyl-3-propenoic acid

Literature Cited

(1) Burges, M. A., Hurst, H. M., and Walkden, S. B., Geochim. et Cosmochim. Acta (1964), 28, 1547.
(2) Flaig, W., *in* "Coal Science" (1966), Adv. in Chem. Series No. 55, Amer. Chem. Soc., p. 58.
(3) Given, P. H., Brown, J. K., Lupton, Vera, and Wyss, W. F., Proceedings of Institute of Fuel Conference, "Science in the Use of Coal" (1958), Sheffield, p.A-38.
(4) Blom, L., Edelhausen, L., and van Krevelen, D. W., Fuel (1957), 36, 135; (1959), 38, 537.
(5) Vastola, F. J., Pirone, A. J., Given, P. H., and Dutcher, R. R., *in* "Spectrometry of Fuels", ed. Friedel, R. A., Plenum, New York (1970), p. 29.

RECEIVED February 10, 1978

6

Oxidation of Coal by Alkaline Sodium Hypochlorite

SUJIT K. CHAKRABARTTY

Fuel Sciences Division, Alberta Research Council, Edmonton, Alberta, Canada

The use of halogens in alkaline solution to effect the oxidation of ketones (1) and of hypochlorite as a bleaching agent are well-known, but the application of alkaline sodium hypochlorite to oxidize coal has only been recently reported. The close proximity of the standard potentials of halogens in various oxidation states and the ease of disproportionation into species with varying degrees of oxidizing power are responsible for rendering aqueous halogen solutions versatile oxidants (2). For this reason, various mechanisms can be encountered in the course of any reaction between a substrate and a halogen in aqueous media. With sodium hypochlorite (at pH above 10, 99 per cent of the available active oxidant is OCl^-), the reaction rates depend on several variables, e.g. concentration of the oxidant, pH, buffer-constituents, ionic strength, temperature, presence of catalyst, etc. Under defined reaction conditions, the hypochlorite anion can perform very selective oxidations; the haloform reactions, the oxidation of enolizable ketones to carboxylic acids, the oxidation of active methine, methylene and methyl groups to ketone or carboxyl functions, the replacement of active hydrogen by halogen, decarboxylation and decarbonylation are some of the well-studied reactions used in preparative organic synthesis (1c,2). Since coal contains a fair amount of labile hydrogens, hypochlorite is expected to initiate at least one of these reactions.

The cleavage of aromatic rings by hypochlorite anion is a pertinent question for coal-oxidation studies. In the presence of a catalyst, particularly (3), RuO_4, OsO_4, RhCl and $IrCl_3$, sodium hypochlorite degrades benzene rings; 3-phenylpropionic acid is oxidized to succinic acid (94 per cent yield); phenylcyclohexane can be oxidized to cyclohexane carboxylic acid (25 per cent). Without catalyst, phenols and naphthols (4) are known to react with sodium hypochlorite at moderate temperature. The reaction of picric acid with hypohalite to give halopicrin (5) is another remarkable reaction that results in the complete fragmentation of a benzene ring into six single carbon

0-8412-0427-6/78/47-071-100$05.00/0

units under very mild conditions. In all these cleavage reactions, initially an extensive halogenation of the benzene ring occurs until the aromatic resonance stabilization is sufficiently weakened. Low pH conditions favour halogenation and addition of hypohalite acid.

Landolt and co-workers (6) undertook oxidation of aromatic compounds at constant pH. In 4 to 6 hours in a nitrogen atmosphere, naphthalene, with an excess of sodium hypochlorite at 60-70°C was oxidized at pH 8.5-9 to phthalic acid (70 per cent). At pH above 11, the yield of cleavage products decreases considerably and unoxidized naphthalene along with some chlorinated products were recovered. Although 2-methylnaphthalene reacted similarly to naphthalene at pH 8.5-9, 1-nitro-2-methylnaphthalene reacted to give only 10 per cent yield of 3-nitro-4-methylphthalic acid and 80 per cent recovery of starting compounds. Phenanthrene was found to be less reactive than naphthalene, and anthracene was inert.

The oxidation of methyl-β-naphthyl ketone with sodium hypochlorite at pH above 10 gives 88 per cent yield of 2-naphthoic acid (7). But further treatment of 2-naphthoic or 2,3-napthalene dicarboxylic acid with excess sodium hypochlorite (presumably at pH lower than 10) resulted in cleavage of an aromatic ring; phthalic, trimellitic and/or pyromellitic acids were obtained as cleavage products (4).

From this discussion it is quite evident that sodium hypochlorite is a versatile oxidant. The degree of selectivity for this oxidant would depend on how precisely the reaction conditions are controlled. Following the procedure developed by Newman and Holms (7) for synthesis of 2-naphthoic acid from methyl-2-naphthyl ketone, several coal samples were treated with hypochlorite anion by Chakrabartty and co-workers (8). Three moles of OCl^- were used for each gram-atom of carbon in coal. It was observed that subbituminous coals would readily react under this condition but bituminous coals with dry, ash-free carbon greater than 82 per cent required previous reaction with nitronium tetrafluoroborate in acetonitrile solvent. The subbituminous coals and derivatized bituminous coals reacted equally well with hypochlorite and gave water soluble products (structurally far less complex than so-called humic acid) at 40-60° in 2-3 hours and at 20-28° in 2-3 days. The products of oxidation were carbon dioxide and carboxylic acids. Table 1 lists the yields from different coals reacted at 60° with initial pH = 12.

The water-soluble products were isolated, were methylated with diazomethane, and were analyzed by gel permeation chromatography (GPC), gas chromatography (GLC), mass spectrometry (MS), and ^1H and ^{13}C nmr analysis. Acetic, propionic, succinic, glutaric, adipic, benzene- and toluene- (with 2 to 6 carboxyl as well as with some nitro groups) carboxylic acids were isolated and identified. The average molecular weight of the more complex products having polycondensed aromatic and/or heteroaromatic (mostly substituted by numbers of carboxyl groups), were between 600 and 800. More than 90 per cent of carbon in coal could be accounted for from this product analysis.

TABLE I. Yield of oxidation products from different coals

Rank	Carbon Dioxide		Carboxylic Acids as COOH-Group	
%C,dmf.	mmole/g-Carbon	%Coal-Carbon	meq/g-Carbon	% Coal-Carbon
76.1	43.2	51.8	13.6	16.3
80.5	43.4	52.1	14.6	17.5
83.1	40.3	48.4	16.4	19.7
85.0	38.3	46.0	14.1	17.0
86.4	38.1	45.7	21.3	25.6
90.2	34.5	41.4	14.0	16.8

With the help of ^{13}C nmr measurement, molecular weight data, and elemental analysis, it was possible to calculate the recovery of carbon in different carbon functional groups (9). For a typical bituminous coal (d.a.f. C - 88%), 18 per cent of total carbon survives oxidation as a single benzene ring, 6 per cent as 2, 3 or 4 condensed aromatic (and/or heteroaromatic) rings, 7 per cent as methyl or methylene and 20 per cent as carboxyl groups (Table II).

From various reactions some ketones and ethers also were observed occasionally as products. The most revealing data of these studies are that, irrespective of the rank of the coal, approximately two thirds of the total carbon is oxidized to carbon dioxide and carboxyl functions. It is pertinent to know the nature of carbon functional groups in coal that undergo this oxidation so readily.

Mayo and Kirshen (10) studied the oxidation of the pyridine-insoluble residue from Illinois No. 6 coal with the objective of attaining maximum yields of soluble product with minimum consumption of oxidant and minimum loss of carbon as carbon dioxide. Because the unreacted coal and the soluble acids compete for the oxidant, repeated oxidations with small proportion of NaOCl were employed with extractions and separations of soluble acids between steps. In one experiment 80 per cent of the original carbon was accounted for as follows: 13.6 per cent in undissolved residue, 59.4 per cent in coloured acids soluble in aqueous bicarbonate, 7.1 per cent in lighter coloured acid readily soluble in water and 19.9 per cent in carbon dioxide. The coloured acids had a number average molecular weight of about 900 and a neutral equivalence of 352. The water soluble

TABLE II. Percentage of coal-carbons recovered as water-soluble products, computed from nmr and combustion data

Product-Fraction	C_{arom}	$C_{carboxyl}$	C_{alkyl}	C_{total}
#1 mono-nuclear	12.0	8.4	4.2	24.6
#2 mono-nuclear	5.8	4.9	2.1	12.8
#3 polynuclear*	0.8	0.8	trace	1.6
#4 polynuclear*	3.1	2.4	trace	5.5
#5 polynuclear	2.9	3.9	0.6	5.9
Total	23.7	20.4	6.9	50.4

*Probable skeletal structural types of the compounds in Fr 3 and Fr 4 are as follows:

Mol Wt 623 $(C_{11.7}H_{2.5}O_{2.5}N_{0.7}S_{0.1})$ $(COOMe)_7(Me)$

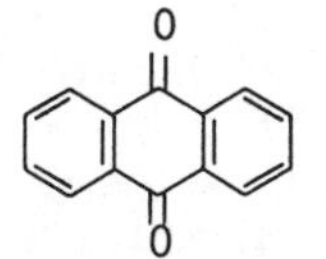

X =O,NH,S

Mol Wt 770 $(C_{14}H_{2.75})(COOM_e)_6(COOH)_4$ $C_2H_9N_{1.8}S_{0.12}$

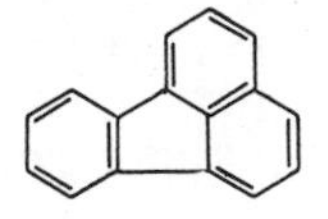

acids had M.W. about 200 and contained 10-15 per cent chlorine. The structural analysis of the products is yet to be completed. Chakrabartty and Dorn (9) identified chloro-alkyl groups in hypochlorite oxidation products by ^{13}C-nmr measurement.

Until a detailed and reliable structural analysis of the intermediate products isolated at the early stages of oxidation is available, it is quite difficult to formulate how hypochlorite reacts with coal. The presence of chloro-alkyl groups in the oxidation products suggests a 'Haloform-type' reaction occurring on certain parts of the coal polymer. Isolation of aliphatic mono- and dicarboxylic acids provides further indication of carbanion-type reactions on non-aromatic structure. But the chloroalkyl carbon and the carbon in the form of aliphatic carboxylic acids constitute a very minor part of the total carbon. From the structural standpoint, it is more pertinent to know how the carboxyl function on the surviving aromatic rings is formed. If one assumes that cleavage of aromatic rings, probably catalyzed by minerals in the coal, is the major reaction, then hypochlorite oxidation data become significant in understanding coal structure. To produce mellitic acid, a structure like (1) in coal may be oxidized (Figure 1). Similarly, the largest polycondensed system (2) should follow the same oxidation path to produce the more complex acid (3) which may further degrade to (4) and finally to mellitic acid. However, the major difficult with this mechanism arises from the fact that bituminous coals are inert to hypohalite oxidation and react only after being nitrated. Nitro groups deactivate aromatic rings for electrophilic attack and suppresses ring-cleavage reactions (see, Landolt and co-workers(6)). Thus it may be presumed that the activity of bituminous coals after nitration arises from non-aromatic structures and aromatic ring-cleavage is not a major reaction.

From this premise, Chakrabartty and co-workers suggested that the reactivity of coal towards hypochlorite oxidation would depend on the accessibility of acidic protons. In low-rank coal, the Brönsted acidity on carbon probably arises from the oxygen functional groups. With the decrease in oxygen content and the increase in molecular weight (?) as the rank increases, progressive loss of Brönsted acidity occurs. However, once the necessary activating group, e.g. $-NO_2$, has been incorporated, the acidity is restored and the coal becomes reactive. The proton on a tertiary carbon is more acidic than on a secondary or primare carbon. Keeping with the elemental composition of coal, particularly the atomic H/C ratio, a bridged tricycloalkane configuration was postulated (8(d)) to accommodate the Brönsted acid sites in coal.

Though nitronium tetrafluoroborate is a well-known nitrating agent (11), recent work by G. A. Olah et al (12) have indicated that this reagent can oxidize benzylic alcohols and silyl-derivatives of secondary alcohols to carbonyl compounds.

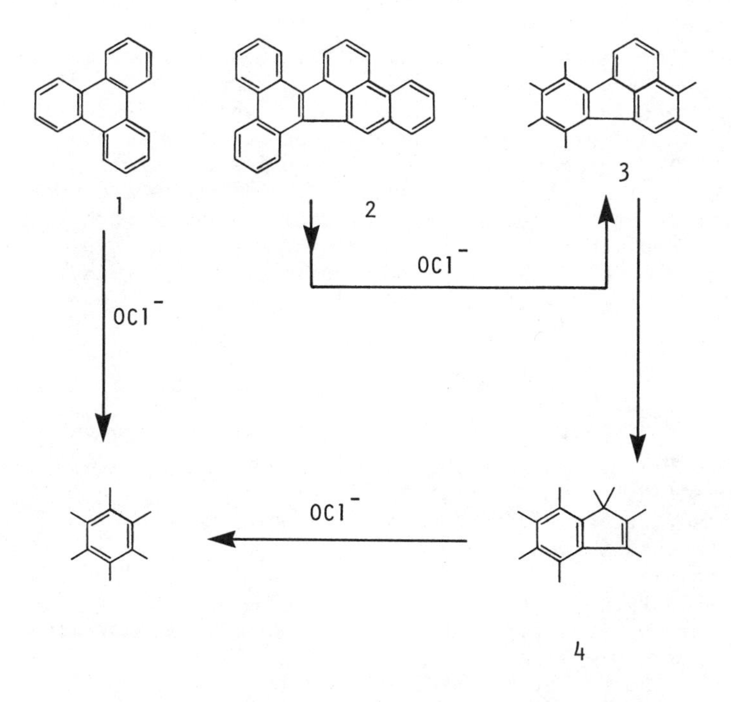

Figure 1. [·—· COOH and/or –Cl/–OH]

The nitrolysis of alkanes, as well as cleavage of ethers could be affected by this reagent very smoothly. Consequently, the reaction of coal with this reagent may introduce, in addition to nitro group, various other oxygen functional groups such as ketone, aldehyde and alcohol. (The total acetylable groups determined after nitration was found to increase considerably.) From this background, it can also be argued that the mechanism of hypochlorite oxidation of coal, after being treated with nitronium tetrafluoroborate, may be similar to phenol oxidation, and a substantial amount of carbon dioxide is generated from cleavage of aromatic rings.

The oxidation studies with hypohalite have been reported on whole coals and few hand-picked vitrain samples. The difference in reactivity between macerals have yet to be studied. Reaction in non-aqueous media can enlarge the scope of applicability of this reagent, e.g. selective oxidation of heterocyclic components by two-phase reactions. All coals are very susceptible to halogenation. With hypohalite, under both high and low pH conditions, halogenation of the substrate is a primary reaction. Consequently, halogenation of coal vis-a-vis hypohalite oxidation would be a useful topic for structural studies.

Literature Cited:

1. a. Fuson, R. C. and Bull, B. A., Chem Rev 15, 275 (1934)
   b. Holst, G., Chem Rev 54, 169 (1954)
   c. Fieser, L. F. and Fieser, M., Reagents for Organic Synthesis, John Wiley & Sons, New York, Vol 1 (1967), Vol 11 (1969)
2. Chakrabartty, S. K., in Oxidation in Organic Chemistry, Part D., Chapter 5, Ed. W. S. Trahanovsky, Academic Press, New York (1978)
3. Wolf, S., Hasan, S. K. and Campbell, J. R., J. Chem. Soc. D(21), 1420 (1970)
4. a. Moye, C. J. and Sternell, S., Austral. J. Chem., 19, 2107 (1970)
   b. Landolt, R. C., Fuel 54, 229 (1975)
   c. Mayo, F. R., ibid., 273 (1975)
5. Birch, A. J., Moye, C. J., Richards, R. W. and Vanek, Z., J. Chem. Soc. 3586 (1962)
6. Angert, J. L., Gatton, S. L., Reilly, M. T. and Landolt, R. G., Fuel 56, 224 (1975)
7. Newman, M. S. and Holms, H. L., in Org. Syn. Coll. Vol. 2, 428 (A. H. Blatt, Ed.) Wiley, New York (1943)
8. Chakrabartty, S. K. and Kretschmer, H. O.
   a. J. Chem. Soc. Perkin 1, 222 (1974)
   b. Fuel 51, 160 (1972)
   c. ibid, 53, 132 (1974)

8. Chakrabartty, S. K. and Berkowitz, N.
   d. ibid., 53, 240 (1974)
   e. ibid., 55, 362 (1976)
   f. Nature, 261 (5555), 76 (1976)
9. Chakrabartty, S. K. and Dorn, H., to be published
10. Mayo, F. R., Huntington, T. and Kirshen, N.
    Paper No. 14, Preprints 1976 Coal Chemistry Workshop Standford Research Institue
    Mayo, F. R. and Kirshen, N., in Quatarly Reports on "Homogenous Catalytic Hydrocracking Process for Conversion of Coal to Liquid Fuels, Basic and Exploratory Research". US Energy Research and Development Administration Contract No. E(49-18)-2202 for periods ending on (a) Jan. 1976; (b) April 1976; (c) July 1976; (d) October 1976; (e) January 1977 and (f) April 1977
11. Fieser, L. F. and Fieser, M., Reagent for Organic Synthesis, Vol I, John Wiley & Sons, Inc., 1967
12. a. Olah, G. A. and Ho T-L, Synthesis 609 (1976)
    b. Olah, G. A. and Ho T-L, Synthesis (Press)
    c. Olah, G. A. and Lin, H. C., J. Amer. Chem. Soc. 93, 1259 (1971)
    d. Olah, G. A. and Ho T-L, J. Org. Chem. 42 (18), 3097 (1977)

RECEIVED February 10, 1978

7

Oxidative Degradation Studies of Coal and Solvent-Refined Coal

RYOICHI HAYATSU, RANDALL E. WINANS, ROBERT G. SCOTT, LEON P. MOORE, and MARTIN H. STUDIER

Chemistry Division, Argonne National Laboratory, Argonne, IL 60439

The bulk of the organic matter in coals consists of a macromolecular material of complex and variable composition. Many workers have attempted to degrade coals to smaller molecules which could be identified and interpreted in terms of coal structure. Oxidation has been one of the more important degradation methods. To date a number of oxidizing agents have been used. Among these were HNO_3, HNO_3-$K_2Cr_2O_7$, $KMnO_4$, O_2, H_2O_2-O_3 and NaOCl, all drastic oxidants. Because these reagents in general result in extensive ring degradations, with benzene carboxylic acids the only aromatic compounds identified, they have been of limited usefulness. Our approach has been to examine the products both from drastic oxidants and more selective ones designed to break up the macromolecules into identifiable units with a minimum of chemical change so that units indigenous to coals can be identified. We have found aqueous $Na_2Cr_2O_7$ to be selective and have reported a number of polynuclear aromatic units which resulted from the dichromate oxidation of a bituminous coal and which we believe to be indigenous to the coal (1,2). We have explored a number of oxidizing agents using the samples listed in Table I. In addition to four coals, a solvent refined coal (SRC) and two completely abiotic samples, the synthetic Fischer-Tropsch polymer, and the polymer from the Murchison meteorite, were used for comparison and to test the oxidation methods.

Results and Discussion

Identification of products: In general the products of oxidation were chiefly carboxylic acids which were esterified with diazomethane to increase their volatility for easier analysis by time-of-flight mass

0-8412-0427-6/78/47-071-108$05.00/0

Table I

Elemental Analysis of Samples (maf %)

No.	Sample	C	H	N	S	O (by diff)
1	Lignite (Sheridan Wyoming)	67.3	4.8	1.3	1.2	25.4
2	Bituminous (Illinois Seam #2)	77.8	5.4	1.4	2.1	13.3
3	Bituminous (Pittsburgh Seam #8)	82.7	5.5	1.3	2.8	7.7
4	Anthracite (Pennsylvania PSOC #85)	91.3	3.9	0.6	1.1	3.1
5	SRC (from Pittsburgh Seam #8)	87.2	5.5	1.8	1.2	4.3
6	SRC benzene-methanol ext.*	86.2	5.6	1.8	0.7	5.7
7	Char	84.9	1.7	---	---	---
8	Synthetic Polymer (from Fischer-Tropsch) (3)	80.0	5.1	1.3	0.0	13.6
9	Polymeric Material from Murchison Meteorite (3)	76.1	4.6	2.8	1.3	15.2

* The SRC was fractionated into 3 fractions on the basis of solubility: hexane (4.5%), benzene-methanol mixture (82.3%), pyridine (11.8%) and a small residue (1.4%).

spectrometry (TOFMS) with a variable temperature solid inlet, GC-TOFMS and high resolution mass spectrometry (HRMS) (1-3).

Nitric Acid Oxidation. All samples of Table I except #6 were refluxed with 70% HNO_3 for 16-24 hours yielding clear orange colored solutions. The acid solutions were evaporated to dryness under reduced pressure and the residue weighed. The methylated acids were analyzed by GC-MS, HRMS and the solid probe. Fragmentation patterns and precise mass determination of both molecular ions and fragments, in particular $(M-OCH_3)^+$, were used for identification. In Figure 1 are summarized the data for benzene carboxylic acids (as their methyl ester). The synthetic sample had been prepared by heating CO, H_2 and NH_3 with an Fe-Ni catalyst at 200°C for six months (3). It was a macromolecular material insoluble in organic solvents, HCl, HF and KOH. Despite the drastic nature of the nitric acid oxidation it appears that useful information can be obtained by the procedure. For example the yield of total acids and the number of carboxylic acid groups per benzene ring seem correlated with the degree of condensation of the original material. The spectra from the synthetic sample from the Fischer-Tropsch reaction, and from char are relatively simple consisting primarily of the benzene carboxylic acid esters with from two to six ester groups. The coal samples and coal derived products (except for char) are more complex and contained nitro substituted esters and pyridine derivatives. Note the similarity between the synthetic sample and the char prepared from Illinois bituminous coal by heating under vacuum to 800°C. The abundance of the benzene hexacarboxylic acids for these two suggests a high degree of condensation in the original samples. The similarity in distribution of the oxidation products from the two bituminous coals (the Illinois and Pittsburgh) is striking. The solvent refined coal derived from the Pittsburgh #8 coal is shown later (Figure 3, Table II) to have a higher degree of aromatic ring condensation than its feed coal. Thus it is surprising to see the shift to fewer acid groups for SRC. This may mean that many aliphatic crosslinks were destroyed in the SRC process and evolved as light hydrocarbons.

The results from the meteoritic polymer (not shown in Figure 1, see Reference (3)) were very similar to those of the synthetic polymer.

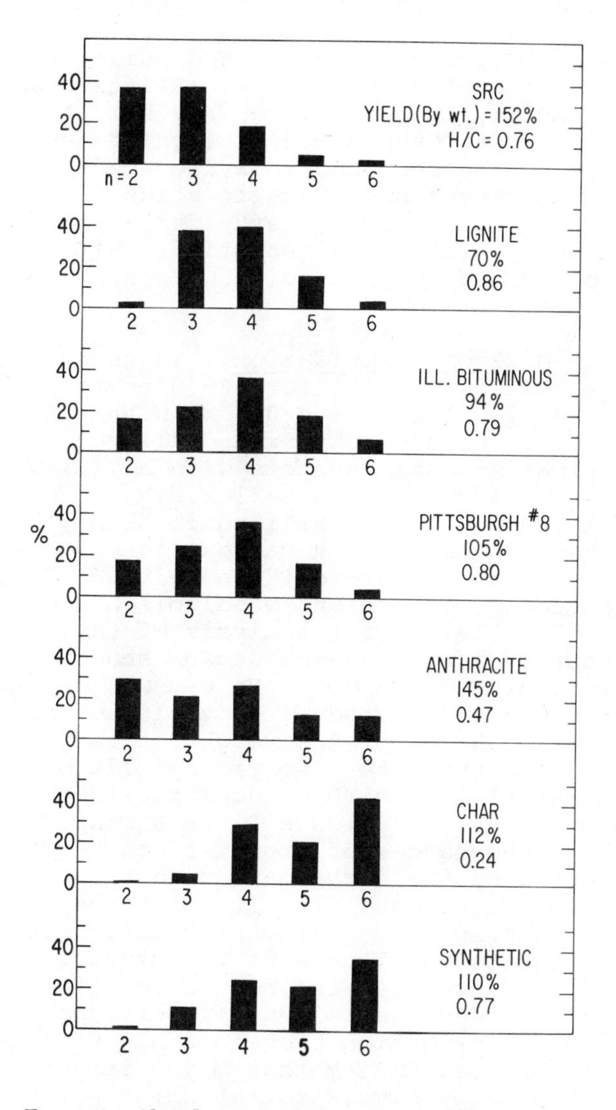

Figure 1. Abundances (%) of benzene carboxylic acids, determined as their methyl esters, produced by nitric acid oxidation. n = *number of* $(COOCH_3)$ *groups per benzene ring.*

15%-HNO_3. Ferguson and Wims have reported that 15%-HNO_3 selectively oxidizes dialkylbenzene to alkylbenzoic acids (4). Refluxing of lignite coal (sample #1) with 15%-HNO_3 for 16 hours gave aromatic acids (58.2 wt% of the original coal) and humic acid-like material (63.5 wt%) respectively. Relative abundances of benzenecarboxylic acids were very similar to those of 70%-HNO_3 oxidation product of lignite (see Fig. 1). In addition to the benzenecarboxylic acids, however, a series of methylbenzenecarboxylic acids (17 wt% of total aromatic acid) and of methylfurancarboxylic acids (3.7 wt%) were positively identified. Nitro derivatives of benzenecarboxylic acids were also found as minor products.

Sodium Dichromate Oxidation. In general 1-3 g of sample was heated at 250°C for 36-40 hours with excess $Na_2Cr_2O_7$ (60-120 ml of 0.4 M-0.6 M) with stirring. This procedure oxidizes side chains and alicyclic appendages to polynuclear aromatic systems with a minimum of degradation of aromatic rings (1,2,5,6). We have found that model compounds are oxidized in high yields (78-95%) to their corresponding carboxylic acids. Samples 1, 2, 3, 4, 6, and 9 were oxidized with $Na_2Cr_2O_7$ with a high degree of conversion to soluble or volatile compounds (70-100%). Mass analysis of the gaseous fraction obtained from the oxidation showed the presence of only CO and CO_2. No H_2, CH_4 or light hydrocarbons were detected. The yields of soluble compounds were 50-70% of the weight of the original samples. The anthracite also yielded 17% of a humic acid type material of high molecular weight soluble in alkaline solution. In Figure 2 are shown graphically the relative abundances of aromatic and heteroaromatic units produced by the dichromate oxidation of three coals (samples 1, 2, and 4). It is obvious that the degree of aromatic condensation increases with rank of coal from lignite to bituminous to anthracite. This oxidation procedure has also been successfully applied to SRL and its feed coal (lignite, North Dakota) (7). The $Na_2Cr_2O_7$-aq oxidation procedure has been questioned because of the possibility that major structural rearrangement with pyrolytic formation of polynuclear aromatic compounds might occur during the reaction at 250°C (8). The fact that no polynuclear aromatic compounds with more than two fused rings were detected in the oxidation products of lignite (sample 1) and that the degree of condensation increases with rank of coal is internally consistent and suggests that condensation during oxidation with $Na_2Cr_2O_7$ is minimal. In a blank

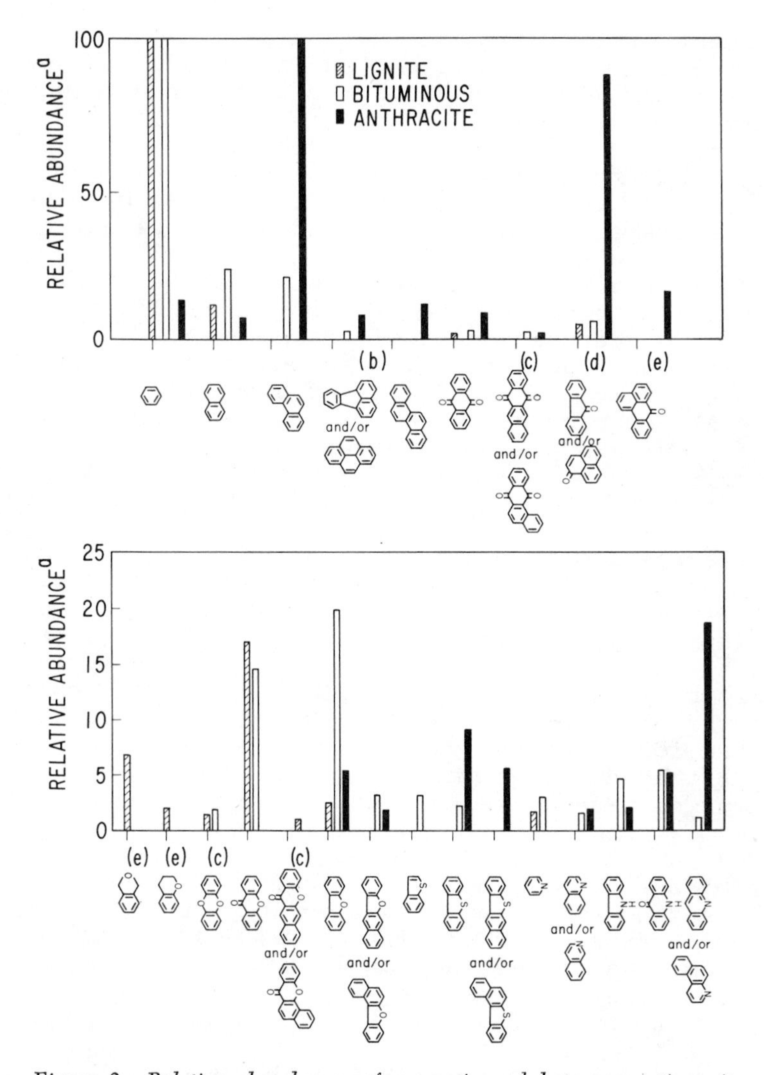

Figure 2. Relative abundances of aromatic and heteroaromatic units produced by sodium dichromate oxidation of lignite, bituminous, and anthracite coals.

Identification based on TOF variable temperature solid inlet, GC-TOFMS and high resolution MS. Aromatic units in the upper graph (left to right): benzene, naphthalene, phenanthrene, fluoranthene/pyrene, chrysene, anthraquinone, naphthacenequinone/benzanthraquinone, fluorenone/phenalenone, benzanthrone. Heteroaromatic units in the lower graph (left to right): phthalan, chroman, dibenzo-p-dioxin, xanthone, benzoxanthones, dibenzofuran, benzonaphthofurans, benzothiophene, dibenzothiophene, benzonaphthothiophenes, pyridine, quinoline/isoquinoline, carbazole, acridone, acridine/benzoquinoline. (a) Determined as methyl esters of carboxylic acids and as nonacidic compounds. The benzene unit is normalized to 100 for the lignite and bituminous coals. Phenanthrene is normalized to 100 for anthracite. Aromatic units with a carbonyl group may have resulted from oxidation of a methylene group. (b) In an earlier paper we reported on the basis of published data (1) "that the pyrene ring would have been extensively degraded." However, we have found it to be rather stable under the oxidation conditions we used. (c) These compounds were identified in the nonacidic fraction only. (d) Because fluorene and bifluorene were observed in other experiments such as pyrolysis we feel that fluorenone is more probable than phenalenone. (e) Identification tentative.

experiment bituminous coal (sample #2) which had been extracted with an organic solvent to remove trapped compounds was heated at 250°C for 40 hours with water. An insoluble residue 96.2% and inorganic salts 2.9% were obtained. Mass analysis of the residue was indistinguishable from that of the original coal and the H/C ratio was unchanged. Furthermore, detailed pyrolysis studies of the coal showed no evidence of significant thermal decomposition until heated above 250°C.

The gas chromatograms of the oxidation products (as methyl esters) from the Pittsburgh coal (sample 3) and its SRC (benzene methanol extract-sample 6) are shown in Figure 3. The numbered peaks are identified in Table II. A greater degree of aromatic condensation of the SRC extract over that of its feed coal is observed. For example, naphthalene and phenanthrene rings are much more abundant in the SRC. The dibenzofuran ring is the most abundant heterocyclic to have survived the SRC process.

Fifteen aromatic ring systems were identified in the $Na_2Cr_2O_7$ oxidation product of the meteorite polymer: benzene, biphenyl, naphthalene, phenanthrene, fluoranthene (or pyrene), chrysene, fluorenone, benzophenone, anthraquinone, dibenzofuran, benzothiophene, dibenzothiophene, pyridine, quinoline or isoquinoline and carbazole (3). These data support the generally accepted idea that the polymeric material in meteorites has a highly condensed aromatic structure (3,9,10) and gives us confidence in the dichromate oxidation procedure.

We have recently developed another selective $Na_2Cr_2O_7$ oxidation. Reaction mixture (3 g SRC, 80 ml of benzene and 60 ml of acetic acid) was stirred at room temperature, and 18 g of $Na_2Cr_2O_7$ in 70 ml of acetic acid was added in 2-3 ml portions over a period of 10-12 hours. Thereafter, stirring was continued at 40°C for 60 hours with 97% of the sample being oxidized. After filtration of the reaction mixture, humic acid-like material (49.4 wt%) was isolated from the insoluble residue by dissolution in methanol and alkaline solution. Aromatic acid fraction (26.5 wt%) and neutral fraction (5.7 wt%) were recovered respectively from the filtrate. The aromatic acid fraction was esterified and was then analyzed by GC-TOFMS, TOFMS-solid inlet and HRMS. The result showed the presence of the same polynuclear aromatic and heterocyclic ring systems which were found in the aq-$Na_2Cr_2O_7$ oxidation product of sample #6 (see Fig. 3a and Table II). Major products were benzenecarboxylic acids (di and tri) and their methyl derivatives, naphthalene

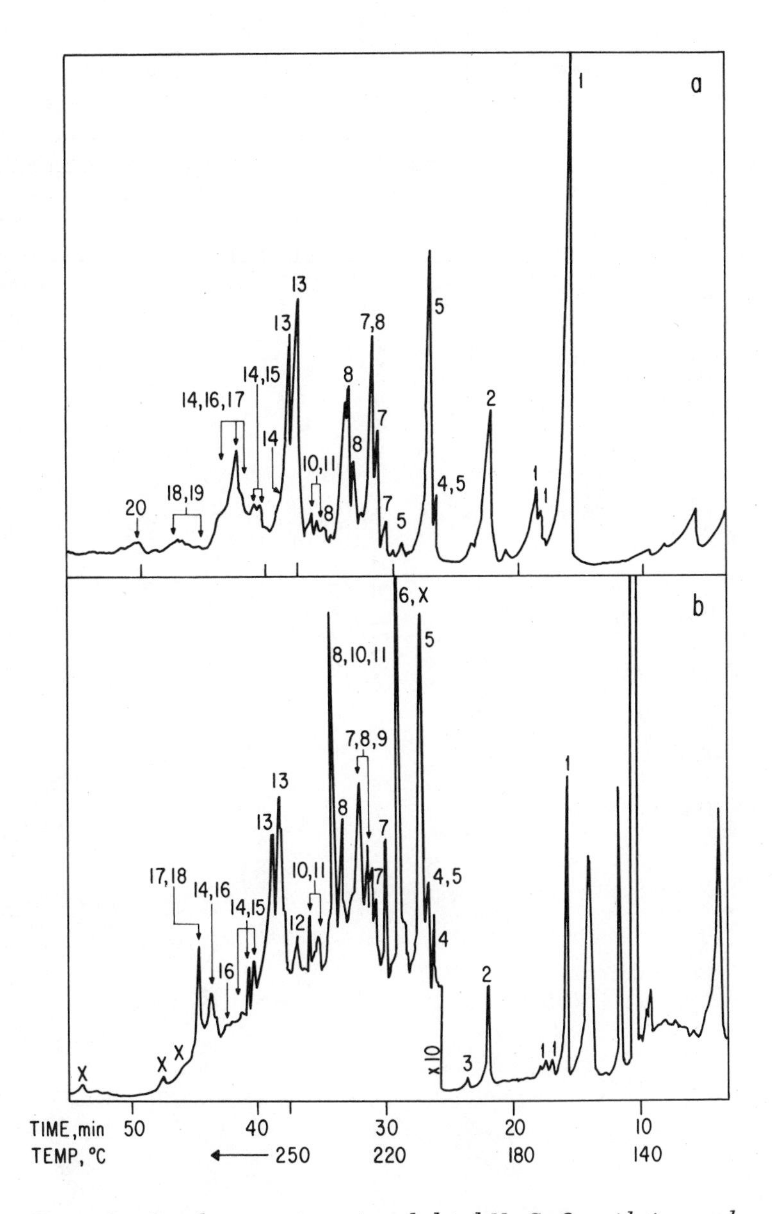

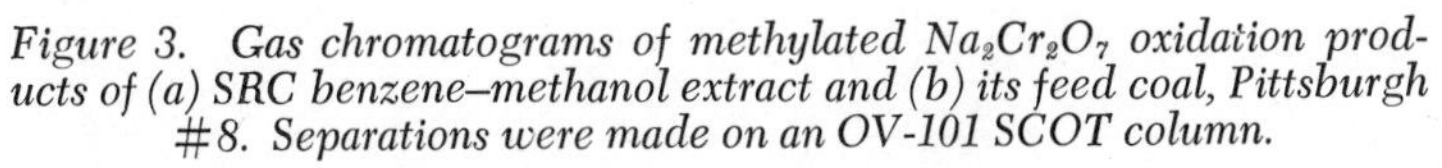

Figure 3. Gas chromatograms of methylated $Na_2Cr_2O_7$ oxidation products of (a) SRC benzene–methanol extract and (b) its feed coal, Pittsburgh #8. Separations were made on an OV-101 SCOT column.

Table II
Aromatic Acids Found in the Oxidation of SRC and Its Feed Coals[a] Identified as Methyl Esters

C Peak Number		Number of $-COOCH_3$	Relative Abundance ($\pm$ 15%)[b]	
			SRC Extract	Coal
1	Benzene	2	100	100
5		3	58	29
10		4	9	10
4	Biphenyl	1	3	9
2	Naphthalene	1	47	30
8		2	82	29
16		3	24	8
		4	*7*	*3*
13	Phenanthrene	1	82	28
20		2	8,*15*	*8*
		3	*5*	*7*
		4	*1* (T)	---
19	Pyrene/Fluoranthene	1	7	*3*
		2	*4*	*2*
11	Fluorenone	1	4	8
		2	*3*	*2*
		3	*3*	---
14	Anthraquinone	1	26	4
		2	*3*	---
7	Dibenzofuran	1	32	21
17		2	9	4
		3	*10*	*4*
15	Xanthone	1	4	9
		2	*6*	*7*
		3	*2* (T)	*3*
3	Benzothiophene	1	---	4
12	Dibenzothiophene	1	*2*	8
		2	---	*5*
	Pyridine	3	---	*5* (T)
	Carbazole	1	*4*	*3*
		2	*3*	*4*
18	Benzoquinoline/Acridine	1	3	7
		2	*5*	*4*
		3	*2*	*2*
6[c]				11
9[d]				22

Table II (Footnotes)

[a]92% of the extract was oxidized and the yield of total acidic and non-acidic less volatile compounds was 59%. For the feed coal, 84% of the coal was oxidized; Yield of totals was 58%.

[b]Benzene dicarboxylic acid methyl ester is normalized to 100.

[c]Peak no. 6 is tentatively identified as trimethoxyxanthone.

[d]Peak no. 9 shows prominent mass ions at 216 and 215. HRMS shows their elemental composition corresponding to $C_{12}H_{10}O_3N$, $C_{12}H_9O_3N$ or $C_9H_{12}O_6$, $C_9H_{11}O_6$. Identification has not been made at this time.

T = identification tentative; --- = not detected; x = peak consists of more than one component which are difficult to identify by GC-MS.

Italics indicate that identification and estimation of relative abundances were made by TOF variable temperature solid inlet and HRMS, because of difficulty of identification by GC-MS.

carboxylic acids (mono and di), phenanthrene mono-carboxylic acid, fluorenone mono-carboxylic acid and dibenzofuran mono-carboxylic acid. Higher molecular weight acid methyl esters were also seen in the spectra of TOFMS-solid inlet and HRMS. In addition to the above observation, diphenyl ether was identified as its mono and dicarboxylic acid methyl esters. This looks like an important finding because the diphenyl ether system has been considered to be present in coal and coal product. However there has so far been little direct evidence. Further oxidation of the humic acid-like material afforded the same aromatic acids.

Photochemical Oxidations: Oxidation by air bubbling through an aqueous HCl solution while irradiating with ultraviolet light from a high pressure mercury lamp was investigated (1-3). In Table III are shown the results obtained with model compounds. From the results with the model compounds it appears that aromatic hydrocarbons are readily oxidized to benzene carboxylic acids. On the other hand N-heterocyclics resist oxidation. In Table IV are listed the aromatic carboxylic acids isolated from a bituminous coal (sample #2) after photo-chemical oxidation. The products from lignite (sample #1) were primarily benzene carboxylic acids with only traces of pyridine tricarboxylic acids and xanthone di- and tricarboxylic acids. Several aliphatic carboxylic acids were identified (after methylation) in the photo-oxidation product of bituminous coal by TOF variable temperature solid inlet and HRMS. They are methyl esters of malonic acid, succinic acid, glutamic acid and saturated monocarboxylic acids (CH_3-$(CH_2)_n$-$COOCH_3$ n = 1-7). The fragments for 3-methyl and 3,3-dimethyl aliphatic carboxylic acid methyl esters were also seen in the mass spectra. A similar observation was made for the oxidation product of lignite.

Hydrogen Peroxide-Acetic Acid: It is probable that in the oxidation experiments described above aromatic units with phenolic groups would have been destroyed. Schnitzer et al. (11) have shown that an acetic acid-H_2O_2 mixture under mild conditions oxidizes humic acids while preserving phenols. Using this procedure we have oxidized lignite (sample #1) and bituminous coal (sample #2) with over 80% conversion to methanol soluble acids and have methylated the acids produced.

Table III

Photochemical Oxidation of Model Compounds[a]

Compound	Unreacted[b] (%)	Major Oxidation Products[c]
p-cresol	0	unident. small species, polymer
anisole	0	unident. small species, polymer
naphthalene	8	phthalic acid
1,4-diMe-naphthalene	5	phthalic acid
2,6-diMe-naphthalene	7	benzene-1,2,4-tricarboxylic acid
indane	3	phthalic acid
fluorene	8	fluorenone
acenaphthene	10	phthalic, naphthalene-1,8-dicarboxylic acids
phenanthrene	24	phthalic acid, phenanthrene-9,10-diketone
pyrene	12	benzene tri- and tetracarboxylic acids
dibenzofuran	32	styrene, salicylic acid
xanthone	78	unidentified chloro compounds
dibenzothiophene	43	styrene, thiophenol
carbazole	78	polymer
N-Et-carbazole	83	polymer
poly(2-vinylpyridine)-polystyrene polymer	8	pyridine-2-carboxylic, pyridine-2-aldehyde, benzoic, malonic, and succinic acids

[a]The oxidation was carried out in 10% aq. HCl solution for 6-10 days.

[b]Values are accurate to ±10%.

[c]After preliminary separation into acidic, neutral, and basic fractions, products were identified by TOFMS with variable temperature solid inlet. Except for the first two samples and the two carbazoles, all identifications were checked by high-resolution MS. For the italicized samples, GC-MS was used as well, after esterification of the product.

All major products were accompanied by lesser amounts of their monochloro derivatives.

Table IV

Aromatic Acids Found in the Photochemical Oxidation of Bituminous Coal: Identified as Methyl Esters[a,g,h]

Nucleus	Number of $-COOCH_3$	Precise Mass $(M-OCH_3)^+$ elemental comp.	observed	Dev. x 10^3	Relative Abundance[b] $\pm$ 15-20%
Benzene	2	$C_9H_7O_3$	163.0388	-0.7	42
	3	$C_{11}H_9O_5$	221.0457	0.8	100
	4	$C_{13}H_{11}O_7$	279.0510	0.6	97
	5	$C_{15}H_{13}O_9$	337.0555	-0.3	29
	6	$C_{17}H_{15}O_{11}$	395.0593	-2.0	4
Cl-benzene[c]	2	$C_9H_6O_3Cl$	196.9990	-1.5	24
	3	$C_{11}H_8O_5Cl$	255.0066	0.6	31
	4	$C_{13}H_{10}O_7Cl$	313.0125	1.1	26
	5	$C_{15}H_{12}O_9Cl$	371.0124	-4.4	2
Fluorenone	2	$C_{16}H_9O_4$	265.0494	-0.6	9
Anthraquinone	2	$C_{17}H_9O_5$	293.0451	0.2	3
	3	$C_{19}H_{11}O_7$	351.0520	1.6	1
Phthalan	2 (T)	$C_{11}H_9O_4$	205.0478	-2.2	2
	3 (T)	$C_{13}H_{11}O_6$	263.0572	1.8	1
Xanthone	2	$C_{16}H_9O_5$	281.0461	1.2	9
	3	$C_{18}H_{11}O_7$	339.0510	0.6	7
Dibenzofuran[d] and/or naphthofuran	2	$C_{15}H_9O_4$	253.0477	-2.3	3
	3	$C_{17}H_{11}O_6$	311.0560	0.6	2
Pyridine	2	$C_8H_5O_3NCl$[e]	197.9974	1.6	3
	3	$C_{10}H_8O_5N$	222.0408	0.5	3
	4	$C_{12}H_{10}O_7N$	280.0488	3.2	3.5
Quinoline and/or isoquinoline	1	$C_{10}H_6ON$	156.0412	-3.7	0.5
	2	$C_{12}H_8O_3N$	214.0468	-3.5	2
Carbazole[f]	1	$C_{13}H_8ON$	194.0598	-0.7	4
	2	$C_{15}H_{10}O_3N$	252.0660	-0.0	2
	3	$C_{17}H_{12}O_5N$	310.0708	-0.6	2
Acridone	1	$C_{14}H_8O_2N$	222.0525	-2.9	5
	2	$C_{16}H_{10}O_4N$	280.0592	-1.8	2
	3	$C_{18}H_{12}O_6N$	338.0635	-2.8	1

Table IV Footnotes

[a]Some preliminary results were reported earlier (1). Identification based on TOF variable temperature solid inlet, GC (Carbowax 20 M)-TOFMS and high resolution MS.

[b]Benzene tricarboxylic acid methyl ester is normalized to 100. Relative abundances were estimated from the GC and an integration of the base peak of each compound during the time that the sample was completely volatilized in the MS.

[c]Monochlorobenzene carboxylic acids were always obtained when coal or model compounds were oxidized in 10% HCl aq. solution. Chlorocarboxylic acids of other aromatics were also observed in very low yield.

[d]Relatively large amounts of dibenzofurans were found to be produced by $Na_2Cr_2O_7$-oxidation (1), while photo-oxidation produced dibenzofuran carboxylic acids in low yield. Perhaps this shows the photochemical procedure destroyed dibenzofuran.

[e]For pyridine dicarboxylic acid, only chloroderivatives were found.

[f]The previous estimates of relative abundances (1) were somewhat high.

[g]The products of lignite photo-oxidation in 5% KOH aq. solution were very similar to those shown in this Table. In addition some mono- and dimethylbenzene carboxylic acids were identified.

[h]Polynuclear aromatic carboxylic acids such as naphthalene and phenanthrene found in the $Na_2Cr_2O_7$ oxidation product (see Fig. 2) are not observed in this product. From our model experiments, we have found that these aromatic compounds are oxidized by the present procedure.

T = identification tentative.

The gas chromatogram of the aromatic portion of the methylated product from lignite is shown in Figure 4. The identification of individual compounds was made by coincident MS and HRMS of the mixture. The methyl esters identified gave the following approximate distribution: 36.1% benzene, 7.6% methylbenzene (not shown in Fig. 4, because of very small peaks), 22.1% methoxybenzene, 15.9% furan and 18.2% dibasic aliphatics. For the bituminous coal the methoxy derivatives were half as abundant. These results suggest that the lignite has twice the phenolic content of the bituminous coal. It is interesting to note that methyl furan tetracarboxylate and other furan derivatives have been identified.

It appears that the coals are acting as catalysts for this oxidation. If 2,6-dimethylnaphthalene is reacted with H_2O_2 under the same conditions as used for the coals, only 2,6-dimethylnaphthaquinone is isolated in essentially quantitative yield. With the addition of a small amount of lignite to the reaction, 5-methyl phthalic acid is obtained as the major product. Also no hydroxylated benzene carboxylic acids were isolated which indicates that this procedure does not hydroxylate aromatic rings. Transition metals are known to catalyze reactions of hydrogen peroxide. Part of the catalytic effect may be due to the mineral matter in the coal.

No polyhydroxyl benzene carboxylic acids have been observed in the products from either coal. It is expected that these species would undergo ring oxidation and subsequent degradation. However, the concentration in coal of these species is probably small. These compounds are expected to be very reactive and probably would not survive the coalification process. We have examined aqueous NaOH extract of the Wyoming lignite. Numerous hydroxylated aromatic hydrocarbons and aromatic acids were identified, but no polyhydroxylated species were detected (12).

Sodium Hypochlorite Oxidation: On the basis of results of oxidation of coal with NaOCl, Chakrabartty et al. have suggested that coal has a non-aromatic "tricycloalkane or polyamantane" structure (13). They pointed out that no evidence for aromatic compounds other than benzene derivatives was found in their oxidation product. The specificity of NaOCl as an oxidant has been questioned (14-17) and is still in dispute. We are attempting to resolve the question and have oxidized samples 6, 9 and 2,6-dimethylnaphthalene

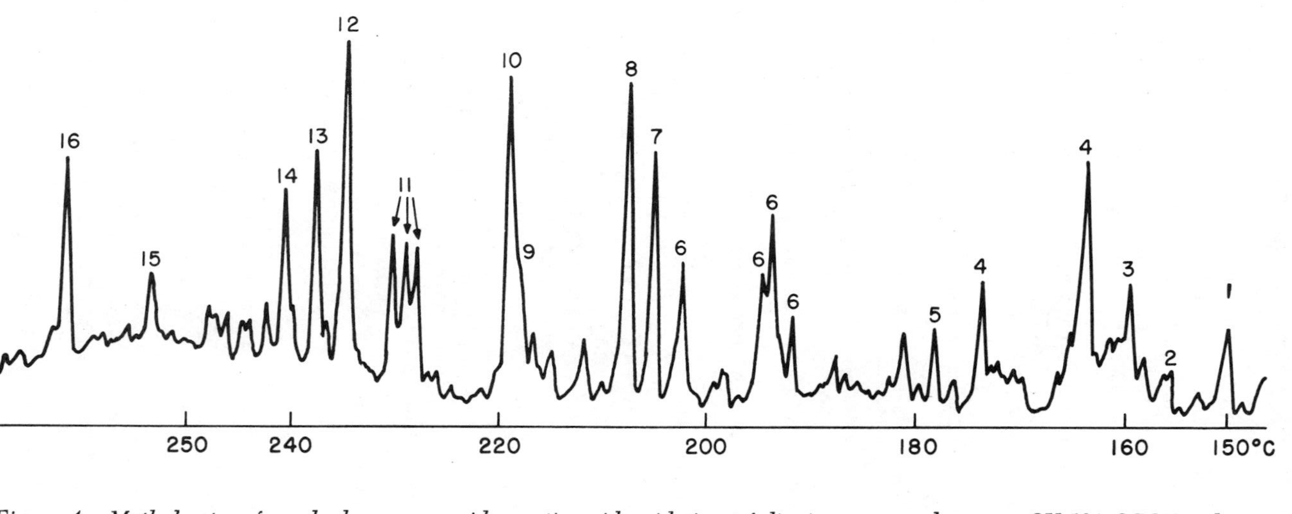

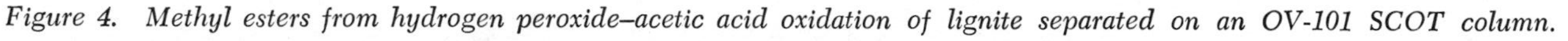

Figure 4. Methyl esters from hydrogen peroxide–acetic acid oxidation of lignite separated on an OV-101 SCOT column.

Numbered compounds: 1. methyl o-*methoxybenzoate; 2. methyl* m- *or* p-*methoxybenzoate; 3. methyl methylfurandicarboxylate; 4. methyl benzenedicarboxylate; 5. methyl carbomethoxyphenylacetate; 6. methyl methoxybenzenedicarboxylate; 7. methyl 1,2,3-benzenetricarboxylate; 8. methyl 1,2,4-benzenetricarboxylate; 9. methyl methoxybenzenetricarboxylate; 10. methyl furantetracarboxylate; 11. methyl methoxybenzenetricarboxylate; 12. methyl 1,2,3,4-benzenetetracarboxylate; 13. methyl 1,2,4,5-benzenetetracarboxylate; 14. methyl 1,3,4,5-benzenetetracarboxylate; 15. methyl methoxybenzenetetracarboxylate; 16. methyl benzenepentacarboxylate.*

under conditions described by Chakrabartty et al. (13, 18). Of particular interest is the oxidation of the polymeric material from the Murchison meteorite (sample 9). Numerous studies have shown this material to be a highly condensed aromatic structure. All samples were nitrated according to the method of Chakrabartty et al. When nitration was followed by NaOCl oxidation at 60°C for 3-4 hours almost no oxidation was observed for nitrated samples 6 (SRC) or 9 (meteorite) while 26% of nitrated dimethyl naphthalene was converted to benzene carboxylic acids and their nitro and/or methyl derivatives. No naphthalene acids were detected. When the reaction was continued for 15 hours, at 65-70°C 57% of sample 9 and 42% of sample 6 were oxidized to water soluble acids. Polynuclear aromatic and heterocyclic acids were not detected in the oxidation products of either sample although both are highly aromatic. Benzene carboxylic acids were the major products (nitro and/or methyl derivatives and a methyl chloro dicarboxylic acid). Apparently polynuclear aromatic systems were destroyed and the product distribution closely resembled that from a nitric acid oxidation. Non-nitrated bituminous coal (sample #2) was also oxidized with NaOCl (60°C 14 hours) with a high degree of conversion to soluble material (88%). The yield of organic acids and humic acid-like material were 9.8 wt% and 49 wt% respectively. Identified major organic acids (as Me esters) were benzene carboxylic acids (from di to penta) and their methyl and/or chloro derivatives. Furan mono and dicarboxylic acids and their methyl derivatives were also found in relatively high abundance. Although they are very minor products, naphthalene mono- and fluorenone mono-carboxylic acids were identified. Further oxidation of the humic acid-like material gave the same organic acids again in a high degree of conversion. Obviously more work is needed to clarify the role of NaOCl in the oxidation of coal.

Various oxidizing agents attack coal and coal products in various ways. To gain insight into the true nature of coals, the use of a number of degradation reagents is necessary.

Acknowledgements: This work was supported by the Department of Energy-Fossil Energy, Division of Coal Conversion-Liquefaction. Chemical and instrumental techniques used were developed with support from the Office of Basic Energy Sciences.

Literature Cited

[1] Hayatsu, R., Scott, R. G., Moore, L. P., and Studier, M. H., Nature (1975) 257, 378; see also Nature (1976) 261, 77.

[2] Hayatsu, R., Winans, R. E., Scott, R. G., Moore, L. P., and Studier, M. H., to appear in Fuel (1978).

[3] Hayatsu, R., Matsuoka, S., Scott, R. G., Studier, M. H., and Anders, E., Geochim. Cosmochim. Acta (1977) 41, 1325.

[4] Ferguson, L. N., and Wims, A. L., J. Org. Chem. (1960) 25, 668.

[5] Friedman, L., Fischel, L., and Schechter, H., J. Org. Chem. (1965) 30, 1453.

[6] Wiberg, K. B. (Ed), "Oxidation in Organic Chemistry", Part A, page 90, Academic Press, New York, 1965.

[7] The results were presented in part. Duty, R. C., Hayatsu, R., Scott, R., Moore, L., Studier, M. H., and Winans, R., 1978 Pittsburgh Conference, Cleveland, Ohio, February 27-March 3, 1978.

[8] Chakrabartty, S. K., and Berkowitz, N., Nature (1976) 261, 76.

[9] Bandurski, E. L., and Nagy, B., Geochim. Cosmochim. Acta (1976) 40, 1397.

[10] Nagy, B., "Carbonaceous Meteorite", Elsevier, Amsterdam, 1975.

[11] Schnitzer, M., and Skinner, I. M., Can. J. Chem. (1974) 52, 1072.

[12] Unpublished data.

[13] Chakrabartty, S. K., and Berkowitz, N., Fuel (1974) 53, 240; see also Fuel (1976) 55, 362.

[14] Angert, J. L., Gatton, S. L., Reilly, M. T., and Landolt, R. G., Fuel (1977) 56, 224.

[15] Mayo, F. R., Fuel (1975) 54, 273.

[16] Ghosh, G., Banerjee, A., and Mazumder, B. K., Fuel (1975) 54, 294.

[17] Landolt, R. G., Fuel (1975) 54, 299.

[18] Chakrabartty, S. K., and Kretschmer, H. O., Fuel (1972) 51, 160.

RECEIVED February 10, 1978

8

Chemistry of Coal Liquefaction

FRANK R. MAYO, JOHN G. HUNTINGTON, and NORMAN A. KIRSHEN

SRI International, Menlo Park, CA 94025

This paper is an extension of an earlier review (1) and a summary of four manuscripts submitted to Fuel (2-5).

Coal is considered to be mostly a crosslinked high polymer, with condensed aromatic aggregates that are difficult to cleave and connecting links that are relatively easy to cleave (breakable single bonds). A range of soluble materials is mixed with the predominantly insoluble material. Associated with these primary bond relations are significant interactions between phenolic groups and pyridine-type bases (1,2). Thus, relatively small polyfunctional molecules, containing both phenolic and basic groups, even if not bound to the network, require complexing solvents, such as pyridine, to dissolve them. Table I shows some

Table I

Comparison of Fractions from Extraction and Solvent-Refining of Illinois No. 6 Coal

	Insoluble Extracted Coal	Soluble in / Insoluble in	Fractions: Pyridine / Toluene	Toluene / Hexane	Hexane	
Coal (78.6% C, 5.1% H, 12.4% O) → Pyridine extraction	78.0	% C	79.5	83.0	84.0	More O
	4.8	% H	5.4	6.3	8.7	
	12.6	% O	10.8	7.3	5.0	
		Increasing C + H, decreasing O →				
Coal → Solvent refining	All soluble	% C	86.1	87.4	89.9	More C + H
		% H	4.9	6.1	6.6	
		% O	5.7	4.0	2.0	
Common designations			Preasphaltenes Asphaltols Polar Compounds	Asphaltenes	Oils	
Average molecular Weight			∿ 1200	∿ 600	∿ 300	

0-8412-0427-6/78/47-071-126$05.00/0

results of fractionating the pyridine extract of Illinois No. 6 coal and a solvent-refined coal (SRC) from the same coal (1,2). About two-thirds of the pyridine-soluble materials will dissolve only in pyridine; they have higher oxygen and nitrogen contents and higher molecular weights than the other soluble fractions. Recent experiments indicate that ethylenediamine will extract from coal much more material, of higher molecular weight.

The insoluble extracted coal in the upper series of fractions in Table I, obtained in about 85% yield, is a preferred material for liquefaction studies, uncomplicated by admixed soluble materials. Progress in breaking bonds can be followed by formation of pyridine-soluble products and probably also by increased swelling of the insoluble material in a chosen solvent. The pyridine-soluble, toluene-insoluble fractions of both coal extract and SRC are preferred substrates for following the upgrading of syncrudes. They have fairly high molecular weights and rather narrow molecular weight distributions, and all the fractions fit a smooth curve for number-average molecular weight against retention time in gel-permeation chromatography (2). Both the extracted coal and the pyridine-solube fraction have elementary analyses (dry, mineral free) and nuclear magnetic resonance (NMR) spectra (3) that are very similar; they apparently differ significantly only in molecular size.

As far as we can determine, the only single bonds in coal that might be broken and assist its liquefaction under mild conditions are the following (Ar = aryl; R = aliphatic or benzyl):

$Ar{-}CH_2{-}Ar$

$Ar{-}(CH_2)_n{-}Ar$

$\left.\begin{array}{l} Ar{-}OAr \\ R{-}OAr \\ R{-}O{-}R \end{array}\right\}$ and their S analogs

The same types of bonds are probably involved in scissions under strongly acidic conditions, as with phenol plus boron trifluoride etherate. Aryl-aryl bonds are too strong to be included, and complete cleavage of rings requires breakage of two bonds in the same ring. The breakdown of coal fractions by refluxing alcoholic KOH indicates that an ester group may have to be added to this list (but see below).

We now consider the solvent-refining process in terms of Table I and the bonds broken. The solvent-refining process consists mainly of conversion of insoluble coal to the pyridine-soluble, toluene-insoluble fraction of SRC (1,2). The net result is loss of about 20% of the original carbon as gases and volatile liquids, loss of three-quarters of the original oxygen (mostly as water with process hydrogen), an increase in aromaticity, and some bond breakage. Because the relations between solubility and molecular weight are the same in the coal extract and SRC series (1,2), the net phenol-base relations have changed little, and so the observed bond breakage appears to be associated with net loss of ethers, some bonds of the Ar-$(CH_2)_n$-Ar type, and perhaps the fracture of some rings after hydrogenation. The conversion of SRC to oil or distillate requires a large hydrogen input and a considerable reduction in molecular weight and heteroatom content. The hydrogen is required in part to cap the fragments left when heteroatoms are removed as H_2O, H_2S, and NH_3, but much of the hydrogen may be required to reduce condensed aromatic systems that cannot be cleaved to partially hydroaromatic systems that can be cleaved, especially by reforming catalysts.

This working hypothesis leads us to suggest cheaper alternative routes to liquefaction. First, there are indications that ether cleavage alone, without removal of oxygen or addition of hydrogen, can render coal soluble (6). The alkylation requirement (6) now seems to be associated more with reducing phenol-base interactions (1,2) by O-alkylation (7) than with C-alkylation (6). As well as the alkali metal route, there should be lower-temperature, lower-pressure routes for ether cleavage, as by acids and bases. Whether ether cleavage alone will convert our coal to preasphaltenes, asphaltenes, or oils, or some of each, as in Table I, is not known.

We have investigated the possibility that selective oxidation will cleave many of the breakable single bonds listed above and that we can distinguish among them chemically. Oxidation of diarylmethanes should give ketones that can be converted to esters with a peracid and then hydrolyzed. Longer aliphatic chains between aryl groups should be oxidized directly to pairs of acids. The alkyl ethers should be oxidized to esters. Hydrolysis of esters should then result in cleavage. Diaryl ethers are easily cleaved by alkali metals. Mild oxidations of extracted coal with subsequent base hydrolysis have indeed produced soluble materials (5), but considerable oxygen is incorporated. Oxidation and NMR studies indicate that the SRC fractions with the most aromatic hydrogen are most reactive, and that this kind of hydrogen is lost preferentially in oxidation (5).

Similar mild oxidations or treatments with alcoholic KOH of the pyridine-soluble, toluene-insoluble fraction of coal extract have resulted in marked reductions in molecular weight and considerable solubility in alcoholic KOH, with only moderate increases in oxygen content (5). The similar effect of the two treatments and the absence of a large additive effect in successive treatments raise doubts about the presence of ester groups, which would be unaffected by oxidation alone.

Investigations of relative reactivites of pure aromatic compounds in competitive oxidations (4) have brought out some difficulties in breaking down coal by mild oxidation. By far the most reactive compounds are anthracene and its alkyl derivatives,

9,10-dihydroanthracene, tetralin, xanthene, and fluorene, all of which oxidize initially to peroxides or ketones without carbon-carbon cleavage. Benzene derivatives in the list of breakable single bonds are one or two orders of magnitude slower in competitive oxidations. Benzyl-type hydrogen atoms next to naphthalene and phenanthrene rings are intermediate in reactivity. These considerations suggest that selective oxidation of breakable single bonds in coal will be difficult without loss of fuel values in oxidizing more reactive groupings, but our moderate success with the pyridine-soluble fraction of coal extract, and the more limited success with extracted coal (5), indicate that the proportions of the very reactive but unbreakable cyclic compounds must be low. The difference between the soluble and insoluble fractions may correspond to the differences between branching and crosslinking, with more oxidation being required to break down the network structure.

These studies have led us to try to dehydrogenate hydroaromatic groups in coal to aromatic groups. We have had some success when ethylene was used as an acceptor for hydrogen in the pyridine-soluble fraction. We hope that this dehydrogenated, still-soluble material will have more tendency to oxidize at breakable single bonds, and increase the possibility that mild oxidation may be used to break down coal more efficiently. The ethane obtained from dehydrogenation could serve as a source of hydrogen in further processing.

The report of Chakrabartty and Berkowitz (8) that coal is oxidized mostly to carbon dioxide and benzene polycarboxylic acids with household bleach (0.8 M aqueous sodium hypochlorite) and their conclusion that bituminous coal consists mostly of benzene rings and polyamantane-like structures aroused considerable interest and opposition and stimulated a more careful study (3). Our work was carried out mostly with pyridine-extracted Illinois No. 6 coal at 30°. It now appears that at pH 11, with an extremely finely divided coal, and repeated treatments with small proportions of hypochlorite, the major product is a mixture of black, bicarbonate-soluble acids of molecular weight above 1000. These acids appear to have become soluble by destruction of some aromatic, possibly phenolic, rings, with little loss of aliphatic carbon or cleavage of our "breakable single bonds" (3). However, oxidation of a coarser substrate leads to more degradation, and oxidation with excess hypochlorite leads to the products reported by Chakrabarrty and coworkers. The black acids contain a high proportion of aliphatic carbon, but such a low proportion of hydrogen to carbon that they must contain a high proportion of tertiary aliphatic carbon atoms, and probably of bridged and condensed aliphatic rings. Consideration of the coal-forming process, however, indicates that the probability of formation of regular, polyamantane, ring structures is unlikely. The soluble black acids represent a new and challenging starting point for investigating the aliphatic material than connects the condensed aromatic structures in coal. The susceptibility of some, but not all, substrates to oxidation by sodium hypochlorite is pH dependent (3).

Acknowledgement

Various portions of this research were supported by the U.S. Energy Research and Development Administration under Contract EF-76-C-01-2202, and by the National Science Foundation under Grants AER74-13359 A02 and AER75-06143.

Abstract

The structure of bituminous coal is discussed in terms of aromatic aggregates and breakable single bonds. The solvent-refining process is interpreted, and low-temperature, low-pressure alternatives are discussed. The latter include combinations of selective oxidation, dehydrogenation, alkali treatment, and various other means of ether cleavage. Aqueous sodium hypochlorite is a remarkable oxidizing agent, converting some of the aromatic aggregates to acids and CO_2, and leaving a soluble remainder as a new material for structure determination. Such investigations of the structure of coal should yield new information on what bonds in coal are most easily broken for liquefaction and how this process might be accomplished most economically.

Literature Cited

1. Mayo, Frank R., Huntington, John G., and Kirshen, Norman A., "Preprints of the 1976 Coal Chemistry Workshop," p. 189, Stanford Research Institute, Menlo Park, (1976).

2. Mayo, Frank R., and Kirshen, Norman A., "Comparison of Fractions of Pyridine Extract and Solvent-Refined Coal from Illinois No. 6 Coal," manuscript submitted to Fuel.

3. Mayo, Frank R., and Kirshen, Norman A., "Oxidations of Coal by Aqueous Sodium Hypochlorite," manuscript submitted to Fuel.

4. Huntington, John G., Mayo, Frank R., and Kirshen, Norman A., "Mild Oxidations of Coal Models," manuscript submitted to Fuel.

5. Huntington, John G., Mayo, Frank R., and Kirshen, Norman A., "Mild Oxidations of Coal Fractions," manuscript submitted to Fuel.

6. Sternberg, H. W., Delle Donne, C. L., Pantages, P., Moroni, E. C., and Markby, R. E., Fuel (1971), 50, 432.

7. Flores, Ruben A., Geigel, Maria A., and Mayo, Frank R., "Hydrogen Fluoride-Catalyzed Alkylations of Pittsburgh Seam (hvb) Coal," manuscript submitted to Fuel

8. Chakrabartty, S. K., and Berkowitz, N., Fuel (1974), 53, 240, and previous papers.

RECEIVED February 10, 1978

9

Early Stages of Coal Carbonization: Evidence for Isomerization Reactions

S. K. CHAKRABARTTY and N. BERKOWITZ

Fuel Sciences Division, Alberta Research Council, Edmonton, Alberta, Canada

At temperatures between 350° and 425°C, the molecular structures that characterize coal are rapidly, and very obviously, transformed into more stable carbon configurations through loss of "volatile matter" (as tar and gas). But little is known about possible configurational changes at lower temperatures. Thermograms of coal (1) indicate specific heat effects as endo- and exotherms from 200°C up, but because enthalpy changes recorded in this manner include sensible heats as well as heats of reaction, it is difficult to assess the nature of the chemical processes which produce the thermograms. On the other hand, low-temperature chemical changes, if such did in fact occur, should be reflected in the "reactivity" of heat-treated coal - and, in particular in its response to oxidation; and if oxidation could be performed so as to yield identifiable products, it should be possible to detect the major configurational changes in the distribution of oxidation products.

It was previously shown (2) that subbituminous coals and derivatized bituminous coals react readily with sodium hypochlorite, and that they are thereby completely converted into water-soluble materials. The principal products are carbon dioxide and carboxylic acids. For a typical bituminous coal, 18 per cent of total carbon survive oxidation as single benzene ring - 6 per cent as 2, 3 or 4 condensed aromatic (and/or heteroaromatic) rings, 7 per cent as methyl or methylene and 20 per cent as carboxyl groups. Acetic, propionic, succinic, glutaric, adipic, benzene- and toluene-carboxylic acids constitute the simple products of oxidations. In all, over 90 per cent of carbon in coal could be accounted for from the product yields.

Although there are some uncertainties about the mechanism of coal-oxidation by hypochlorite, it is reasonable to expect that, under constant reaction conditions, the same mechanism would hold for variously heat-treated coal samples. Accordingly we thought it pertinent to determine whether this technique could be used to monitor low-temperature changes in coal-carbonization. The present paper reports the results of such an

0-8412-0427-6/78/47-071-131$05.00/0

exploratory study.

Experimental and Results

For the purposes of this investigation, two coals - a Western Alberta lvb coal with C = 90%, and a Kentucky hvb coal with C = 85% - were used.

Ten g samples of these coals, each sized to -60 +115 mesh, were preheated in helium for 2 hr at the desired temperature, cooled, and then stored under pure He until required. No significant weight losses or changes in elemental compositions were observed with either coal up to 375°C, but 5-10% weight losses, and slight (0.5-1.2%) increases in carbon contents, with corresponding reductions in oxygen were noted after preheating at 390°-400°C.

For the oxidation experiments, 2 g (preheated) samples were first "activated" by reaction with nitronium-tetrafluoroboate in acetonitrile, and thereafter treated with 125 ml of an aq 1.6N sodium hypochlorite solution at 60°C. The pH of the reaction mixture was maintained at 12 by adding NaOH pellets at regular invervals. When reaction was complete, the mixture was acidified; insoluble matter was filtered off; and soluble carboxylic acids were extracted with ether. The residual solution was freed of water by low-pressure distillation at 40°C, and solid material left behind was extracted with anhydrous methanol.

The ether- and methanol-extracts were then combined, converted to methyl esters by reaction with diazomethane, and separated by gel permeation chromatography on a (Water Associates') Poragel column into two fractions with molecular weights <600 and >600 respectively (fractions A and B). Fraction B (mol wt <600) was further subdivided into "simple" (B,1) and "complex" (B,2) acids by elution chromatography on a Florisil column. (From this column, methyl esters of "complex" acids could only be eluted with 10:1 chloroform-methanol, while esters of "simple" acids could be taken off with pentane, hexane and 2:1 hexane-chloroform.)

The "simple" acids (fraction B,1) were quantitatively analyzed by GC on OV-17/Chromosorb WHP and reference to peak area vs. concentration diagrams for authentic compounds. Where no authentic compounds were available for methyl esters of toluene carboxylic acids, concentrations were computed from response factors derived from those appropriate for the methyl esters of benzene carboxylic acids.

Detailed study of the oxidation products showed that the raw and variously preheated coals furnished substantially identical amounts of carbon dioxide (Figure 1), but that there were significant, though complex, variations in the yields of carboxylic acids (Figures 2-6) and that these yields depended on the nature of the coal as well as on the preheat temperature. Thus, while the lvb coal had to be preheated to at least 350°C

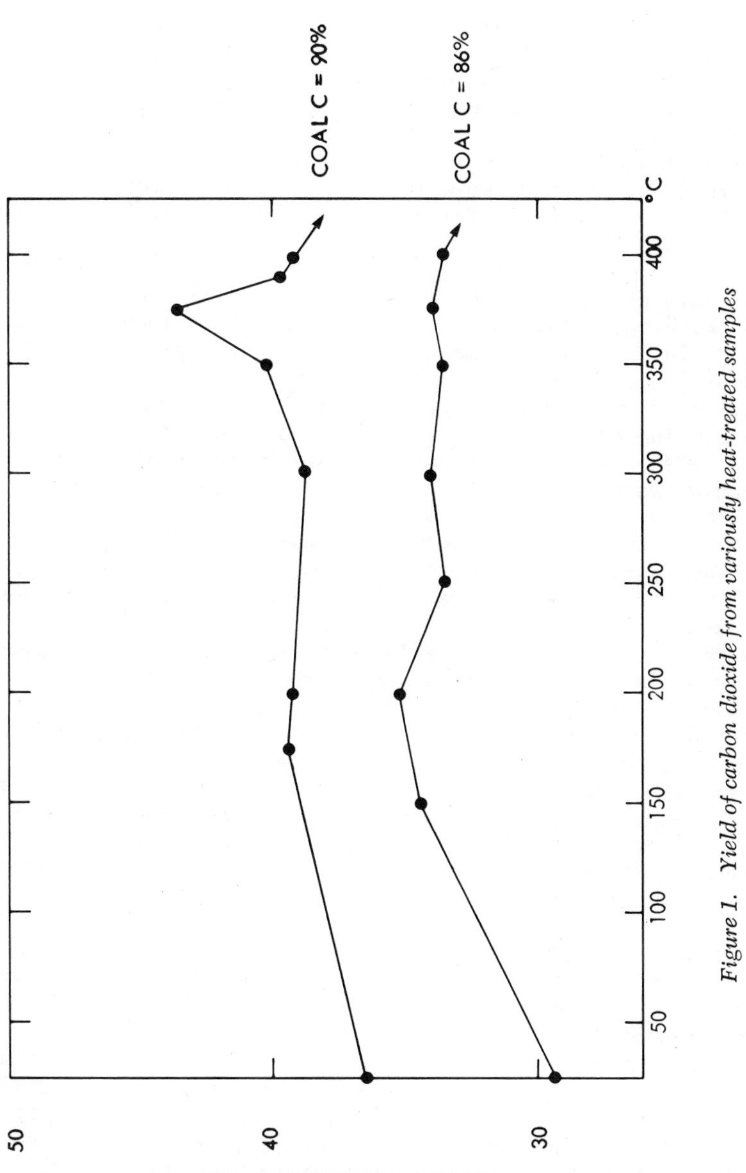

Figure 1. Yield of carbon dioxide from variously heat-treated samples

before it produced greater than the initial amounts of carboxylic acids, the hvb coal needed only to be preheated to 200°C before doing so (Figure 2).

These observations accord well with the enthalpy changes recorded by differential thermal analysis (1), and are, in our view, clearly indicative of thermally induced alteration of the initial carbon configurations in the two coals.

Fraction A - composed of acids with mol. wts. in excess of 600 - presumably represents condensed aromatic fragments of the original and heat-treated samples; and if so, the fact the yields of this fraction increase with preheat temperatures suggests that aromatization begins at temperatures as low as 150°C in hvb coal and at 200°C in lvb coal (Figure 3).

Fraction B,2 - which represents over 50 per cent of the total acid product from each sample and consists of "complex" acids with mol. wts. <600 - can be tentatively identified as originating in coal fragments that produce "pre-asphaltenes" in coal liquefaction and are also interesting (Figure 4). In the case of lvb coal, B,2 yields reach a maximum at 375°C and then decline precipitously at 390°C, but the decline is compensated by an almost equivalent increase in the yield of high molecular weight (>600) material. In contrast, B,2 yields from the hvb coal attain a maximum at 200°C, fall to minimum at 300°C, and then rise to another maximum at 400°C.

The "simple" acids with mol. wts. <600 (Fraction B,1), which were completely characterized and invariably accounted for 22-30 per cent of the total acid products, are evidently produced from easily oxidizable open-structured coal fragments; and from the lvb coal, the yield of this fraction decreased steadily as preheat temperatures rose. However, in the case of the hvb coal, the yield was found to remain constant up to 300°C, and to fluctuate thereafter (Figure 5).

Finally, some note must be taken of the distribution pattern of penta- and hexa-carboxyl benzenes vis-a-vis that of tri- and tetra-carboxyl toluenes. Quite generally, maxima for benzenes are almost coincident with minima and maxima of toluenes. But here again, significant differences between lvb and hvb coal are observed. For the lvb coal, maxima of benzenes lie at 175° and 375°C, while for the hvb coal, they appear at 350° and 400°C. Maximum yields of toluenes were obtained from lvb coal after preheating at 300°C, and from hvb coal after preheating at 150°C (Figures 6 and 7).

Since a methyl group attached to an aromatic ring is most reactive towards hypohalite oxidant, the variation in the yield of toluene carboxylic acids has added significance in terms of the chemistry of low-temperature heating effect. During heat treatment, presumably, a methylene or methine group is converted to methyl group which occupies relatively less active position to survive oxidation.

We believe that these variations in yield and distribution

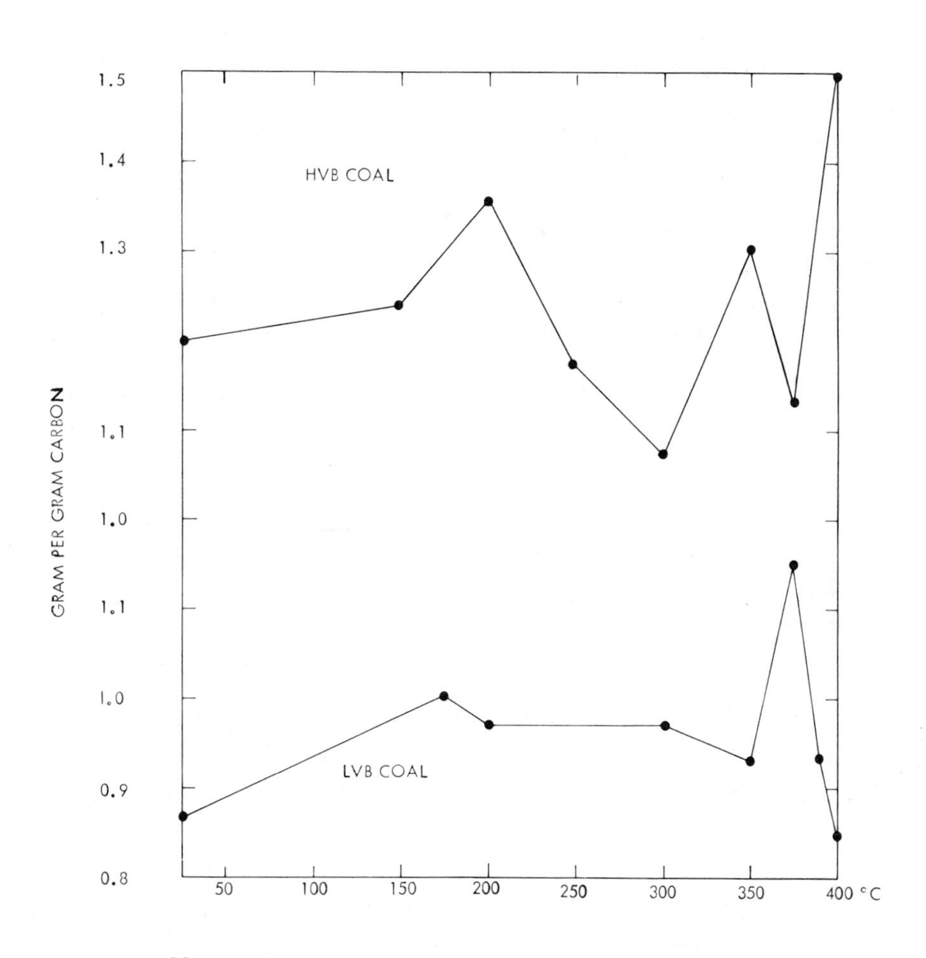

Figure 2. Yield of total carboxylic acids as methyl esters from variously heat-treated samples

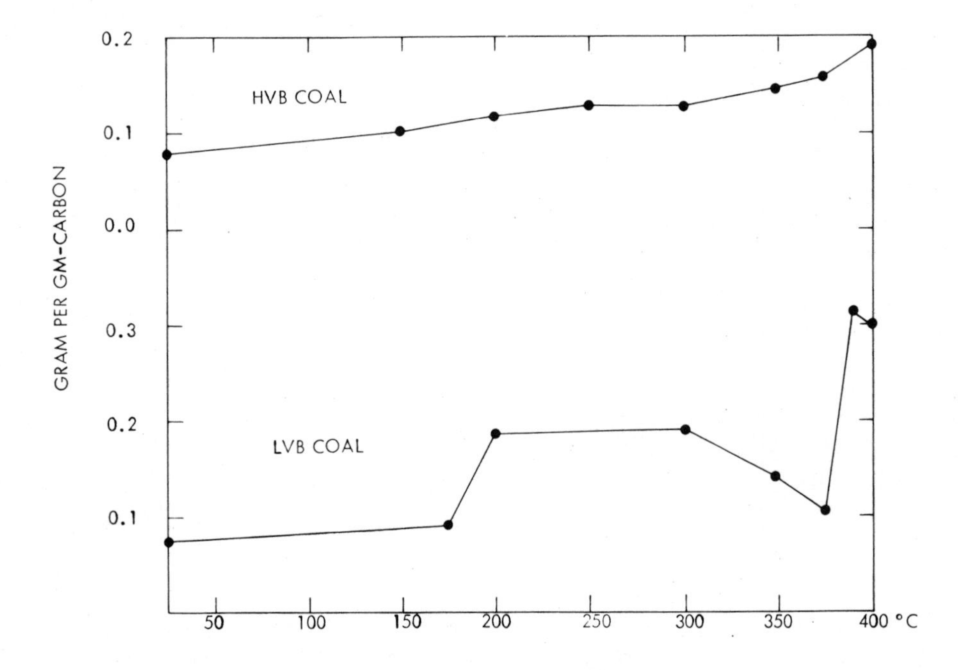

Figure 3. Yield of complex acids as methyl esters, M.W. > 600 from variously heat-treated samples

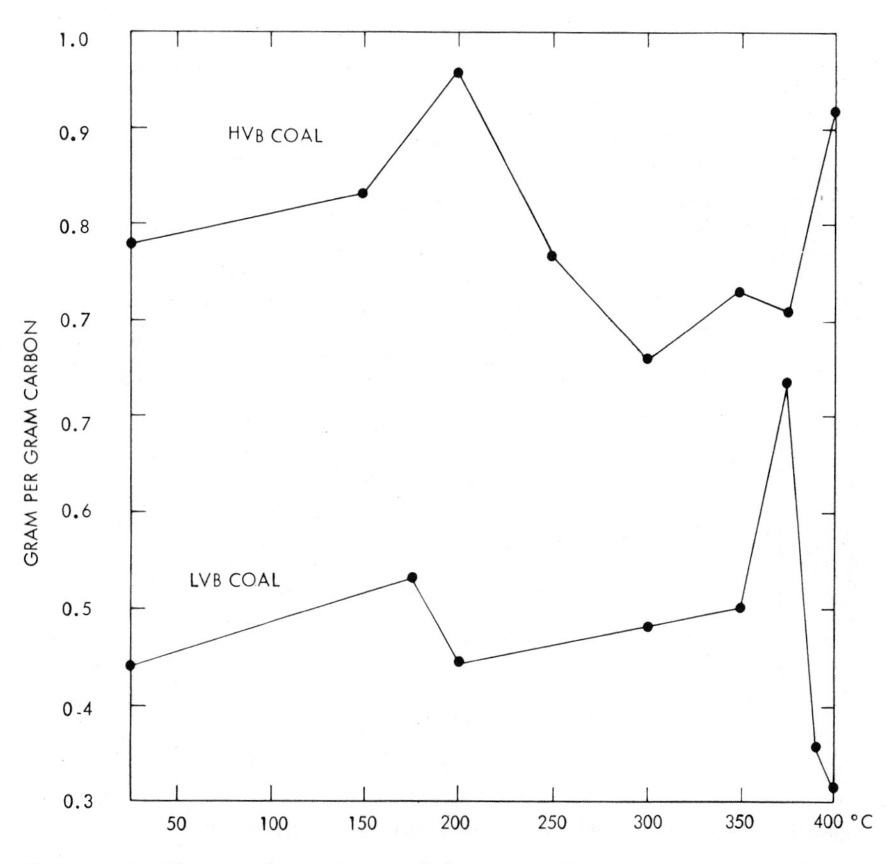

Figure 4. Yield of complex acids as methyl esters, M.W. < 600 from variously heat-treated samples

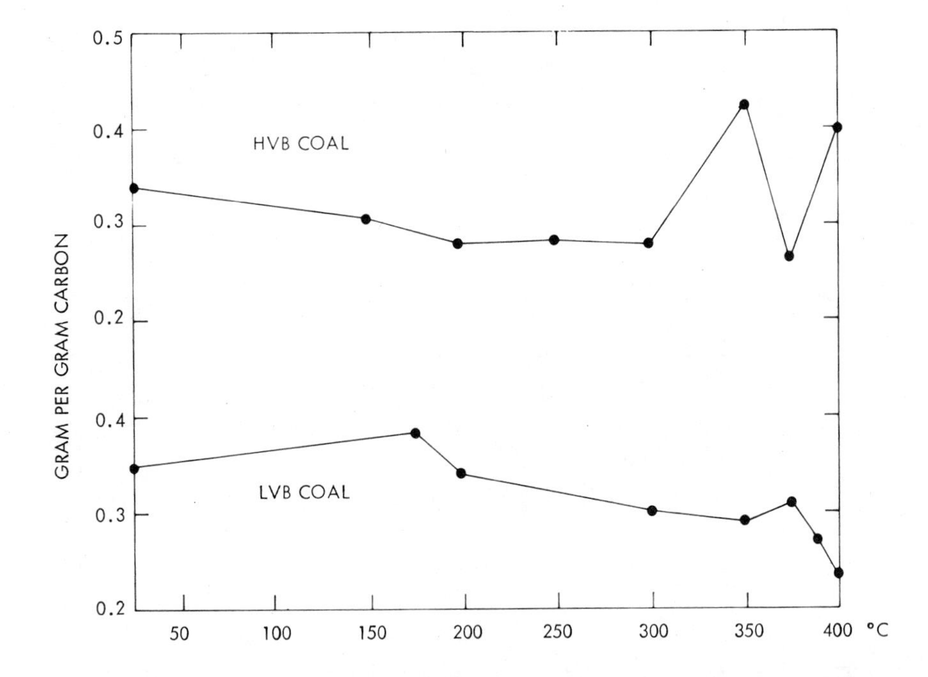

Figure 5. Yield of simple acids as methyl esters from variously heat-treated samples

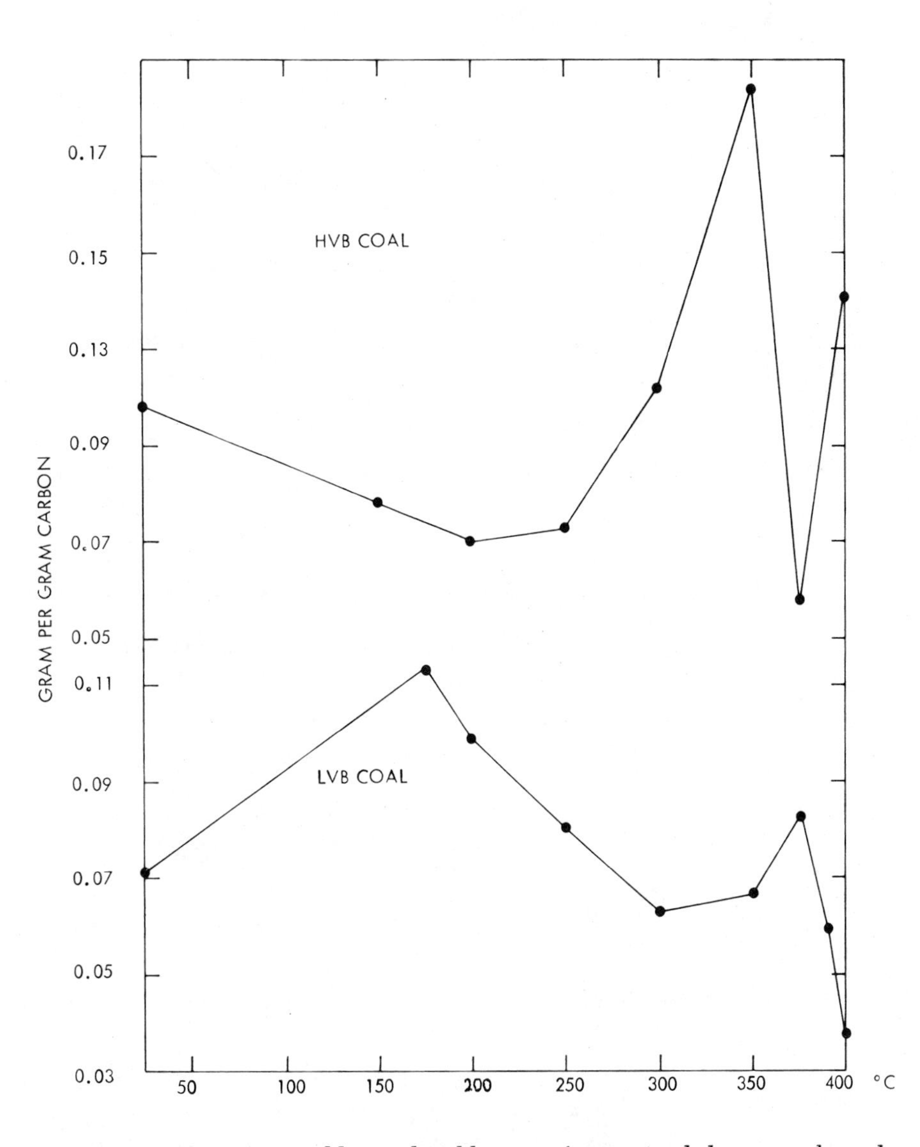

Figure 6. Yield of penta- and hexacarboxyl benzenes from variously heat-treated samples

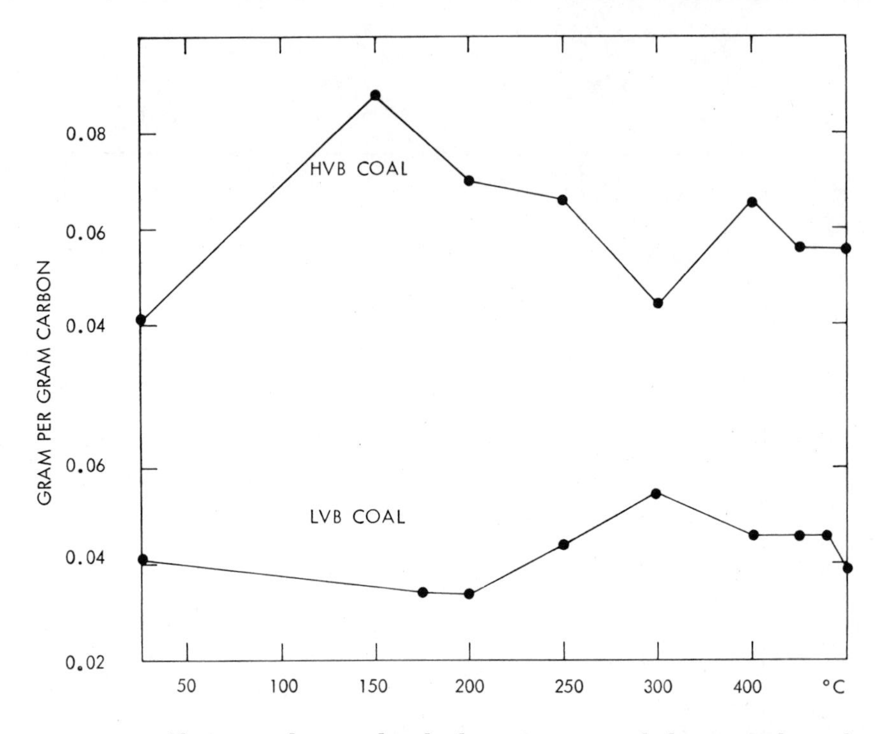

Figure 7. Yield of tri- and tetracarboxyl toluene from variously heat-treated samples

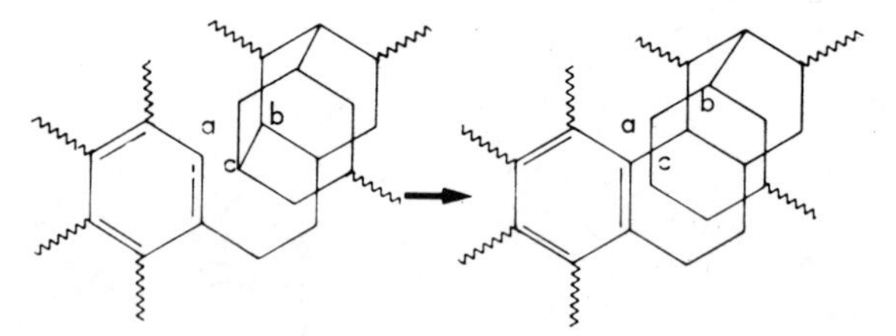

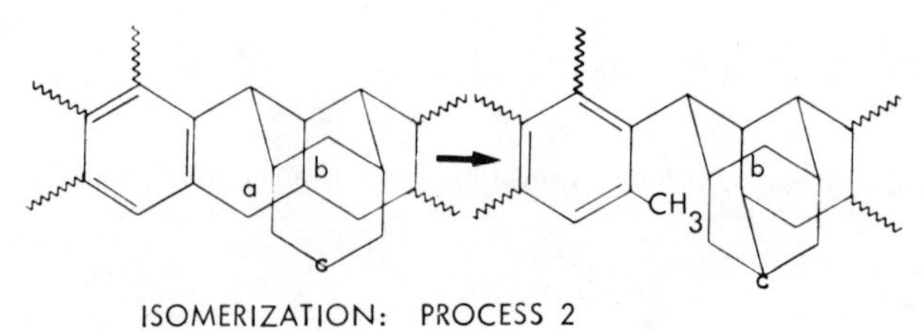

Figure 8.

of carboxylic acids are best understood in terms of low-temperature intra-molecular isomerization reactions which modify the initial carbon-hydrogen configurations of the coal well before it begins to undergo thermal cracking. Of the two specific processes that could be postulated as effecting such isomerization, one is trans-annular bond formation and the other is isomerization of benzylic carbon to methylphenyl derivatives and more complex ring systesm (Figure 8). Either change could readily occur at temperatures as low as 150°C and create structures susceptible to oxidation by sodium hypochlorite.

We note, in this connection, that ether-oxygen also appears to play a major role in the behaviour of coal at elevated temperatures (4); but whether or not isomerization and/or cleavage of ether-linkages also occurs at low temperatures could not be determined in this study, since hypochlorite oxidation easily degrades heterocyclic functional elements under acidic as well as basic conditions.

Abstract

Detailed study of the oxidation products obtained by reaction of preheated coals with sodium hypochlorite shows significant, though complex, changes in the yields of various benzene and toluene polycarboxylic acids. These observations can be accounted for by postulating a series of trans-anular bond formations at 170-200°C and isomerization of benzylic carbon with formation of methyl-phenyl derivatives between 200° and 300°C.

Literature Cited

1. Kirov, N. Y. and Stephens, J. N.; Physical Aspects of Coal Carbonization Research Monograph, U. of New South Wales, Sydney, Australia, Chap. 12 (1967)
2. Chakrabartty, S. K. and Kretschmer, H. O.; Fuel (Lond.) 53. 132 (1974)
3. Whitehurst, D. D. and Mitchell, T. O.; ACS Div. Fuel Chem. Preprints 21 (5), 127 (1976)
   Farcasiu, M., Mitchell, T. O. and Whitehurst, D. D.; ACS Div. Fuel Chem. Preprints 21 (7), 11 (1976)
4. Lazarov, L. and Angelova, G.; Fuel (Lond.) 50, 338 (1971)
   Szladow, A. J. and Ignasiak, B.; Fuel (Lond.) 55, 253 (1976)
   Ignasiak, B. and Gawlak, M.; Fuel (Lond.) 56, 216 (1977)

RECEIVED February 10, 1978

10

Electron Spin Resonance Studies of Coals and Coal-Derived Asphaltenes

H. L. RETCOFSKY, G. P. THOMPSON, M. HOUGH, and R. A. FRIEDEL

U.S. Department of Energy, Pittsburgh Energy Research Center, 4800 Forbes Avenue, Pittsburgh, PA 15213

The discovery of electron spin resonance (esr) absorption in natural carbons by Uebersfeld (1) and Ingram (2) prompted a number of investigators to apply the technique to coal and materials derived from coal. At least three excellent review articles describing the early esr studies of coals have been published (3-5). Although the exact nature of the species responsible for the esr absorption has not been established unambiguously, it is generally thought that the unpaired electrons, at least in non-anthracitic and possibly young anthracitic coals, are associated with organic free radical structures. One group of investigators, however, has proposed that charge-transfer complexes rather than stable free radicals may be responsible for the absorption (6).

During the present investigation, esr spectra were obtained for vitrains and fusains from a large number of coals. For most ranks of coal, samples of both lithotypes were studied. The objectives of the investigation were: 1) to better characterize the immediate chemical environment of the unpaired electrons; 2) to deduce information about the metamorphic changes that occur during vitrinization and fusinization; and 3) to provide needed background information for future studies of the role of free radicals in coal liquefaction. To further pursue the latter objective, esr spectra of asphaltenes, which are considered by many coal researchers to be intermediates in the conversion of coal to liquid fuels, were also obtained. A secondary purpose for examining the asphaltenes was to explore the recent hypothesis (7) that charge-transfer interactions may be important binding forces between the acid/neutral and base components in these materials.

EXPERIMENTAL

Samples. Most of the vitrains and fusains studied were of

0-8412-0427-6/78/47-071-142$05.00/0

high petrographic purity, although several of the vitrains are more appropriately described as vitrain-rich samples. Sixty-three vitrains, including samples from Antarctica, Austria, Canada, Germany, Japan, Pakistan, Peru, the United States, and Yugoslavia, and 30 fusains, most of which were separated from U. S. coals, were investigated. Petrographic characterization and other properties of many of these samples were published previously (8).

The asphaltene samples were derived from products from the Pittsburgh Energy Research Center's SYNTHOIL (9) coal liquefaction Process Development Unit using a recently described solvent separation method (10). The acid/neutral and base components of the asphaltenes were prepared according to published procedures (11).

Spectral Measurements. The esr measurements were made over a period of approximately ten years. The experimental techniques were essentially those published earlier (12) except for minor modifications that were made over the years to facilitate the measurements. All esr measurements were made on evacuated samples ($\sim 10^{-6}$ torr) to prevent line broadening by oxygen in the air. The electrical properties of many of the higher rank samples necessitated that the samples be dispersed in a non-conducting medium to avoid microwave skin effects.

RESULTS AND DISCUSSION

Vitrains and Fusains. ESR data obtained for vitrains and fusains during the present investigation are far too numerous to tabulate here; complete listings of the data are available from the authors upon request. The plots of Figures 1-6 show some of the more significant correlations of the data with coal composition or coal rank.

The concentrations of unpaired electrons in the vitrains, as estimated by comparing the esr intensity of each sample with that of a standard sample of diphenylpicrylhydrazyl, are shown as a function of the carbon contents of the samples in Figure 1. The relationship of Figure 1 is similar to those from earlier studies (3-5, 12), although the scatter of the data points is more pronounced in the present work. These data show that, in general, the concentrations of unpaired electrons increase with increasing coal rank up to a carbon content of approximately 94% after which the spin concentrations decrease rapidly. The initial, crudely exponential increase in spin concentration is generally attributed to the formation of organic free radicals during vitrinization. The free radical electrons are thought to be delocalized over aromatic rings and thus stabilized by resonance. Resonance stabilization of the radicals is greater for the vitrains from higher rank coals since these presumably contain the larger polynuclear condensed aromatic ring systems.

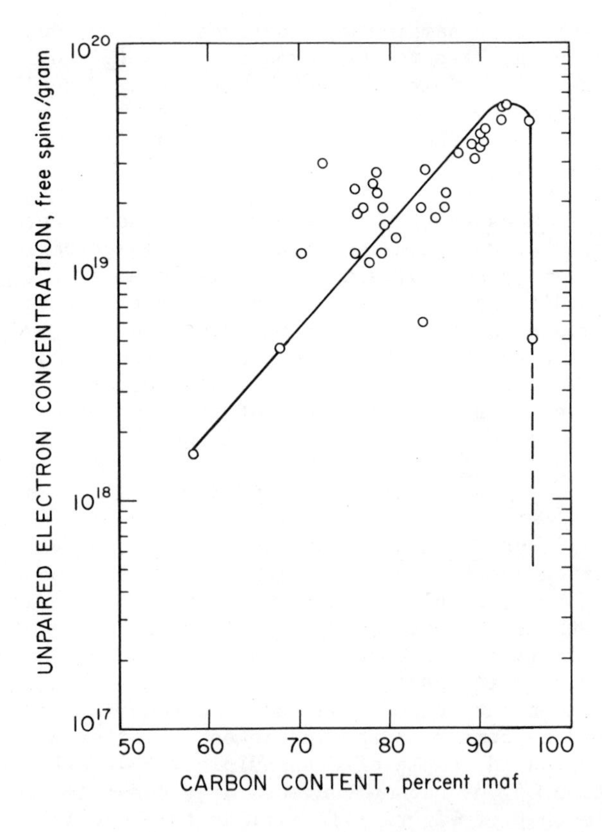

Figure 1. Concentrations of unpaired electrons as a function of carbon content for vitrains from selected coals

The precipitous decrease in spin concentration above 94% C results from the increased conductivity of the samples.

In contrast to the results for vitrains, the spin concentrations of the fusains (Figure 2) exhibit no readily discernable dependence on carbon content. The wide variation in thermal history experienced by fusains during their formation is the most likely explanation for this behavior (13).

The changes that occur in esr linewidths and g values during vitrinization and fusinization are shown in Figures 3-6. The abscissa used in these figures is based on the coalification plots of Schopf (14) with slight modifications by Parks (15) and the present authors. The ranks shown on the plots are those of the parent coals as determined by standard procedures (16).

For the vitrains, an increase in esr linewidth with increasing rank is first observed (Figure 3); this trend is reversed at the low rank bituminous stage. The rate of decrease becomes larger as coalification progresses through the higher rank bituminous stages to the early anthracitic stages. Some of the anthracites and most of the meta-anthracites exhibit very broad lines (not shown in the figure); a linewidth in excess of 60 gauss was observed for one such coal. The linewidth results can be interpreted as follows: Nuclear broadening, i.e., unresolved proton-electron hyperfine interactions, plays an important role in the observed linewidths of peats, lignites, and bituminous coals. The relatively narrow lines observed in the spectra of some of the anthracites probably result from the smaller number of protons in the samples, although exchange narrowing of the esr resonances may also be occurring. The proton line broadening hypothesis is supported by a recent esr study of coals before and after catalytic dehydrogenation (17). The very large linewidths of the highest rank materials are undoubtedly due to the presence of graphite-like structures which form during the latter stages of coalification. The difficulty in differentiating between anthracites and meta-anthracites (18) may be responsible for the apparent lack of predictability of linewidths in vitrains from coals of these ranks.

The esr linewidths for the fusains (Figure 4) are very small, frequently less than one gauss, except for samples from the lowest rank coals. Unlike the results for the vitrains, no evidence was found for the formation of graphitic structures during the latter stages of fusinization. In addition, the gradual decrease in linewidths of vitrains as coalification proceeds from the low rank bituminous stages to the early anthracitic stages (Figure 3) appears as a very rapid change in the fusinization plot (Figure 4). This is in accord with Schopf's (14) representation of fusinization as a process which has an early inception and progresses rapidly in the peat and lignitic stages, after which the metamorphic changes become

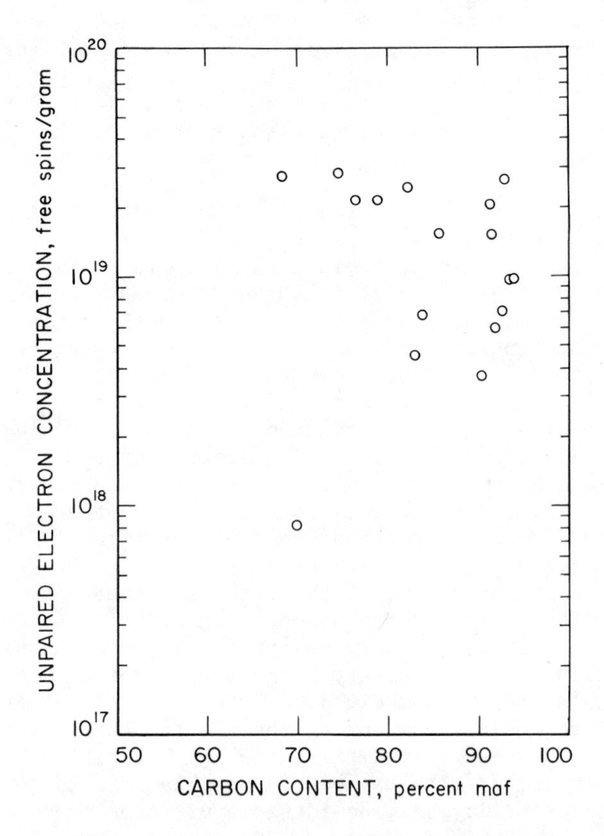

Figure 2. Concentrations of unpaired electrons as a function of carbon content for fusains from selected coals

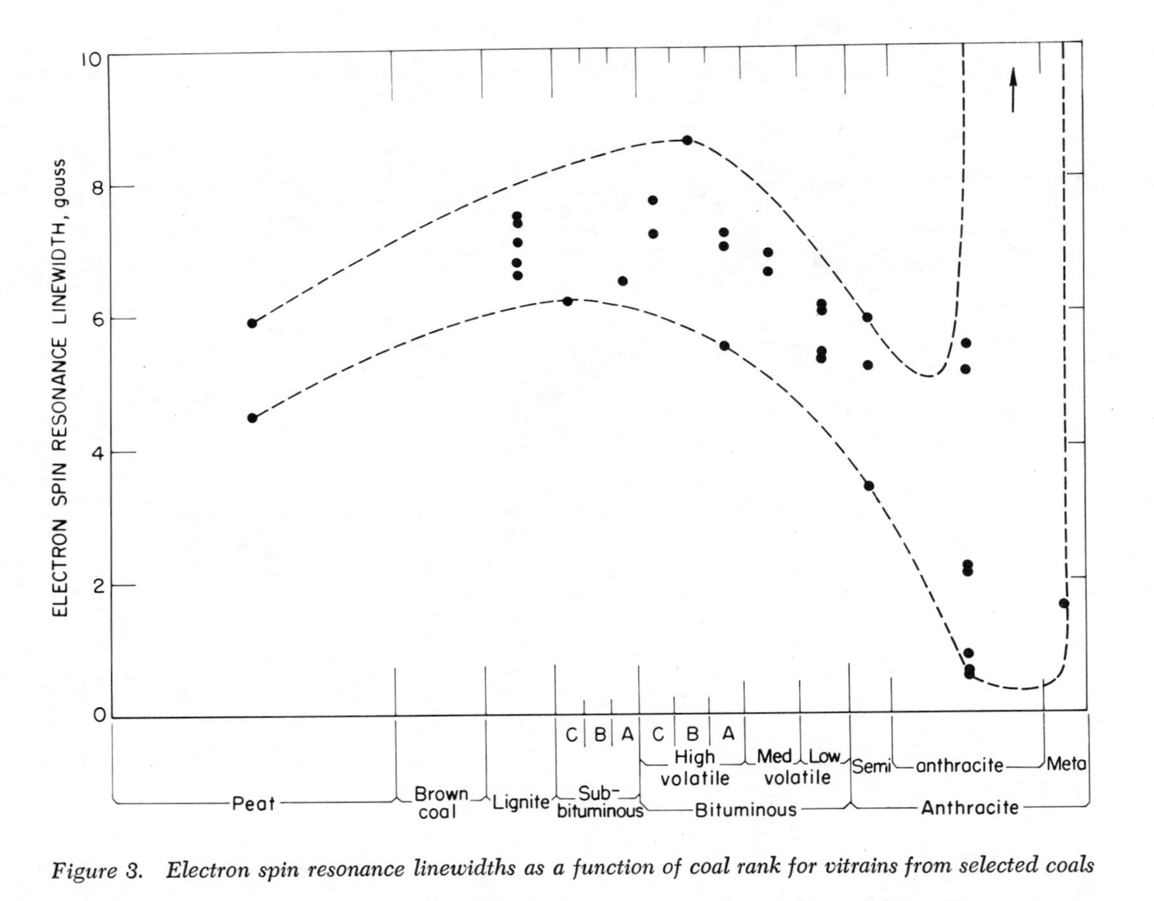

Figure 3. Electron spin resonance linewidths as a function of coal rank for vitrains from selected coals

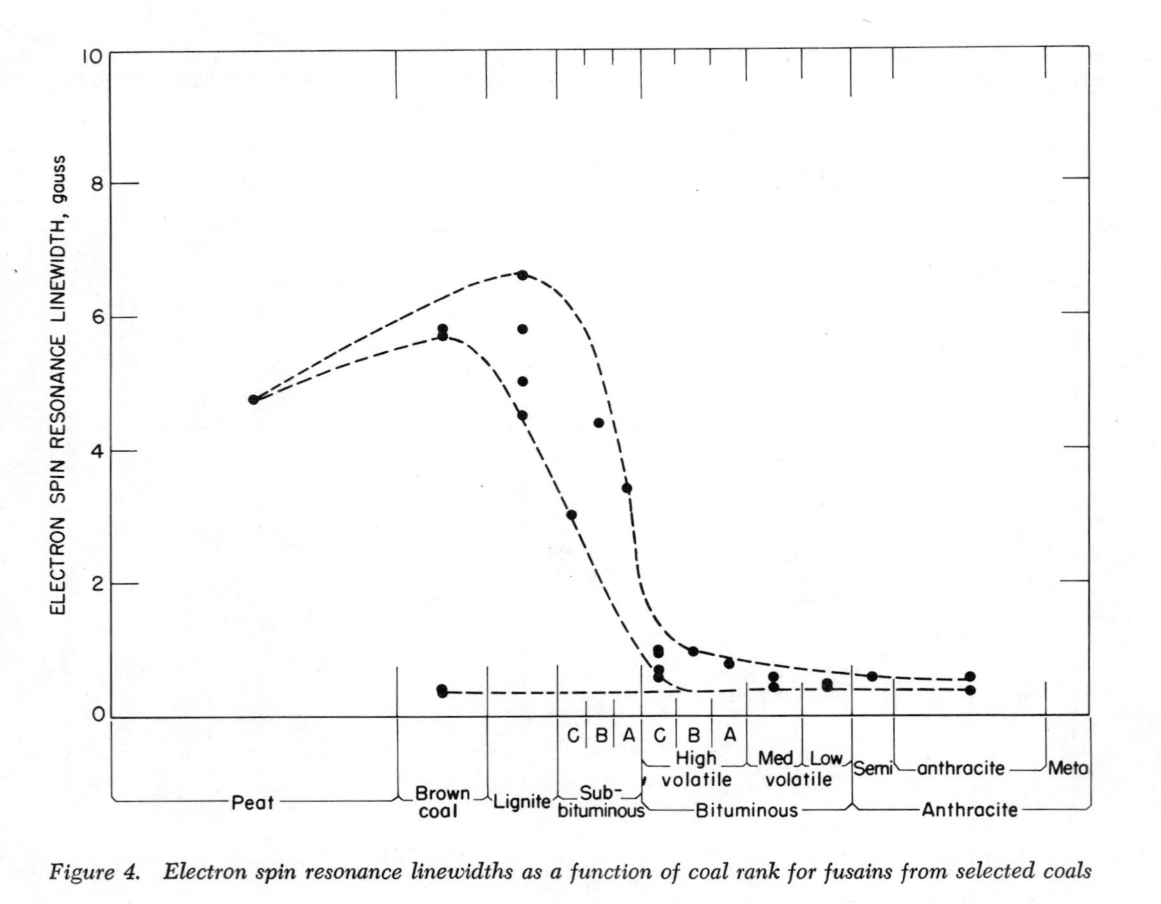

Figure 4. Electron spin resonance linewidths as a function of coal rank for fusains from selected coals

nearly imperceptible. Vitrinization, on the other hand, involves a progressive change throughout all stages of rank development.

The relationship between the g values of the lithotypes and coal rank (Figures 5 and 6) also supports Schopf's theories of vitrinization and fusinization. The large g values found for the vitrains from meta-anthracites is in accord with the final step in vitrinization being the fusing of aromatic rings into graphite-like structures. The g value of each of the vitrains and fusains is higher than that of the free electron and lies in the spectral region expected for simple organic free radicals. The only exceptions are vitrains from the more highly metamorphized coals, one of which exhibited a g value of 2.011.

The fact that esr g values of organic free radicals are greatest for radicals in which the unpaired electron is localized or partially localized on atoms having high spin-orbit coupling constants can be used to explain the g value results for vitrains. Since the heteroatom contents of coals decrease with increasing rank, the high g values for peats and lignites can be interpreted in terms of aromatic radicals with some partial localization of the unpaired electrons on heteroatoms, particularly but not exclusively oxygen. As coalification progresses the g values decrease, suggesting that the radicals become more "hydrocarbon-like." The g values of many of the vitrains from bituminous and young anthracitic coals compare favorably with those exhibited by aromatic hydrocarbon radicals. During the final stages of coalification, the g values become quite large as one would expect if continued condensation of the aromatic rings into graphite structures occurs. The observation of a small, but reproducible, anisotropy in the g value of certain anthracites (Figure 7) suggests that some ordering of the polynuclear condensed aromatic rings is occurring.

Coal-Derived Asphaltenes. To better understand the chemistry of coal liquefaction, an esr investigation of coal-derived asphaltenes was initiated. Preliminary results are presented here. Of particular concern was the temperature variation of the esr intensities of asphaltenes and their acid/neutral and base components (Figure 8). The most significant finding to date is that the weighted average of the temperature dependencies of the two components reproduces the temperature dependence of the total asphaltene (before separation) exceptionally well. This suggests that charge transfer interactions, at least in the Mullikan sense, are relatively unimportant binding forces between the acid/neutral and base components of the asphaltenes.

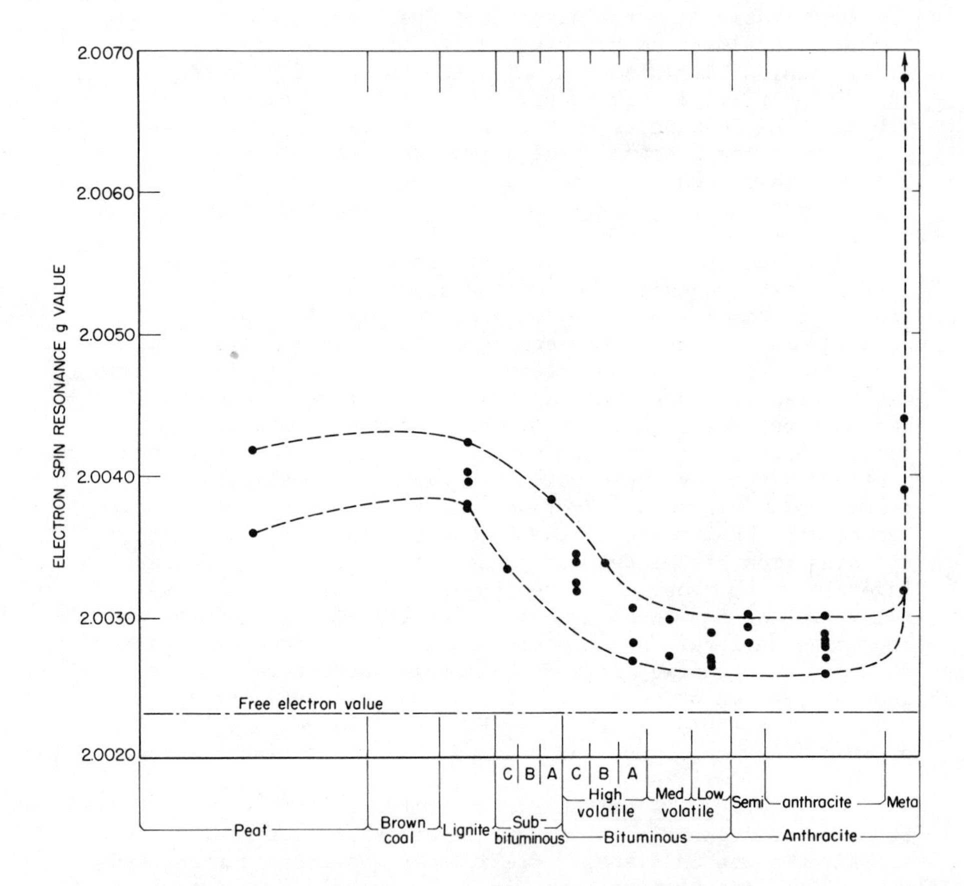

Figure 5. Electron spin resonance g values as a function of coal rank for vitrains from selected coals

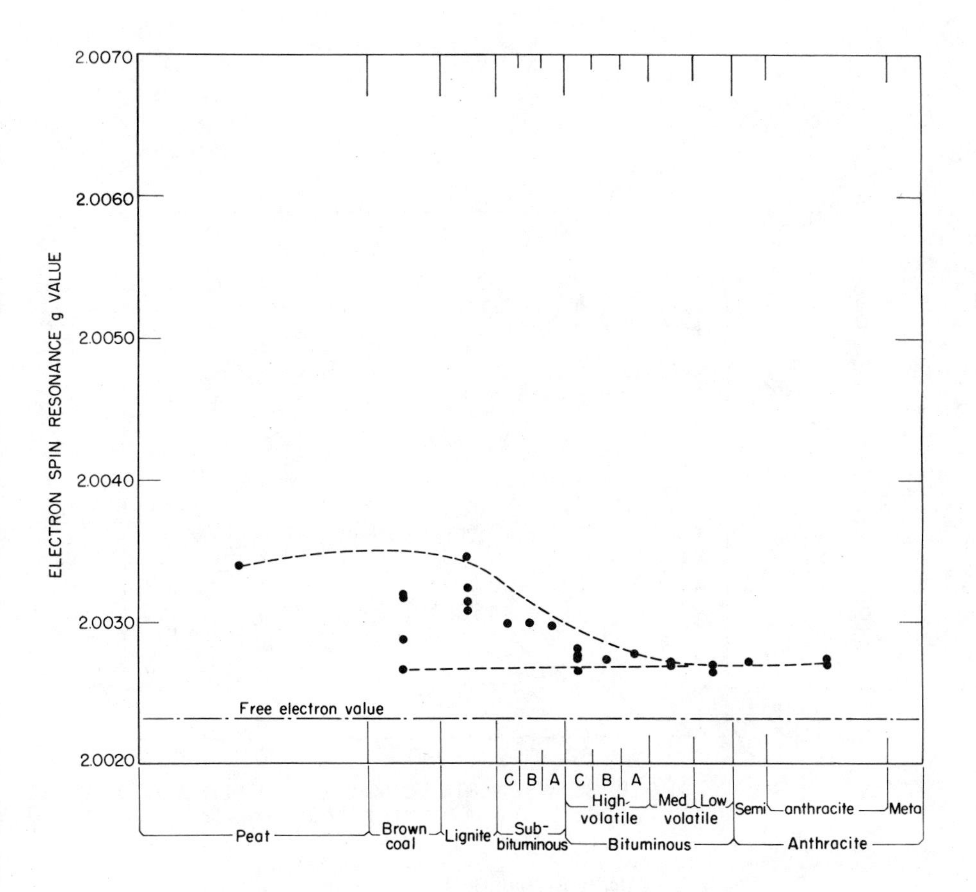

Figure 6. Electron spin resonance g *values as a function of coal rank for fusains from selected coals*

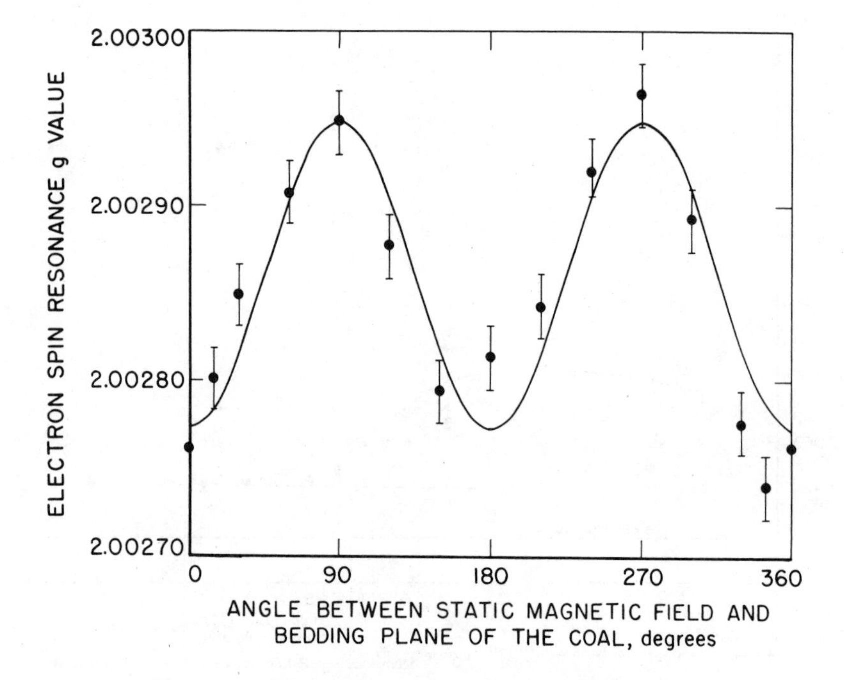

Figure 7. Value anisotropy in Huber Mine anthracite

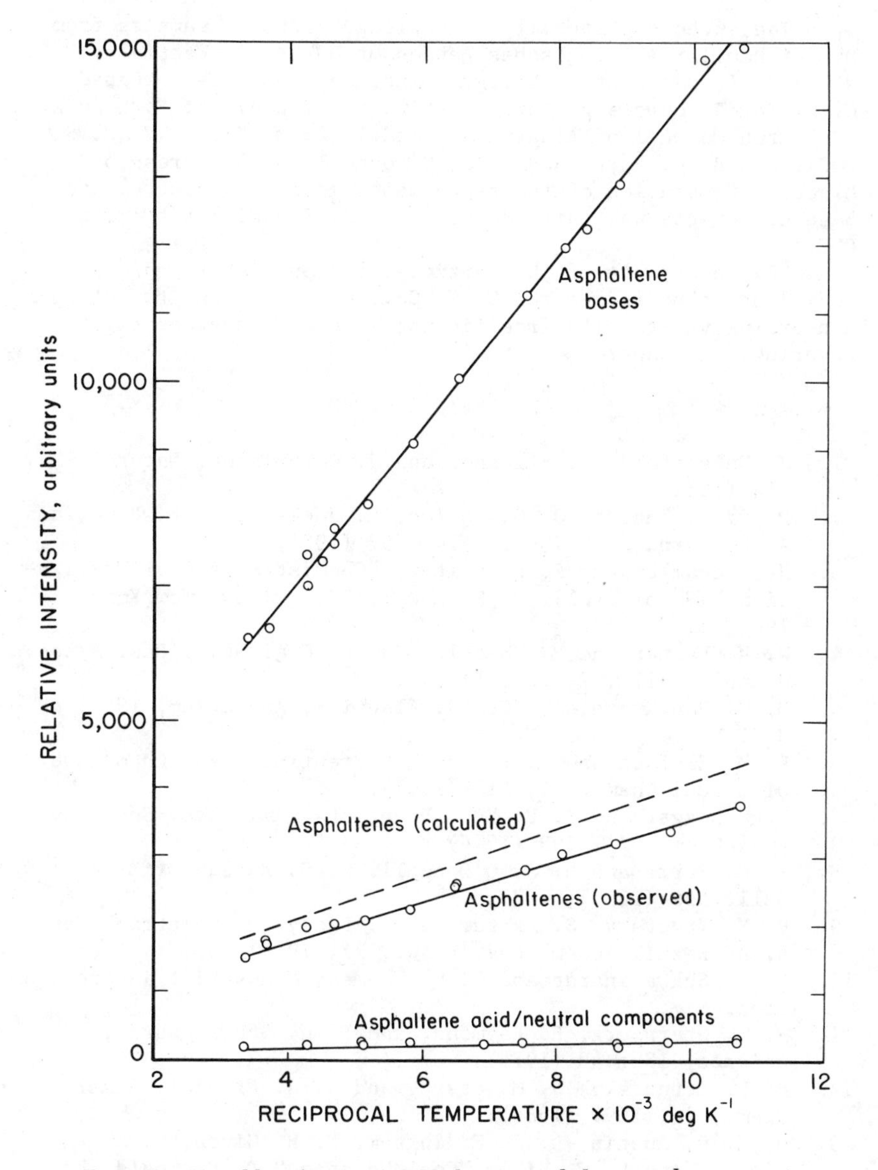

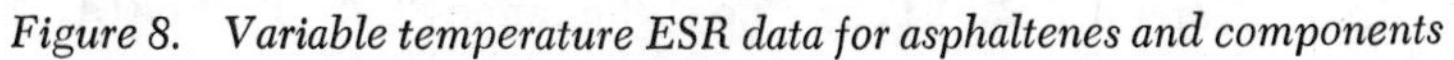

Figure 8. Variable temperature ESR data for asphaltenes and components

ACKNOWLEDGMENTS

The authors gratefully acknowledge gifts of samples from M. Teichmüller (Geologisches Landesamt Nordrhein-Westfalen), B. N. Nandi (Department of Energy, Mines and Resources - Canada), H. Honda (Resources Research Institute - Japan), K. F. Schulz (Research Council of Alberta - Canada), P. M. Yavorsky (U. S. DOE), and B. C. Parks and H. J. O'Donnell (U. S. Bureau of Mines). Separation of the asphaltenes into acid/neutral and base components was carried out by F. K. Schweighardt (U. S. DOE).

LITERATURE CITED

1. J. Uebersfeld, A. Étienne, and J. Combrisson, Nature, 174, 614 (1954).
2. D. J. E. Ingram, J. G. Tapley, R. Jackson, R. L. Bond, and A. R. Murnahgan, ibid., 174, 797 (1954).
3. H. Tschamler and E. De Ruiter, "Chemistry of Coal Utilization," Suppl. Vol., H. H. Lowry, Ed., Wiley, New York, 1963, p. 78.
4. W. R. Ladner and R. Wheatley, Brit. Coal Util. Res. Assoc. Monthly Bull., 29, 202 (1965).
5. D. W. Van Krevelen, "Coal", Elsevier, Amsterdam, 1961, p. 393.
6. R. M. Elofson and K. F. Schulz, Preprints, Am. Chem. Soc. Div. Fuel Chem., 11, 513 (1967).
7. I. Schwager and T. F. Yen, Preprints, Am. Chem. Soc. Div. Fuel Chem., 21, 199 (1976).
8. B. C. Parks and H. J. O'Donnell, U. S. Bureau of Mines Bull. 550 (1956), 193 pp.
9. P. M. Yavorsky, S. Akhtar, J. J. Lacey, M. Weintraub, and A. A. Reznik, Chem. Eng. Prog., 71, 79 (1975).
10. F. K. Schweighardt and B. M. Thames, Manuscript in preparation.
11. H. W. Sternberg, R. Raymond, and F. K. Schweighardt, Science, 188, 49 (1975).
12. H. L. Retcofsky, J. M. Stark, and R. A. Friedel, Anal. Chem., 40, 1699 (1968).
13. D. E. G. Austin, D. J. E. Ingram, P. H. Given, C. R. Binder, and L. W. Hill, "Coal Science," R. F. Gould, Ed., Amer. Chem. Soc., Washington, D. C., 1966, p. 344.
14. J. M. Schopf, Econ. Geol., 43, 207 (1948).
15. B. C. Parks, "Chemistry of Coal Utilization," Suppl. Vol., H. H. Lowry, Ed., Wiley, New York, 1963, p. 6.
16. Standard Specifications for Classification of Coals by

Rank (D 388-66), 1975 Annual Book of ASTM Standards, ASTM, Philadelphia, 1975.

17. H. L. Retcofsky, G. P. Thompson, R. Raymond, and R. A. Friedel, Fuel, 54, 126 (1975).
18. M. Mentser, H. J. O'Donnell, and S. Ergun, Fuel, 41 153 (1962).

RECEIVED March 6, 1978

11

Photochemistry Hydrogen Atoms as a Structural Probe of the Surface of Coal[1]

GILBERT J. MAINS, MUTHU SP. SUNDARAM, and JOSEPH SOLOMON[2]

Department of Chemistry, Oklahoma State University, Stillwater, OK 74074

Considerable interest in the interaction of hydrogen atoms with coal surfaces has been evident for over twenty-five years (1), resulting in a review of the literature (2) in 1965 and recent research by N. Berkowitz, et al (3, 4). Despite this intense interest the data are conflicting, and range all the way from reports of no reaction at ambient temperature to complete gasification. Some of these discrepancies arise from the discharge technique used to generate the H atoms; coal dust probably found its way into the discharge itself in some instances. Photochemical production of H atoms was mentioned and dismissed as being too inefficient (2). Because of the experimental discrepancies and their probably cause, the system H · atoms + coal was reinvestigated with photochemically generated H atoms.

Experimental

In work reported elsewhere in this symposium it was found that cryocrushing was the grinding method of choice if the surface of the coal was to represent bulk coal more accurately. The grinding techniques are reported elsewhere (5) and will not be described here. The bituminous coals were dried at 200°C in an oven through which N_2 gas was passed continuously. The subbituminous Wyoming-Wyodak coal was dried in a vacuum oven at 100°C. Illinois No. 6, Pittsburgh Seam, Utah-Emery, and Wyodak coal samples were then cryocrushed, sieved, and stored in a nitrogen environment. Only the -53/+38 micron fractions were used in these studies.

---

[1]Based in part on data to be submitted by Muthu Sp. Sundaram as a Ph.D. Dissertation.

[2]On sabbatical leave, Philadelphia College of Pharmacy and Science, Philadelphia, Pa.

0-8412-0427-6/78/47-071-156$05.00/0

Fifteen grams of coal were mixed with an equal weight of ground glass, +50 micron sizes, and placed on a frit, F_1, Figure 1, for dispersal in the quartz reactor. The mercury saturator was heated to near 100°C and ambient condenser water used to produce saturation. The reactor walls were heated by passing current through the nichrome heaters. The six, meter-long, germicidal lamps, 36 watts each, were cooled with rapidly flowing, filtered air and turned on. Next, a stream of H_2 sufficient to disperse the coal dust (about 300 liters/min) was initiated. Traps T_2 and T_3, packed with 1/4" glass beads, were precooled in liquid nitrogen for ninety minutes before the experiment and were by-passed initially. When the temperature in the quartz reactor reached 200°C, the liquid-N_2-cooled traps were opened and the gas by-pass closed. Throughout this period and the remainder of the experiment, the finer coal dust, which tended to accumulate on the Millipore filter, F_4 on Figure 1, was recycled using a mechanical vibrator. At the end of an hour, during which the temperature sometimes dropped as low as 185°C, the flow was discontinued and the traps were isolated for analyses.

In the actinometry experiments, the identical procedure was followed except that the coal was omitted and the carrier gas contained 2% ethylene.

The traps were disconnected from the apparatus and the excess H_2 was pumped away at liquid nitrogen temperature. The residual gases, after the traps were warmed to ambient temperature for several hours, were sampled with a 25-cc sample loop and injected on to a gas chromatographic column at 60°C, 1/8" x 10', packed with $\underline{n}$-C_{18}/Porasil C, in a Perkin-Elmer Model 990 gas chromatograph equipped with a thermal conductivity detector using 8% H_2 in He carrier gas. The resultant gas chromatograms are reproduced in Figures 2 through 5, inclusive. Tentative product identifications, based on elution times, are included on Figure 2.

Results and Discussion

Because the vapor pressure of Hg was high, probably near a millitorr, most of the 2537A resonance radiation was absorbed within 0.2 cm. of the quartz wall. However, the resultant excited $^3P_1$Hg atoms, $\tau_0 \approx 10^{-7}$ sec, were essentially completely quenched by the atmosphere of H_2 and thus generated two H atoms per photon in this 240 cm^3 zone. The extent to which the rapidly moving H atoms attained a uniform concentration throughout the vessel is not known, nor is the rate at which they were removed at the quartz reactor surface by recombination. When 2% ethylene was added to the flow stream in the absence of coal and the resultant butane, ethane, ethylene mixture analyzed, a rate of H atom formation of 2.90×10^{18} atoms/sec in the reactor was indicated. If these were uniformly distributed, 5.4×10^{14} atoms/cc/sec was the stationary state rate of hydrogen atom production and an overall

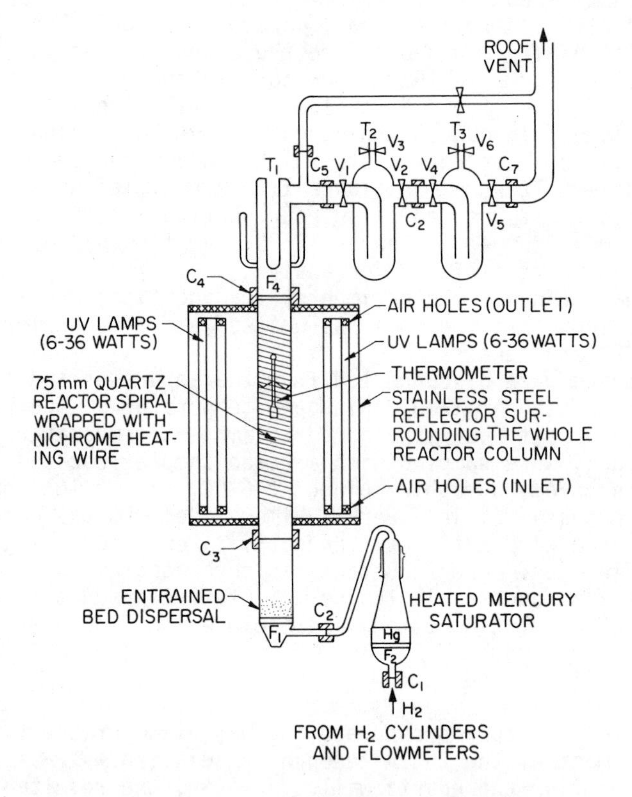

Figure 1. Photoreactor: C = couplings, F = frits/filter, V = stopcocks, T = traps. Products collected in T_2 *and* T_3.

steady state concentration of about $2x10^{13}$ H atoms/cm^3 can be calculated, which compares well with those generated by discharge techniques.

Although no coal dust deposited on the walls of the quartz reactor, it did collect on the Pyrex thermometer, the Millipore filter paper, F_4, and in the lower un-irradiated zone. This deposition precluded a quantitative determination of the particle concentration. Swirling coal dust in the reactor zone is visible to acute eyes and is inferred readily by light scattering when He-Ne laser beam is directed through the reactor. A coarse (and, probably, high) estimate of the particle concentration can be made if one assumes that 0.01% of the coal is dispersed in the flowing H_2 at any given time; making this admittedly poor assumption, we calculate $6x10^6$ coal particles/cc in the reactor. This suggests about Avogardro's Number of collisions per second between atomic H and coal dust, and the yields suggested in Figures 2 to 4 are consistent with the formation of 10^{14}-10^{15} product molecules per second. This suggests a collision efficiency of the order of 10^{-9} to 10^{-8} which is below the $7x10^{-3}$ value observed by Snelson (6) for graphite; the estimated coal particle concentration must be far too high and/or the profile of atomic hydrogen not uniform across the reactor. Based on the overall product yield estimates and the actinometry for the entire reactor, the product efficiency per hydrogen atom must be in the 10^{-2} to 10^{-3} range. Clearly, further experimentation, especially product identity and quantification, is required to decrease the uncertainty in the yield per H atom.

Examination of Figures 2, 3, and 4 shows that the gas chromatographic traces are extremely similar for the hydrogenation products from Illinois No. 6, Pittsburgh Seam, and Utah-Emery coals. Although all three are indeed bituminous coals, they are physically very different and one would be surprised if they had identical surface compositions. The "fingerprints" do show subtle differences but, in view of the uncertainties just discussed, one is struck more by their similarity than anything else. (At this writing, product identification is based on elution times. It is hoped that GC-MS identification will be made on every peak later.) Either the surface structures which can be "cracked off" by H atoms at 185°-200°C are the same for these three coals or all the products represented by peaks have common precursors which are liberated from the coals in the experiment, e.g. some combination of aromatic free radicals. Worth mentioning is one experiment in which He carrier gas was used instead of H_2. Only very small yields of benzene (~17 minute elution time) and the product with 35-minute retention time were found in the traps, presumably photo-detached from the coal surface. The yields were far too small to account for all the other products by hydrogen atom cracking of these compounds. Blank experiments using H_2 but leaving the ultraviolet lights off gave no detectable products in an hour. While considerable experimentation remains to be done,

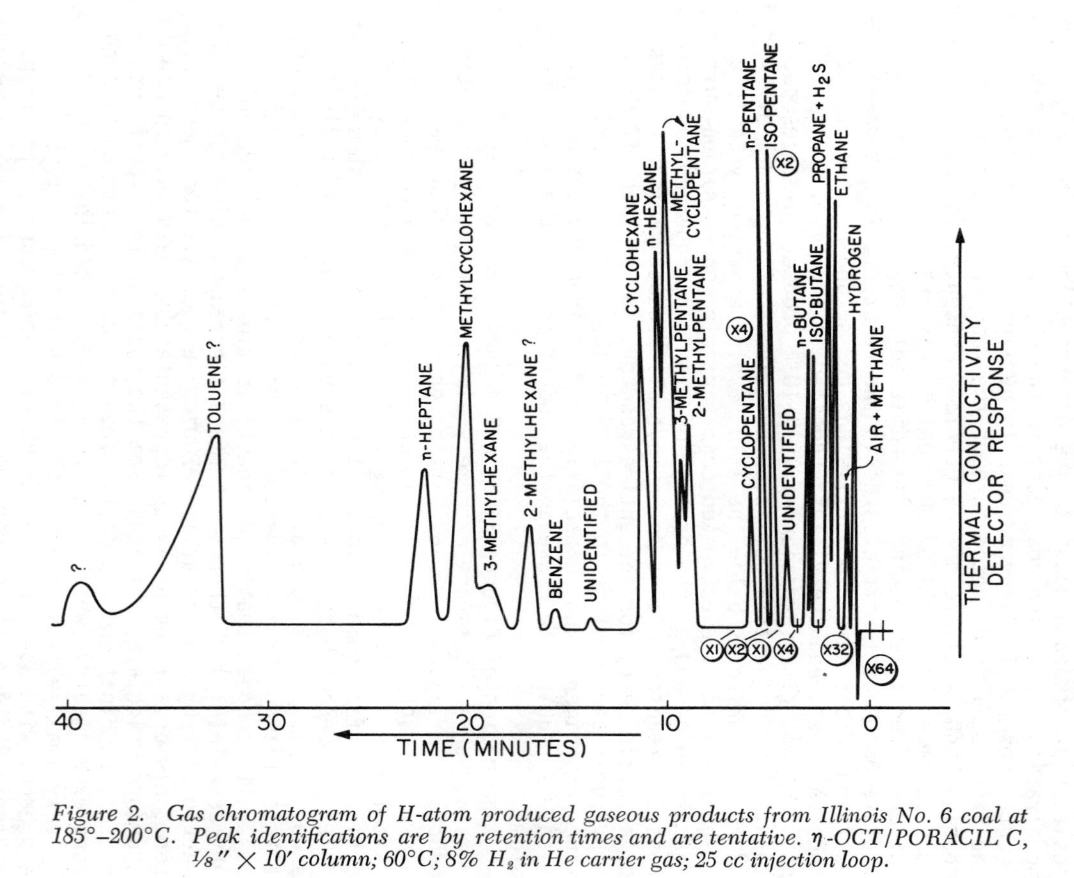

Figure 2. Gas chromatogram of H-atom produced gaseous products from Illinois No. 6 coal at 185°–200°C. Peak identifications are by retention times and are tentative. η-OCT/PORACIL C, 1/8″ × 10′ column; 60°C; 8% H_2 in He carrier gas; 25 cc injection loop.

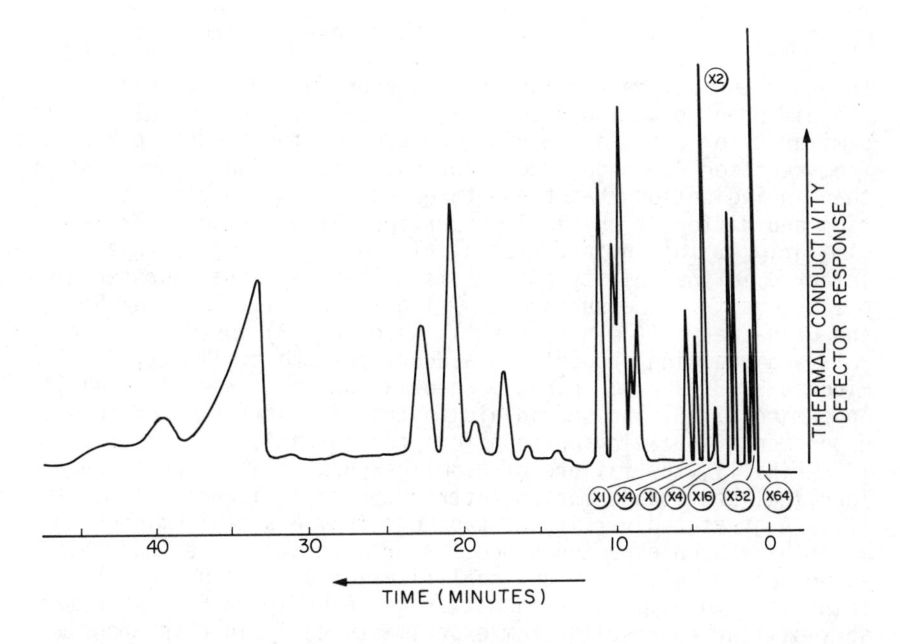

Figure 3. Gas chromatogram of H-atom produced gaseous products from Pittsburgh Hi-Seam Coal at 185°–200°C. η-OCT/PORACIL C, 1/8" × 10' column; 60°C; 8% H_2 in He carrier gas; 25 cc injection loop.

the similarity of the products from the three bituminous coals suggest the following mechanism:

$$Hg + h\nu \rightarrow Hg^* (^3P_1) \quad (1)$$

$$Hg^* + H_2 \rightarrow Hg + 2H \quad (2)$$

$$H + coal \rightarrow Precursors + ? \quad (3)$$

$$H + Precursors \rightarrow C_2, C_3, C_4, C_5, C_6 \quad (4)$$

hydrogenation products

In other words, the similarity of the product distributions from the different bituminous coals is proposed to be a result of the similarity of secondary H atom cracking of the product precursors from reaction (3) under the identical reactor conditions, rather than an indication that these three coals, with different histories and different physical properties, have surfaces that are similar. At this point, the only likeness in surface that can be inferred is the ability of H atoms to liberate the hydrogenation precursors from the surface of Illinois #6, Pittsburgh Hi-Seam, and Utah-Emery bituminous coals. Reaction (3) must be similar in rate and reaction products to account for the similarity of Figures 2, 3, and 4. Kinetic information about reaction (4) is appearing (7, 8) and should aid in the interpretation of future experiments involving more uniform concentrations.

Finally, the failure of Wyoming-Wyodak coal to react under identical conditions must be attributed to a slowness of reaction (3). A greater diversity of products from a sub-bituminous coal might have been expected since its organic structures had been subjected to less stringent coalification conditions and it contains more volatile organic matter. Since Figure 5 looks remarkably similar to results from experiments early in this program using Illinois #6 coal at ambient temperature, one might expect a greatly enhanced ESR spectrum such as was observed then. That is, a result of H atom bombardment is the production of nonvolatile free radicals. Further work is clearly warranted.

Nonetheless, it has now been demonstrated that photo-produced H atoms do interact with the surface of bituminous coals at 200°C to yield C_2 to C_8 hydrocarbons. However, the authors feel that further experiments, some altering the steady state H atom concentration and adding suspected precursors, are required to elucidate the mechanisms. Also, an accurate method for determining coal particle concentrations in the photoreactor needs to be found before reliable rate data can be forthcoming. Finally, other sub-bituminous coals need to be examined to see if the Wyodak results, reproduced several times, are common to this rank of coal. Whether H atoms will prove an effective coal surface

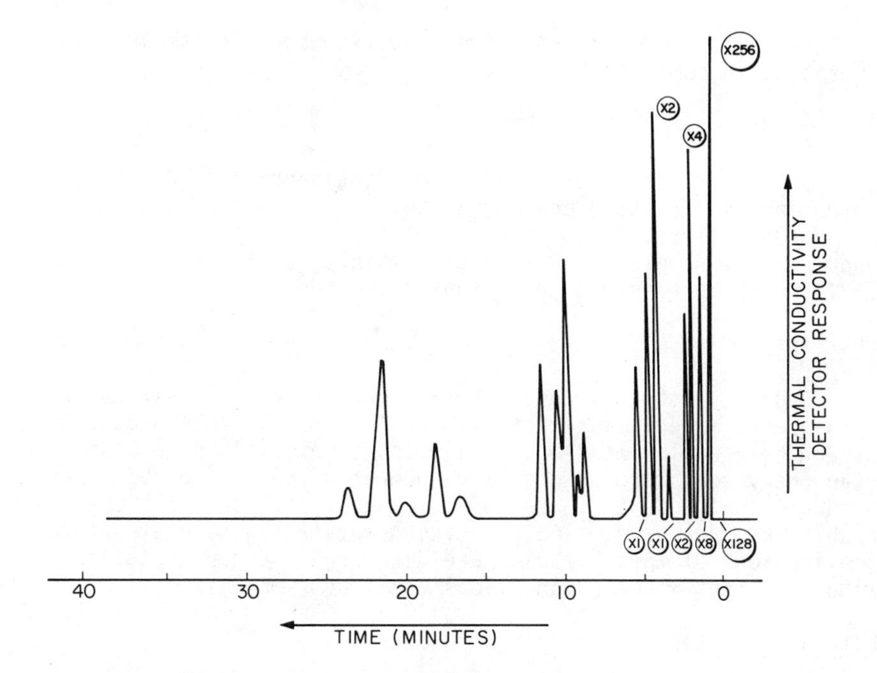

Figure 4. Gas chromatogram of H-atom produced gaseous products from Emery–Utah Coal at 185°–200°C. η-OCT/PORACIL C, 1/8" × 10' column; 60°C; 8% H_2 in He carrier gas; 25 cc injection loop.

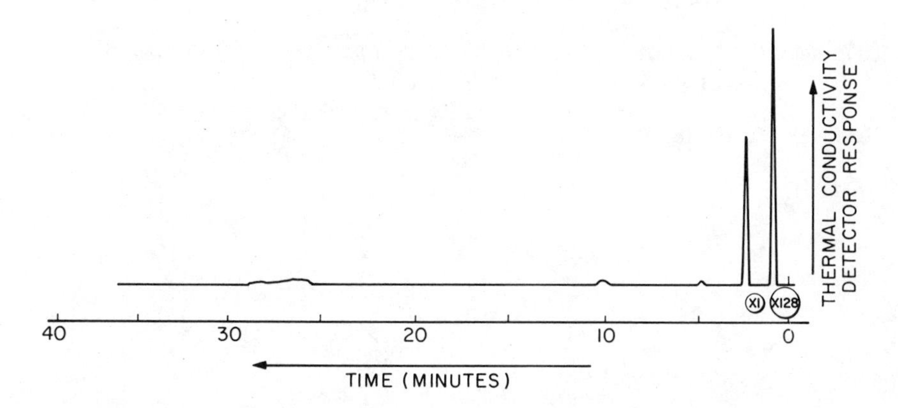

Figure 5. Gas chromatogram of H-atom produced gaseous products from Wyodak–Wyoming Coal at 185°–200°C. η-OCT/PORACIL C, 1/8" × 10' column; 60°C; 8% H_2 in He carrier gas; 25 cc injection loop.

probe remains to be proven. These preliminary results are encouraging in some respects.

Acknowledgments

We wish to thank Dr. Fred Radd, Continental Oil Company, who enthusiastically encouraged us to begin coal research, and Mr. Ed Obermiller, CONOCO Coal Development, for both encouragement and the coal samples. Last, but certainly not least, we thank E.R.D.A. for a research grant that supported these experiments.

Abstract

Hydrogen atoms, produced by the mercury photosensitization of H_2, were made to interact with coal dust, -53 to +38 microns, at 200°C in a flow reactor. Illinois No. 6, Pittsburgh Seam, and Utah-Emery coals produced a large number of saturated hydrocarbon products in the C_2 to C_8 range. Wyoming-Wyodak coal was considerably less reactive. The kinetic, quantitative, and structural implications of these results are discussed. Experimentation with different reactor conditions continues.

Literature Cited

(1) Avaremko, V. J., J. Phys. Chem. (U.S.S.R.), (1946), 20, 1299.
(2) Pinchin, F. T., Brit. Coal Util. Res. Assoc., Monthly Bulletin, (1965), 29, 105.
(3) Sanada, Y., Berkowitz, N., Fuel, (1969), 48, 375.
(4) Kobayashi, K., Berkowitz, N., Fuel, (1971), 50, 254.
(5) Solomon, J. A., Mains, G. J., Fuel, (1977), 56, 302.
(6) Snelson, A., A.C.S. Div. Fuel Chem. Proc., (1973), 18, 101.
(7) Kim, P., Lee, J., Bonnano, R., Timmons, R., J. Chem. Phys., (1973), 59, 4593.
(8) Amano, A., Horie, O., Hanh, W., Int. J. Chem. Kinetics, (1976), 8, 321.

RECEIVED March 13, 1978

12

Isotopic Studies of Thermally Induced Reactions of Coal and Coal-Like Structures

CLAIR J. COLLINS, BEN M. BENJAMIN, VERNON F. RAAEN, PAUL H. MAUPIN, and W. H. ROARK (*1*)

Chemistry Division, Oak Ridge National Laboratory, Oak Ridge, TN 37830

We recently (2) reported that under conditions of coal conversion (tetralin, 400°) several diarylalkanes undergo carbon-carbon cleavage, and that the scission of carbon-carbon bonds must therefore be considered as an important process in asphaltene formation (3). We also reported (2) that vitrinite (from Illinois No. 6 coal) was a "better hydrogen transfer agent" than tetralin itself for the hydrogenolysis of 1,1,2-triphenylethane to diphenylmethane and toluene. We have now extended these studies to establish a) that vitrinite is indeed a better hydrogen donor than tetralin toward several organic structures; b) that tetralin, in addition to its function as a hydrogen donor, can undergo certain other reactions with coal and coal-like structures which involve both carbon-carbon bond formation and bond cleavage.

A Comparison of Tetralin and Vitrinite as H-donors

When 1,2-diphenyl-1-*p*-tolylethane is heated at 400° (either in glass capillaries or in stainless steel tubes) with an excess of tetralin, the major products are toluene and phenyl-*p*-tolylmethane. The same products are obtained when 1,2-diphenyl-1-*p*-tolylethane is heated at 400° in the presence of an excess of vitrinite (handpicked from Illinois No. 6 coal). Given in Table I is a comparison of the extent reaction — as determined by g.c. analysis of the products — after various contact times with tetralin or with vitrinite.

Another diarylalkane which is easily decomposed in the presence of excess tetralin or excess vitrinite is 1,3-diphenylpropane. The major products in both cases are toluene and ethylbenzene, although a multiplicity of minor products are produced. Also given in Table I are comparisons of the extent reaction of 1,3-diphenylpropane (400° for 30 minutes) a) with excess tetralin; b) with excess tetralin and vitrinite; and c) with excess vitrinite. The extent reaction in each case was estimated from the g.c. trace.

0-8412-0427-6/78/47-071-165$05.00/0

Table I

A Comparison of Tetralin and Illinois No. 6 Vitrinite as Hydrogen Donors

Reactants	Conditions	Percent Reaction
1,2-Diphenyl-1-*p*-tolylethane	400°, 5 min[a] tetralin	2%
"	400°, 30 min[b] tetralin	94%
"	400°, 5 min[a] vitrinite	50%
1,3-diphenylpropane	400°, 30 min[b] tetralin	23%
"	400°, 30 min[b] tetralin & vitrinite	43%
"	400°, 30 min[b] vitrinite	65%

[a]The oven was at 400°, but the warm-up time is 15 minutes, thus the actual temperature was considerably less than 400°.

[b]30 min included warm up time.

Since the reactions were monitored by g.c., which would detect neither nonvolatile polymeric material, nor high molecular weight products of reaction with vitrinite, it is possible that the vitrinite is acting not as a hydrogen donor, but merely as a catalyst, and that the source of the hydrogen for the hydrogenolyses comes from the 1,2-diphenyl-1-*p*-tolylethane or from the 1,3-diphenylpropane. To circumvent this problem, we heated benzophenone to 400° for one hour a) in the presence of excess tetralin, and b) in the presence of excess vitrinite. The major products are diphenylmethane and water, with traces of toluene and benzene. The reaction in tetralin proceeded to the extent of only 12%, whereas in the presence of vitrinite 35% reaction had occurred. (6,7)

Reactions of Tetralin other than Hydrogen Donation

Tetralin-1-^{14}C reacts with Wyodak coal at 400° (1 hour) to the extent that the pyridine-insoluble residue contains chemically bound carbon-14 equivalent to 5% tetralin by weight. Further, when the residue was reheated in normal tetralin (400°, one hour) the reisolated solvent contained no measurable amount of either tetralin-^{14}C or of naphthalene-^{14}C. There were, however, traces of labeled alkylated naphthalenes, which were identified by g.c. retention times as 1- and 2-substituted methyl- and ethylnaphthalenes. These products undoubtedly arise as a result of free radical intermediates. We therefore investigated the possibility that methyl- or ethylnaphthalenes could be produced by the reaction of tetralin with structures containing aromatic moieties separated by two or more methylene groups, or with aryl alkyl ethers. Both types of structure (4,5) are known to be present in different kinds and ranks of coal.

We heated several diarylalkanes and aryl alkyl or aralkyl ethers to 400° in tetralin for varying periods of time. Many of these reactions yielded measurable quantities of methyl- and ethylnaphthalenes in addition to other products. Typical are the reactions of 1,3-diphenylpropane and of phenetole, both of which were investigated with carbon-14—labeled species. The products were analyzed a) by gas chromatography combined with radioactivity monitoring of carbon-14—labeled products; b) by gas chromatography equipped with mass spectrographic analyzers; and c) by isolation of specific products using preparative g.c. followed by nmr analysis (Varian XL-100 Spectrometer). Given in Tables II and III are the major products obtained — together with appropriate yields — from the reactions of 1,3-diphenylpropane and phenetole, respectively, with tetralin.

Table II

Major Products and Yields Obtained[a] on Heating 1,3-Diphenylpropane with Tetralin One Hour at 400°

Product	Yield
Toluene	28%
Ethylbenzene	19
1- and 2-(2-Phenylethyl) tetralins	8
1,4-Diphenylbutane	5
1- and 2-Methylnaphthalenes	3
Styrene	1.5
1,3-Diphenylpropene	1.5
Methyldihydronaphthalenes 1,2-Diphenylethane 1- and 2-(2-Phenylethyl) naphthalenes Other	34

[a]Based on 1,3-diphenylpropane consumed.

The 1- and 2-methylnaphthalenes were isolated and identified by nmr analysis. Their genesis from the reaction of 1,3-diphenylpropane-2-^{14}C (^{14}C=*) and tetralin was determined as follows:

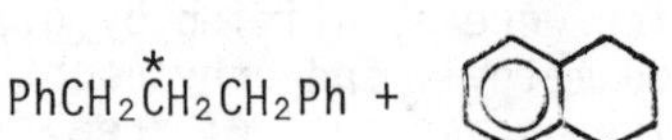 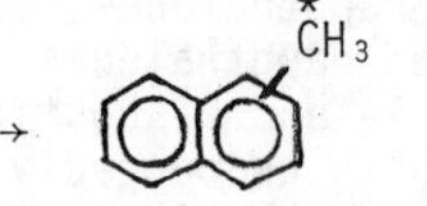 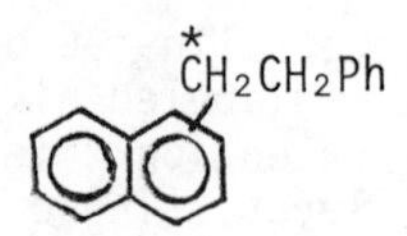

Table III

Major Products and Yields Obtained on Heating Phenetole with Tetralin Eighteen Hours at 400°

Phenol	37%
Methylnaphthalenes	13%
Toluene	7
Ethylbenzene	7
Ethylnaphthalenes	7
Methyltetralins	4
Ethylphenol	3
Ethyltetralins	3
Ethylmethylbenzene	From Decomposition of Tetralin
Methylindane	From Decomposition of Tetralin
Butylbenzene	From Decomposition of Tetralin

The mixture of ethylnaphthalenes was identified by g.c. retention times and radioactivity assay by means of the g.c. radioactivity monitor. Traces of methylindane and of butylbenzene were always present after reactants were heated with tetralin. That these latter two products were derived from tetralin was demonstrated by the fact that they contained carbon-14 when tetralin-^{14}C was used in the reaction. In like manner, labeled phenetole and tetralin were subjected to the conditions of reaction with the following results:

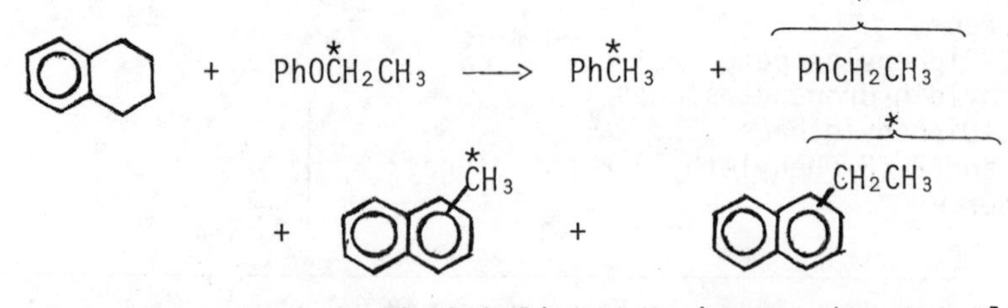

It is clear from the isotopic labeling experiments that tetralin has entered into the reaction both with 1,3-diphenylpropane and with phenetole. The results are nicely accommodated by the postulation of radical intermediates. A possible mechanism for the reaction of 1,3-diphenylpropane is indicated in Table IV.

TABLE IV

Possible Mechanism for the Reaction of Tetralin with 1,3-Diphenylpropane

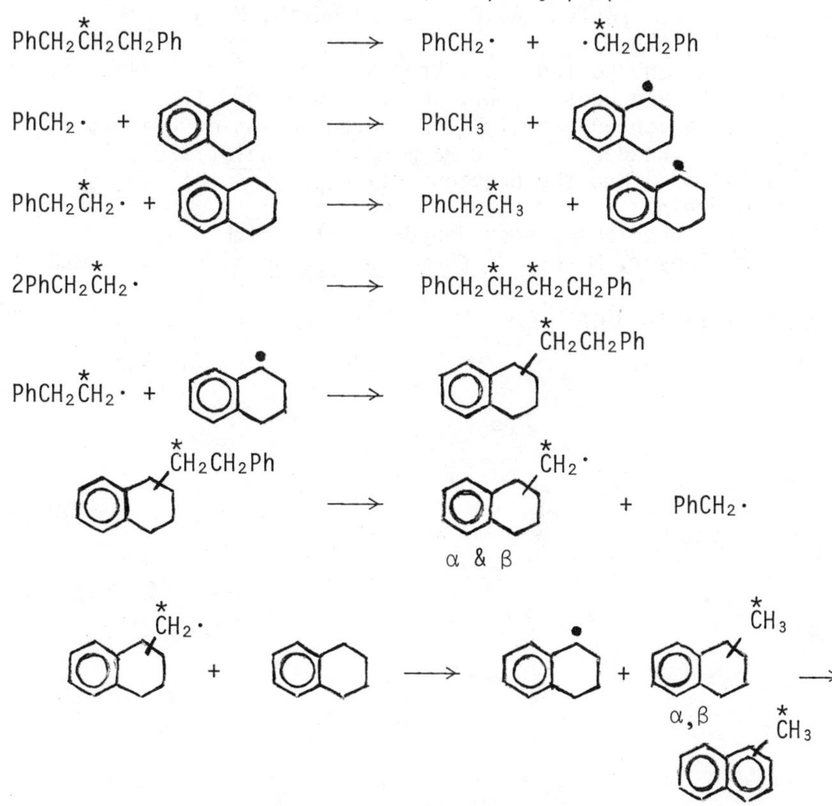

Acknowledgements: We acknowledge with thanks the assistance of Mr. L. L. Brown in running the nmr spectra, and of Dr. W. T. Rainey, Mr. E. H. McBay and Mr. D. C. Canada of the Analytical Chemistry Division, for capillary g.c. and mass spectrographic analyses of several of the hydrogenolysis products. We thank the Sahara Mining Co., Harrisonburg, Ill., for a generous sample of Illinois No. 6 coal; thanks are also due Drs. L. A. Harris and A. S. Dworkin for a field trip to the mine site.

Literature Cited

1. Research sponsored by the Division of Basic Energy Sciences of the Department of Energy under contract with the Union Carbide Corporation.
2. Collins, C. J., Raaen, V. F., Benjamin, B. M., and Kabalka, G. W., Fuel (1977), 56, 107.

3. See also Shozda, R. J., Depp, E. A., Stevens, C. M., and Neuworth, M. B., J. Amer. Chem. Soc. (1956), 78, 1716; Depp, E. A., Stevens, C. M., and Neuworth, M. B., Fuel (1956), 35, 437.
4. Herédy, L. A., Kostyo, A. E., and Neuworth, M. B., Fuel (1965), 44, 125.
5. Brücker, R. and Kölling, G., Brennst. Chem. (1965), 46, 41; Kölling, G. and Hausigk, D., ibid. (1969), 50, 1.
6. For information on "Catalylic Dehydrogenation of Coal" see Reggel, L., Wender, I., and Raymond, R., Fuel (1973), 52. 162-163 (1973) and the previous six papers in this series cited therein.
7. Coals have previously been dehydrogenated with p-benzoquinone, Peover, M. E., J. Chem. Soc. (London) (1960), 5020.

RECEIVED February 10, 1978

13

Supercritical Solvents and the Dissolution of Coal and Lignite

JAMES E. BLESSING and DAVID S. ROSS

SRI International, 333 Ravenswood Avenue, Menlo Park, CA 94025

The unique solvent properties of supercritical fluids suggest their use in coal extraction as a novel scheme for isolating syncrude-like materials. One advantage such a process might offer is the ease of separating solvent from extract. We have studied the degree of coal dissolution possible with a number of solvents in the supercritical state and examined the importance of system parameters, such as solvent type, density, and temperature, on the success of extractions.

Shortly after we began our work, Bartle, Martin, and Williams of the National Coal Board of Britain reported a 17% yield of low-ash, high-H/C material from the extraction of coal with supercritical toluene at 350°C (1). Since then, Maddocks and Gibson have reported greater yields, with up to one-third extraction of an Illinois No. 6 coal with toluene at 400°C (2). They estimated that their process would be economically competitive with the COED and SRC operations.

This paper reviews the fundamentals of supercritical extraction, discusses our data in terms of theoretical expectations, and draws some conclusions regarding the role of extraction *per se* in obtaining products from coal.

Background

A "supercritical" fluid is one that is above its critical temperature (T_c), the point beyond which a phase boundary no longer exists between gas and liquid. In the supercritical region, the density of a fluid is a continuous function of its pressure, no distinction exists between gas and liquid, and the fluid has no surface tension.

One hundred years ago, Hannay and Hogarth observed the dissolution of KI in supercritical ethanol (3). Yet, until now, little practical use has been made of supercritical extractants. Paul and Wise have described the theoretically based expectations of the use of supercritical fluids as solvents or extractants, both

0-8412-0427-6/78/47-071-171$05.00/0

generally and with some emphasis on coal dissolution (4). Given below are some of the properties of supercritical fluids, discussed in their monograph:

- At low densities, gases have no solvent power, and the concentration of a compound solute in the gas phase is described by its partial pressure. However, at a given temperature, the solvent power of any gas increases dramatically as its density approaches that of liquids.
- For a given pressure, a gas is at its highest density near its critical temperature, where it is least ideal.
- For a given gas density, the concentration of a solute in a gas increases with increasing temperature due to increased solute volatility, but solvent pressures rise rapidly as the temperature exceeds T_c and the solvent gas becomes more ideal. Theory predicts that the solvent power of such a gas is primarily a function of its physical properties and is relatively independent of its chemical structure and functionality.
- Because a gas is generally less viscous than a liquid, it can better penetrate porous substrates, such as coal.
- Though the solvent power of a dense gas may not be high compared with liquids, the gas is more easily separated from materials like coal, and solvent recoveries can therefore be better.

The conclusions of Paul and Wise thus suggest that supercritical extraction is a promising procedure for coal conversion. The results of our research verify the applicability of these principles to coal extraction, but also point to the importance of processes other than simple extractions in the production of coal products.

Experimental Procedures

A variety of experiments were performed using several solvents over a range of conditions to extract samples of Illinois No. 6 coal and North Dakota lignite. The coals are characterized in Table I.

All experiments were done in batch mode, in a 300 cm^3, 316 SS AE MagneDrive autoclave. Most experiments were run for 90 min at 335°C. The run procedure is summarized in Figure 1.

The workup procedure is shown in Figure 2. Each reaction product is separated into a filtrate and a solids fraction. In every case, the filtrates were fully pyridine soluble. The pyridine solubilities of the solids were determined by stirring 0.5 g solid in 50 ml pyridine for 1 hr at room temperature, and

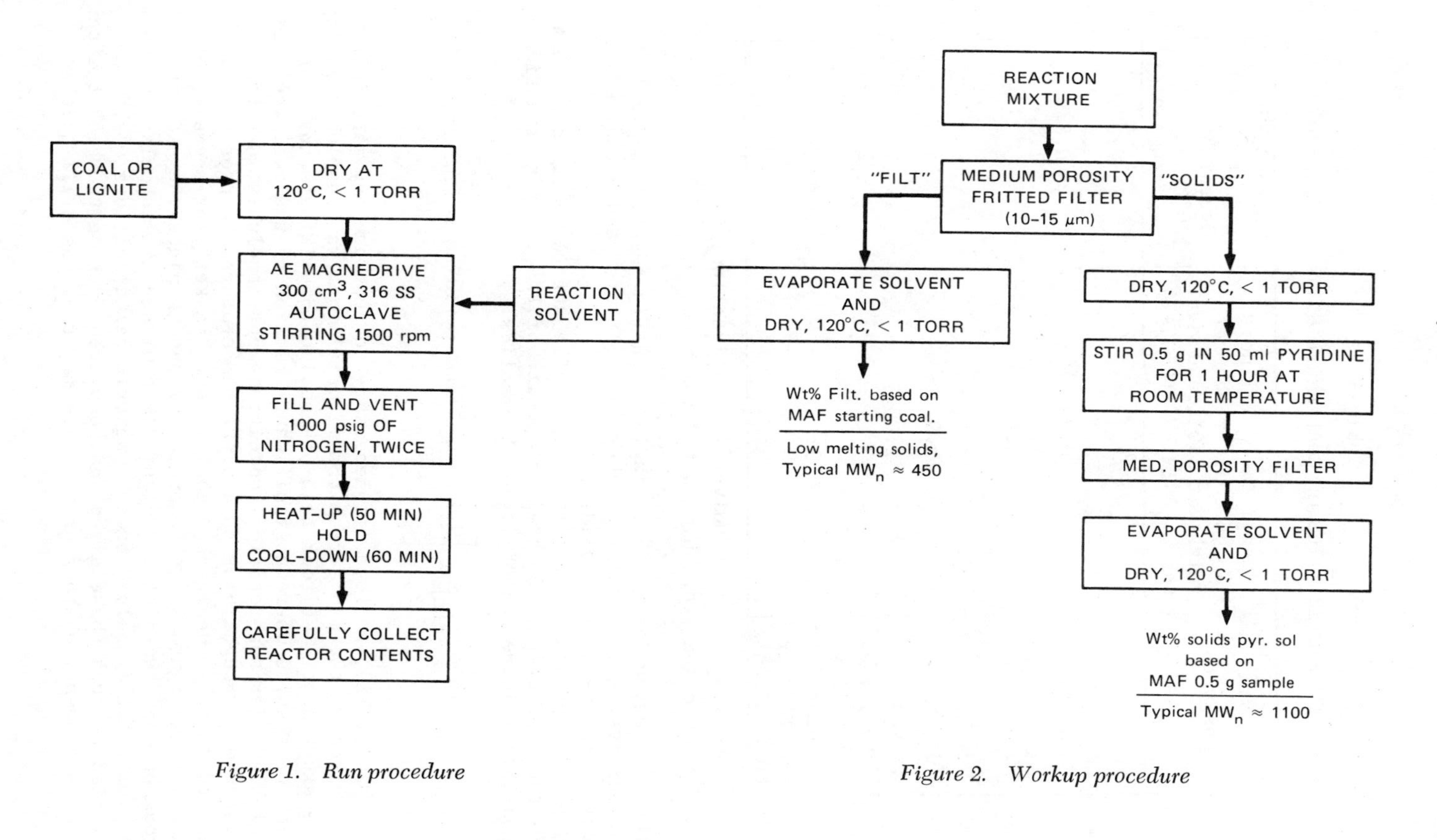

Figure 1. Run procedure

Figure 2. Workup procedure

Table I

CHARACTERISTICS OF COAL AND LIGNITE SAMPLES

Analyses	Beneficiated Illinois No. 6 Coal Dried Overnight at 120°C and < 1 torr. ASTM HVC (PSOC-26)[a]	North Dakota Lignite Dried Overnight at 120°C and < 1 torr (PSOC-246)[a]
%C	77.2	62.0
%H	5.1	4.5
Molar H/C	0.79	0.87
%N	1.7	1.0
$\%S_{org}$	2.1	0.7^{b}
$\%S_{inorg}$	∿ 0	
$\%O_{\Delta}$	11.9	14.8
%Ash	2.0	17
% Pyridine solubility	13	2
% Solubility in all reaction solvents	< 1	< 1

[a] Pennsylvania State University designation.
[b] Organic and inorganic sulfur combined.

then filtering (Figure 2). Included in Figure 2 are typical number-average molecular weights for filtrates and pyridine soluble portions of the solids.

The pyridine solubility (PS) of the entire reaction product is the appropriate algebraic composite of the filtrate fraction and the pyridine solubility of the solids:

$$PS \equiv \left(\%\ \text{Filt}\right) + \left[\left(\frac{100\% - \text{Filt}}{100}\right)\left(\begin{array}{c}\%\ \text{pyridine solubility}\\ \text{of solids}\end{array}\right)\right]$$

Initially, we found it necessary to separate any materials dissolved in the media from the material that was insoluble <u>during the experiment</u>. We took this precaution to eliminate any confusion of results where a significant fraction of the coal would be soluble under supercritical conditions but insoluble when the system was brought back to ambient temperature and pressure. Accordingly, we designed a coal filter "basket" for these experiments (see Figure 3). The procedure was to place the starting coal between two sintered-glass discs within the autoclave so that any material dissolved under supercritical conditions would be carried through these discs and recovered outside the basket after the experiment, whether or not this material was still soluble.

Results and Discussion

Initial Extractions. We performed several experiments using the extraction apparatus of Figure 3. Table II compares the results of these experiments with those of experiments run identically but without the extraction apparatus (i.e., with the coal dispersed in the reactor). The filtrate yields are comparable, and in the basket runs, little if any solid material is recovered outside the basket.* Thus, no more material is soluble in the supercritical media at 335°C than in the solvent at ambient conditions after the run. In light of these results, all subsequent work was done without the extraction apparatus. The filtrate yields are much greater than the solubles of the untreated coal in these same solvents at ambient conditions (Table I).

The soluble material, therefore, either was physically entrapped in the coal and required supercritical treatment to liberate it, or was not initially soluble, but a result of some chemical change in the coal during supercritical treatment. We believe the latter to be the case, as will be discussed later.

The high pyridine solubilities obtained with isopropanol are due to hydride donation chemistry, and are discussed elsewhere (5).

The experiments described above might be considered as continuous extractions carried out within a batch reactor. There can be limitations imposed on results acquired in batch type experiments, perhaps the most important of which is the lack of fresh solvent continuously entering the reactor and removing dissolved material. This constraint could result in a saturation of the solvent, thereby limiting extraction yields. To determine if we were encountering solvent saturations, we performed two experiments in 0.62 g/cm^3 benzene at 335°C for 3 hr, one with 1 g of coal and the other with 5 g. The amount of filtrate found in the reaction solvent was 12% of the starting coal in both cases. Furthermore, the reaction solids of both runs were found to be completely insoluble in fresh solvent after reaction, indicating that these runs were not saturation-limited.

These batch-reaction results, though not identical with continuous extraction results, lead to conclusions that may apply to continuous unit operations as well as batch operations.

---

*The extraction results for the i-PrOH/lignite/basket run are significantly low, perhaps because the lignite tended to agglomerate under these conditions, thereby restricting the circulation of i-PrOH through the basket.

Table II

THE EFFECTIVENESS OF SUPERCRITICAL EXTRACTION OF ILLINOIS NO. 6 COAL AND NORTH DAKOTA LIGNITE AT 335°C FOR 90 MINUTES[a]

Medium $(g/ml, T_c)$[b]	Mode	Dissolved in the Medium (% Filtrate)	Solids Recovered Outside Basket (%)	PS (%)	Mass Recovered[e] (%)
Coal:					
iPrOH (0.3,253)	Basket	11	0	45	97
iPrOH (0.3,253)	No Basket	11	--	40	97
Benzene[c] (0.6,289)	Basket	12	1	14	95
Benzene[c] (0.6,289)	No Basket	13	--	18	92
MeOH[d] (0.4,240)	Basket	12	2	--	97
MeOH (0.4,240)	No Basket	8	--	--	94
Lignite:					
Toluene (0.2,320)	Basket	3	0	--	87
Toluene (0.2,320)	No Basket	4	--	--	88
iPrOH (0.1,253)	Basket	5	0	--	86
iPrOH (0.1,253)	No Basket	12	--	--	85

[a] 5 g coal or lignite is used. Reaction volume is ≈ 280 ml.

[b] Density of media during run, and critical temperature of solvent (°C).

[c] 1 g coal is used, for 3 hr.

[d] 1 g coal is used.

[e] Percent of starting coal or lignite recovered.

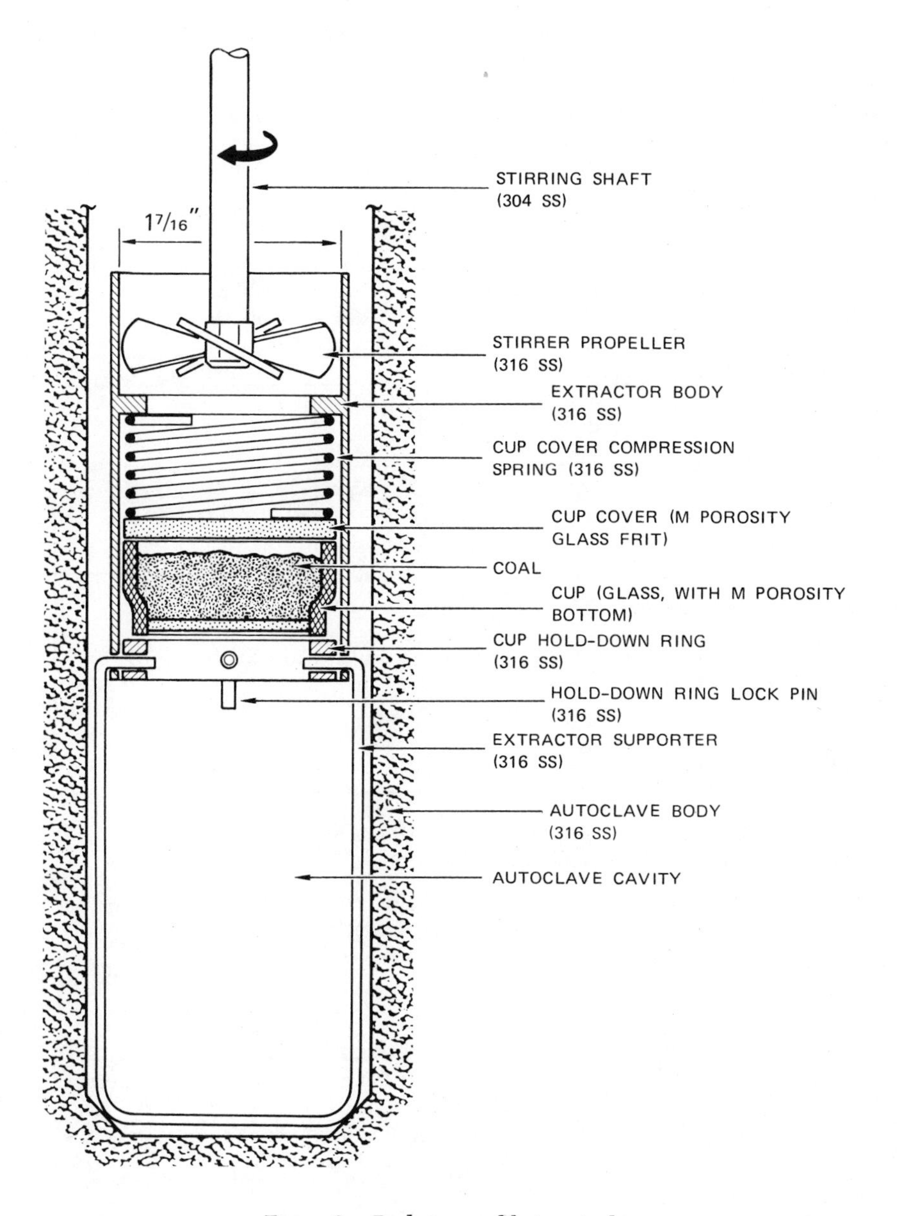

Figure 3. Basket assembly in autoclave

Effects of the State, Temperature and Density of the Medium. We performed one set of experiments to determine if increased dissolution requires a solvent that is strictly supercritical. In these runs, toluene was used at fifteen degrees above and below its critical temperature (T_c = 320°C). To account for simple thermal effects, we used benzene at these same temperatures. Since the T_c of benzene is 288°C, benzene was supercritical in both runs. The results of these experiments are presented in Table III.

Table III

EXPERIMENTS ON ILLINOIS NO. 6 COAL WITH TOLUENE BELOW AND ABOVE ITS CRITICAL TEMPERATURE FOR 90 MINUTES

Experimental Conditions				Solubility (%)	
Solvent	Solvent Density (g/ml)	Temp[a] (°C)	psig	Filt	PS
Toluene (subcritical)	0.65[b]	305	2400	8	13
Toluene (supercritical)	0.65	335	3400	13	24
Benzene (supercritical)	0.70	305	3400	8	13
Benzene (supercritical)	0.70	335	4100	10	20

[a] The critical temperatures for toluene and benzene are 320°C and 288°C, respectively.

[b] The liquid has expanded to fill the entire reactor and thus this density is the same as in the supercritical experiment.

The results show that toluene at 15°C above its critical temperature extracts significantly more material from the coal than it does at 15°C below its critical temperature. The comparable runs using benzene, however, show parallel increases. Thus, a supercritical condition has no significant effect on dissolution. Simple thermal effects must be primarily responsible for these increased solubilities.

To investigate further the effects of temperature on these yields, we performed one experiment with toluene at a higher temperature. The results for three temperatures are compiled in Table IV.

Table IV

EFFECTS OF VARYING
TEMPERATURE ON 0.65 g/ml
TOLUENE/COAL TREATMENTS FOR 90 MIN

Reaction Temperature (°C)	Filt (%)	PS (%)	Reaction Pressure (psig)
305	8	13	2400
335	13	24	3400
375	17	28	4800

Clearly, the filtrate yield is strongly temperature-dependent. The amount of toluene soluble material obtained from the 375°C reaction exceeds the pyridine solubility of the untreated coal, and the pyridine solubility of the entire product is more than twice that of the untreated coal. Thus, thermal processes are probably breaking down the coal to soluble materials.

Considerations of bond strengths and thermal cleavage kinetics do not predict much thermal activity at 335°C. For example, given the relatively weak C-C bond in 1,2-diphenylethane (57 kcal/mol), the half-life for its thermal cleavage at 335°C

$$PhCH_2\text{-}CH_2Ph \rightarrow 2PhCH_2\cdot$$

is 160 hr (6). On this basis, little thermal fragmentation should occur in 90 min at this temperature. On the other hand, the half-life value is for a gas-phase, low-density system and may not strictly apply here.

When coal itself is heated in nitrogen at 335°C for 90 min, its pyridine solubility declines slightly. In the presence of solvent, however, the pyridine solubility of a product coal increases, as we have just seen. The results of a brief series of experiments in which only the density of the solvent was varied are presented in Table V. This table reveals a clear increase in both the filtrate yield and the product pyridine solubility with increasing solvent density, as when the temperature was varied. Thus, a dense solvent must play some part in the production of increased filtrate yields and pyridine solubilities at these temperatures. The solvent may be acting simply as a solvent, where the solvent power is changing with density, or it may be reacting with the coal, or both.

Table V

EFFECTS OF VARYING SOLVENT DENSITY IN TOLUENE/COAL TREATMENTS AT 335°C FOR 90 MIN

Solvent Density (g/ml)	Filt (%)	PS (%)	Filt Molar H/C	Reaction Pressure (psig)
0.21	5	9	0.98	600
0.60	11	18	0.93	2150
0.65	13	24	—	3400

The Hildebrand solubility parameter δ is a measure of the cohesive forces in a solvent and has been considered in terms of polar and nonpolar contributions (7). The application of the solubility parameter to coal processing has been discussed by Angelovich et al., who concluded that solvents with a nonpolar solubility parameter of about 9.5 are most effective in coal dissolution (8). The polar/nonpolar effects should diminish above the critical temperature of a solvent as it becomes more like an ideal gas.

Giddings et al. (9) found a correlation between the solvent capabilities and the δ values of a number of supercritical fluids at liquid densities. Their expression for the parameter, the one we are using, is

$$\delta = 1.25\, P_c^{1/2}\, \rho_r/\rho_\ell \quad ,$$

where P_c is the critical pressure of the medium in atmospheres, ρ_r is its reduced density ($\rho_{critical}$), and ρ_ℓ is the reduced density of liquids, taken to be about 2.66. Thus δ is a linear function of the experimental density, all other variables in the equation being constant for any given solvent.

Product pyridine solubility versus δ is plotted in Figure 4. The data for all the solvents appear to fall about a line.* Regardless of their structural differences, all these compounds largely perform in accordance with their solvent capabilities. Thus, the media are acting simply as solvents and are apparently not chemically active. Since the pyridine solubility of the starting coal is 13%, the use of these solvents at lower densities is actually counterproductive.

---

*A comparable run with Tetralin and coal yields a product that is about 50% pyridine soluble.

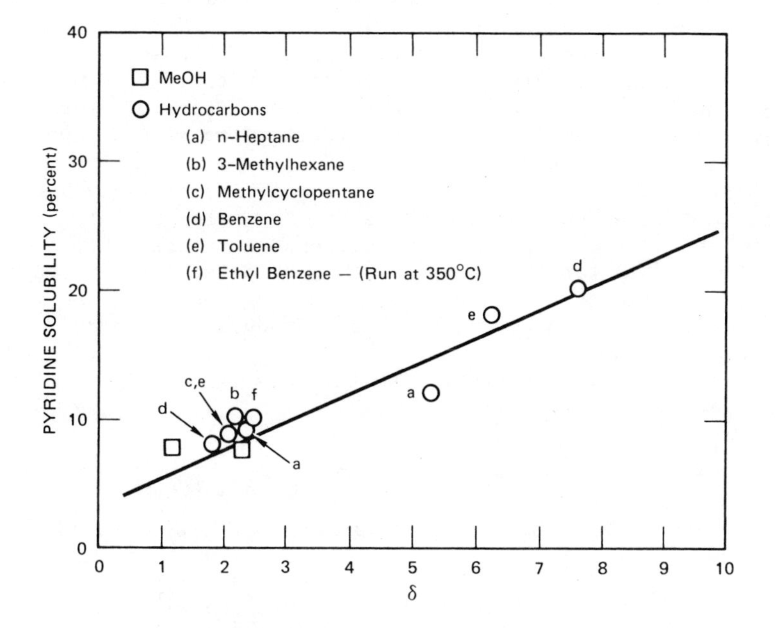

Figure 4. Product pyridine solubilities vs. δ (reaction conditions: 90 min, 335°C, approx. 500–5000 psig)

Conclusion

The conclusions presented here are based on the results of our experiments with coal. On the basis of some limited research with lignite, we suggest that these conclusions also apply to lignite. The yields for lignite, however, were generally lower than those for coal under comparable conditions.

The correlation of pyridine solubility of the coal products with the Hildebrand solubility parameter coupled with the temperature dependence of the product yields leads us to suggest that the soluble product materials result from an initial, thermally induced fragmentation of the coal involving the action of a dense solvent. Thermochemical considerations suggest that the rate of thermal fragmentation is too slow to account for the results. As stated, however, these thermochemical calculations are for a gas phase system, at densities several orders of magnitude lower than those used in our experiments.

Wiser has suggested that coal thermolysis rates may be significantly enhanced in the presence of a solvent (10). Although neither theoretical nor independent experimental justification exists for this suggestion,* our data, and particularly, our finding of a correspondence between the density of the medium and the conversion to pyridine-soluble products, are best explained that way.

Thus, a model consistent with the data is

$$\sim\text{C-C}\sim \xrightarrow[1]{\text{solvent participation}} 2\ \sim\text{C}\cdot \xrightarrow[2]{} \text{conversion to stable products}$$

C-O bonds can be considered similarly. Step (2) may involve either

(i) Hydrogen-transfer from a hydrogen-rich portion of the coal,

$$\sim\text{C}\cdot + \text{RH} \longrightarrow \text{R}\cdot + \sim\text{CH}$$

(ii) Disproportionation of the radical species,

$$2\ \sim\dot{\text{C}}\text{H-CH}_3 \longrightarrow \sim\text{CH=CH}_2 + \sim\text{CH}_2\text{-CH}_3$$

or

(iii) β-scission, splitting off a small, relatively stable free radical (R=H, alkyl, benzyl)

$$\sim\dot{\text{C}}\text{-C-R} \longrightarrow \sim\text{C=C} + \text{R}\cdot$$

---

*For the simple thermal homolysis, R-R → 2R·, no evidence exists that the rate is significantly enhanced by the presence of solvent. Moreover, if the process is strictly one in which no charge separation occurs in the transition state, there is no theoretical expectation of a significant solvent effect.

(An alternative mechanism is discussed in the Appendix.)

Supercritical solvents, therefore, clearly can provide moderate yields of syncrude-like materials from coal, and these yields are not primarily due to any unique characteristics of supercritical conditions. Since high solvent densities are desirable, solvents that are liquid at liquefaction temperatures could prove at least as effective as those that are supercritical, and at lower pressures. Liquids, however, are subject to the limitations of surface tension and higher viscosities, which diminish their usefulness in this scheme. Additionally, solvents that are liquid at liquefaction temperatures are very difficult to separate from coal products.

Thus, supercritical solvents offer solvent fluidity, a relatively wide range of usable types of compounds, and easily obtainable high solvent recoveries in the extraction of low molecular weight materials from coal. An understanding of the importance of the thermal processes involved in the treatment of coal with hot, dense solvents and the principles of supercritical extraction, as enumerated at the outset of this paper, could lead to an effective use of supercritical solvents in coal and lignite processing.

Appendix

An alternative scheme involving charge separation is suggested for coal thermolysis:

$$\text{coal} + \text{coal} \rightleftharpoons (\text{coal}^{+\cdot} \ldots\ldots \text{coal}^{-\cdot}) \xrightarrow{\text{solvent}} \text{solvent separated pair}$$

$$\text{coal}^{+\cdot} \xrightarrow[\text{coal or H-donor}]{} \text{coalH}_2^{+\cdot}$$

As unlikely as this suggestion may seem at the outset, the proposition is consistent with the following:

- As discussed in the text, an increase in density of the media increases the degrees of coal conversion. The conversion process, in turn, entails irreversible changes in the coal, and thus is not just a simple solvent-solute interaction.

- Wiser states that coal pyrolysis in the absence of Tetralin is second order in coal, and when Tetralin is present the process is first order in both Tetralin and coal (10).

- Tetralin readily donates hydrogen to electron-poor systems at 50 to 160°C. Typical H-acceptors are quinones. The reaction is accelerated by electron-withdrawing substituents on the quinone, is accelerated by polor solvents, and is unaffected by free radical initiators (11).

- Radical cations readily accept H_2 from hydrocarbon donors (12).

- The established acid-base character of coal-derived asphaltenes (13) suggests that charge separation, i.e., donor-acceptor complexes, either are present in coal itself or form thermally.

- Poly-condensed aromatic structures, like those in coal, are known to form readily both radical cations and radical anions (15).

This suggested model could provide practical insight into the action of catalysts in conversion processes. One might consider, thus, the use of catalysts that promote C-C scission by radical cation intermediates.* The implications of this scheme await the results of further research into the H-donor process and coal conversion chemistry.

Acknowledgment

We acknowledge the support of the Department of Energy for this work on Contract EF-76-C-01-2202.

Literature Cited

1. Bartle, Keith D., Martin, Terence G., and Williams, Dereck F., Fuel (1975), 54, 226.
2. Maddocks, R. R., and Gibson, J., Chem. Eng. Prog. (June 1977), 73, 6, 59-63.
3. Hannay, J. B., and Hogarth, J., Proc. Roy. Soc. (London), Ser. A (1879), 29, 324-26.
4. Paul, P.F.M., and Wise, W.S., "The Principles of Gas Extraction," Mills and Boon Limited, London, 1971.

*Trahanovsky and Brixius have shown that at temperatures below 100°C, Ce(IV) promotes the cleavage of $PhCH_2$-CH_2Ph, yielding oxidation products by way of a radical cation intermediate (16). It would be of interest to carry out the reaction with H-donor solvents present.

5. Ross, D. S., and Blessing, J. E., "Isorpopyl Alcohol as Coal Liquefaction Agent," Fuel Division Preprints for the 173rd National Meeting of the Amer. Chem. Soc., New Orleans, LA, (March 1977). Manuscript in preparation.
6. Benson, S. W., and O'Neal, H., National Standard Reference Data Series — NBS 21, U.S. Government Printing Office, Washington, D.C., 1970.
7. Blanks, R., and Peausnitz, J., Ind. Eng. Chem. Fundamentals (1964), 3, 1.
8. Angelovich, J., Pastor, G., and Silver, H., Ind. Eng. Chem. Process Des. Dev. (1970), 9, 160.
9. Giddings, J., Myers, M., McLaren, L., and Keller, R., Science (4 October 1968), 162, 67.
10. Wiser, N., Fuel (1968), 47, 475.
11. Braude, E. A., Jackman, L., and Linstead, R., J. Chem. Soc. (1954), 3548, 3564, 3569.
12. Doepker, R., and Ausloos, P., J. Chem. Phys. (1960), 44 (5), 1951; Kramer, G., and Pancirov, R., J. Org. Chem. (1973), 38, 349.
13. Sternberg, H., Raymond, R., and Schweighardt, F., Science (4 April 1975), 188, 49.
14. Franklin, J., Dillard, S., Rosenstock, H., Herron, J., and Draxl, K., "Ionization Potentials, Appearance Potentials, and Heats of Formation of Gaseous Positive Ions," National Standard Reference Data Series — NBS 26, U.S. Government Printing Office, Washington, D.C., 1969.
15. Compton, R., and Huebner, R., in "Advances in Radiation Chemistry," Burton, M., and Magee, J., Ed., Vol. 2., 1970 p. 281; Christophorou, L., and Compton, R., Heath Physics (1967), 13, 1277.
16. Trahanovsky, W., and Brixius, W., J. Amer. Chem. Soc. (1973), 95 (20), 6778.

RECEIVED February 10, 1978

14

Homogeneous Catalytic Hydrogenations of Complex Carbonaceous Substrates

J. L. COX, W. A. WILCOX, and G. L. ROBERTS

Battelle, Pacific Northwest Laboratories, Richland, WA 99352

Hydrogenation of unsaturated organic compounds with homogeneous catalysts has been known and practiced for some time. Such hydrogenations have been of both an academic and commercial interest. Some of the more extensively studied catalysts include $Co(CN)_5^{-3}$ (1), RhCl $(P\Phi_3)_3$ (2), $Ir(CO)Cl(P\Phi_3)_2$ (3,4), $IrH(CO)(P\Phi_3)_3$ (3,4), $OsHCl(CO)(P\Phi_3)_3$ (3,4) and Ziegler-type catalysts (5,6). These catalysts, except for the Ziegler-type, have not been observed to hydrogenate aromatics. In fact, very few homogeneous catalysts have been reported that will hydrogenate aromatics. Wender, et al. (7) have shown that polynuclear aromatics are partially hydrogenated with $Co_2(CO)_8$ Efimov, et al., (8,9) have observed rapid hydrogenation of polynuclear aromatics in the presence of a rhodium complex of N-phenylanthranilic acid (NPAA), formulated as $(RhNPAA)_2$. This rhodium catalyst is more active than the dicobalt octacarbonyl and shows a greater hydrogenation activity toward polynuclear aromatics than the Ziegler catalyst. Holly et al. (10) investigated the use of this rhodium complex and other homogeneous catalysts for coal liquefaction, concluding that such catalysts do not appear to offer a viable route to coal liquefaction. Muetterties and Hirsekorn (11) have reported the hydrogenation of benzene to cyclohexane at 25°C and 1 atm pressure in the presence of η^3-allylcobalt phosphite; η^3-$C_3H_5Co[P(OCH_3)_3]_3$.

Here we report the results of homogeneous catalytic hydrogenation of complex unsaturated substrates including coal and coal-derived materials.

Hydrogenations

Using organic soluble molecular complexes as catalysts, a number of hydrogenations of various organic substrates (a Hvab coal, solvent refined coal (SRC) and COED pyrolysate) were performed. The analysis of these feed materials is contained in Table I. The hydrogenations were carried out in a 300 cc stirred autoclave by mixing coal with carrier solvent containing solubi-

0-8412-0427-6/78/47-071-186$05.00/0

lized catalyst under prescribed conditions. Upon completion of the run the gases were measured with a calibrate wet test meter and analyzed by routine gas chromatography procedures. The solid carbonaceous residue was separated from carrier solvent by filtration, then thoroughly washed with benzene and finally dried in a vacuum oven.

TABLE I. Analysis of Feed Materials

	Coal[(a)] As Received	Coal[(a)] Vacuum Dried	SRC	COED	Hydrogenated Coal Run 25
Moisture	1.1	0.0	0.0	0.0	1.8
Ash	14.5	14.7	0.3	0.1	27.0
H	4.8	4.6	5.6	7.3	4.4
C	68.8	68.3	87.7	85.0	55.0
O	6.0	6.1	4.0	7.2	5.9
N	1.2	1.2	2.2	1.1	0.2
S	4.6	4.6	0.5	1.3	4.0

[(a)]Both samples were -200 mesh.

All catalysts except the Ni-Ziegler are commercially available and were used without further purification. The Ni-Ziegler was prepared under a nitrogen atmosphere by reacting 4 moles of triethylaluminum with 1 mole of nickel naphthenate in anhydrous n-heptane. The activity of this catalyst was first tested with benzene before proceeding to more complex substrates. Hence, 10 ml benzene were hydrogenated in 40 ml n-heptane containing 4×10^{-3} moles of the Ni-Ziegler catalyst for 1 hour at 150°C and 1000 psig H_2 (ambient temperature). Even though the hydrogenation covered a 1 hour period the autoclave pressure rapidly dropped to 650 psig once 150°C was reached, signaling rapid hydrogenation of the benzene. Liquid product analysis by gas chromatography revealed 99% conversion of the benzene to cyclohexane.

Hydrogenation conditions and results for coal and coal-derived materials are summarized in Table II. The change in atomic hydrogen-carbon ratio (Δ) is the principal criterion for comparing catalyst activity and extent of hydrogenation. Since no attempt has been made to account for the lighter hydrocarbons that were removed with the carrier solvent by filtration the hydrogenation criterion is very conservative. The Δ value has been obtained by subtracting the experimentally determined atomic hydrogen-carbon ratio of carbonaceous substrate from that of the product. The carbon-hydrogen analysis was performed on a Perkin Elmer model 240 elemental analyzer. Since a small amount

TABLE II. Summary of Homogeneous Catalytic Hydrogenations of Carbonaceous Substrates

Run No.	Catalyst/Feed[a]/Solvent	Temperature/Pressure[d]/Time[b]	H	C	At.H/C	Δ(atomic H/C)[c]
31	No catalyst/15g C/decalin	300/2880/2	4.64	68.6	0.806	-0.003
17	14 mmole $Co_2(CO)_8$/30g C/decalin	200/2950[e]/2	4.69	67.4	0.829	0.020
21	7 mmole $Co_2(CO)_8$/15g C/decalin	300/3080[e]/2	5.16	69.0	0.891	0.082
20	7 mmole $Co_2(CO)_8$/15g C/decalin	400/3400[e]/2	5.26	71.8	0.873	0.064
18	13 mmole $Fe_3(CO)_{12}$/30g C/decalin	200/2830[e]/2	4.56	65.9	0.824	0.015
19	7 mmole $Ni[(PhO)_3P]_2(CO)_2$/15 g C/decalin	220/2720[e]/2	4.94	70.7	0.833	0.021
25	7 mmole Ni-Ziegler/15g C/decalin	200/2770/2	4.48	55.8	0.957	0.148
36	8 mmole Ni-Ziegler/15g C/heptane	200/1300/22	5.88	63.6	1.10	0.291
38	8 mmole Ni-Ziegler/10g SRC/THF	200/1200/23	7.29	77.1	1.13	0.377
40	5.7 mmole Ni-Ziegler/17.2g COED/THF	200/3850/21	8.28	74.8	1.32	0.290

[a] Feed materials include: Consolidation coal (C), 4.64%H, 68.3%C, At.H/C = 0.809; Solvent Refined Coal (SRC), 5.55%H, 87.7%C, At.H/C = 0.753; FMC pyrolysate (COED), 7.32%H_2, 85.0%C, At.H/C = 1.03.

[b] Variables temperature, pressure and time reported as °C, psig and hr., respectively.

[c] Δ, is the change in atomic H/C ratio between substrate and product.

[d] Pressures are those at reaction temperature and due to hydrogen and solvent unless otherwise stated.

[e] Gas composition of 25%CO, 75%H_2, used in hydrogenation.

of unremoved solvent in the product can seriously affect the interpretation of results, extreme precaution was taken to ensure its complete removal. This was accomplished by washing with a volatile solvent (tetrahydrafuran) followed by vacuum drying with a nitrogen bleed at 110°C for at least 24 hours. In order to check thoroughness of solvent removal a dried sample showing H/C of 1.10 was further dried and reanalyzed. There was essentially no change in the H/C (i.e., 1.10 versus 1.09).

It is apparent from Table II that the NI-Ziegler catalyst is more active than $Co_2(CO)_8$, $Ni[(PhO)_3P]_2(CO)_2$ and $Fe_3(CO)_{12}$. In Run 25 a Δ of 0.148 for the Hvab coal was observed over a 2-hour reaction time at 200°C and 2770 psig. This change in atomic H/C ratio from hydrogenation corresponds to a hydrogen usage of only 0.85% (w/w) of coal. Even in Run 36 where a Δ of 0.291 was effected at 200°C and 1300 psig H_2 after 22 hours only 1.7% (w/w) H_2 is consumed in the hydrogenation. These hydrogenations may be compared to Run 31 where a slight decrease in Δ, -0.003, was observed in the hydrogenation of the Hvab coal with no catalyst for 2 hours at 300°C and 2880 psig. In contrast to these hydrogen consumptions about 2% H_2 (w/w maf coal basis) is used in the SRC process, 2.5% in Synthoil and 4% for H-Coal.

Homogeneous catalytic hydrogenations were also conducted on solid products from the SRC and COED coal liquefaction processes. The analyses of these substrates are contained in Table I. Examination of the hydrogenation results summarized in Table II reveals that the ease of hydrogenation under these homogeneous catalytic conditions is SRC > COED > Hvab, although some reservation must be made since the hydrogenations were not made under identical conditions. That the coal-derived substrates are more readily hydrogenated than the coal is not too surprising since they are liquids at reaction temperatures ($\geq$200°C) and quite soluble in carrier solvent permitting effective catalyst-substrate interaction. Diffusional resistances to hydrogenation are also expected to be less for these materials than the solid coal.

Product gas analysis on each experimental hydrogenation run revealed predominantly reactant gases. In Runs 17, 18, 19, 20 and 21 the product gas consisted of $\geq$98% H_2 and CO, while Runs 25, 31, 36, 38 and 40 showed at least as great a concentration of hydrogen. Except for Run 20, CO_2 and CH_4 contributed $\leq$0.2% to the gas balance. The gas composition of Run 20 was different from the others in that low but noticeable concentrations of light hydrocarbons including CH_4, C_2H_6, and C_3H_8 were found. The observed light hydrocarbons in this run are undoubtedly due to the higher temperature (400°C versus $\leq$300°C) employed, which was high enough to affect hydrogenolysis. The gas analysis of Runs 25, 36, 38 and 40 employing the Ni-Ziegler catalyst were unique in that appreciable concentrations (1 to 2%) of ethane were observed. The ethane is attributed to loss of ethyl groups from the Ni-Ziegler catalyst or decomposition of excess triethylaluminum which was used in its preparation.

Even though there is no kinetic data to support a proposed mechanism for the observed homogeneous catalytic hydrogenation of aromatics the following mechanism is consistent with the chemistry of analogous systems. In this proposed mechanism L represents coordinated ligands and solvent and M is the transition metal. This proposed mechanism depicts that generally accepted for the hydrogenation of olefins (including cyclohexene) by a number of group VIII metal complexes.

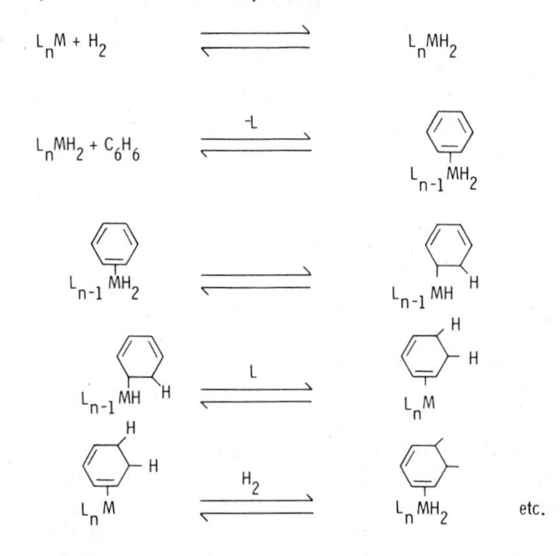

We know for instance that molecular hydrogen will undergo oxidative addition reactions with a number of transition metal complexes as shown in the first equation (12). The metal migrations during hydrogenation may occur through a π-allyl complex. Finally, intermediates similar to those postulated in this mechanism have been supported by chemical evidence and include the two following compounds (13).

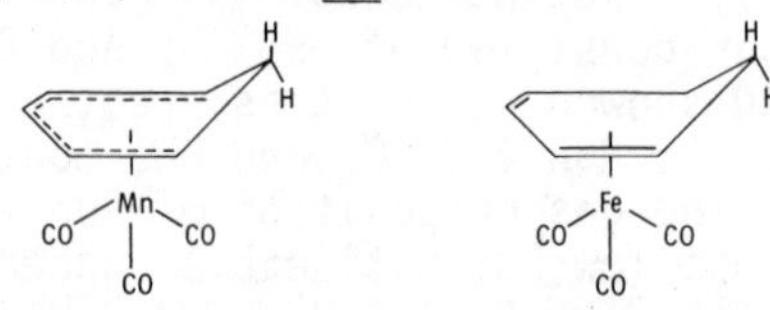

Even though the above mechanism has been written with benzene as the substrate it is relatively easy to write analogous mechanisms for yet more complex aromatic substrates.

Product Characterizations

The products from the hydrogenation of the coal and coal-derived materials were all solid glossy black materials at room temperature, except in Run 20 where the temperature reached 400°C and a black viscous liquid was obtained. The carbon and hydrogen analyses of these hydrogenation products are contained in Table II. With respect to these complex carbonaceous substrates hydrogenation is often falsely taken to be synonymous with liquefaction. The polymeric structure of coal (14,15) implies that complete saturation with hydrogen would not alter its physical state. An appropriate analogy would be the hydrogenation of polystyrene to produce the solid saturated polymer.

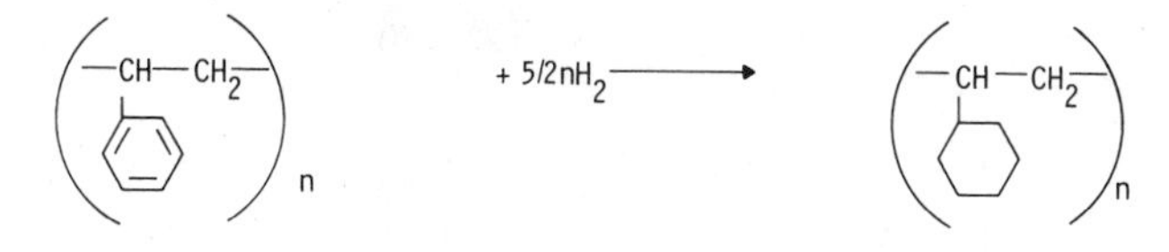

Even though polyethylene has an H/C ratio of 2.0, it is also solid. Hence with respect to coal liquefaction, undoubtedly the hydrogenolysis reactions and not hydrogenation are responsible for converting the solid to a liquid by conventional processing schemes. The liquefaction that occurred in Run 20 was undoubtedly due to hydrogenolysis as a consequence of the high temperature (400°C) employed. Again using polystyrene for analogy, the hydrogenolysis may be represented by the following reaction in which the product ethyl benzene is clearly a liquid.

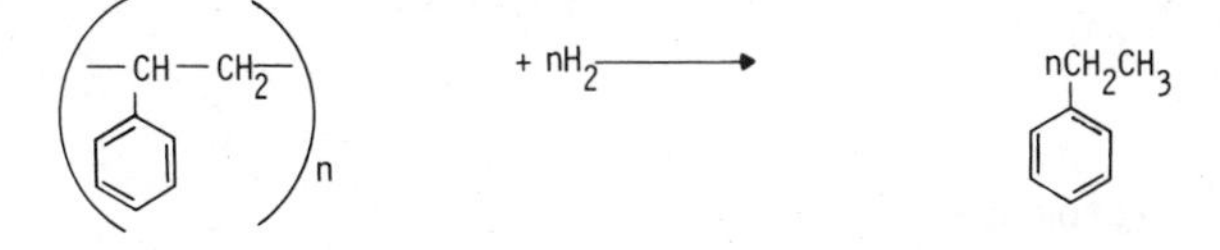

Coal hydrogenolysis has been contrasted with its hydrogenation by carrying out a material balance estimation on the Synthoil process as shown in Table III. These results show that one net result of catalytic hydrogenolysis is a decrease in the product (char + crude oil) H/C ratio (0.765 versus 0.883), despite the addition of 9000 scf H_2 per ton of coal (maf). The decrease in the H/C ratio is a direct consequence of the hydrogenolysis reaction which breaks structural bonds in the solid substrate and adds hydrogen to the fragments. Hence, a large hydrogen consumption is seen in removing oxygen and sulfur from the coal as H_2O and H_2S, respectively. Considerable hydrogen is also consumed by the formation of light hydrocarbon gases (C_1 - C_3) as a result of hydrogenolysis reactions. On the other hand, hydrogen consumed by hydrogenation goes directly to increase the substrate H/C ratio. This is why the H/C ratio is relatively high in the product for the amount of hydrogen consumed in the homogeneous catalytic hydrogenations. Of course, from a practical viewpoint it is highly desirable, if not necessary, to at least remove the sulfur which is invariably accompanied by oxygen removal and light hydrocarbon gas production.

TABLE III. Synthoil Material Balance

Materials Charged	lb-moles		lb
Coal[a] (atomic H/C = 0.883)		2000	
C	1117.90		1416
H	104.20		105
O	22.50		360
S	3.00		96
N	7.50		105
Hydrogen[b] (H)	50.04	50.5	
		2050.5	
Materials Out			
C_1-C_3 Hydrocarbons[c]		100	
C	6.52		78.3
H	21.53		21.7
Heteroatoms		467.8	
H_2O[d]	21.25		382.8
H_2S[e]	2.5		85.0
Product[f] (atomic H/C = 0.765)		1482	
C	111.4		1338
H	85.3		86
O	1.2		20
S	0.5		16
N	1.6		22
		2049.8	

[a] maf basis.

[b] 50.04 lb-moles H equivalent to 9000 scf H_2.

[c] Amount to 5 wt% of maf coal charged with estimated composition of 60% C_1, 30% C_2 and 10% C_3.

[d] 340 lb O + 42.8 lb H..

[e] 80 lb S + 5.04 lb H.

[f] Crude oil + char.

Complete hydrogenation of the unsaturated carbon-carbon bonds in the complex carbonaceous substrates is estimated to result in a H/C ratio of at least 1.5. The maximum value we have obtained is 1.32 for the COED pyrolysate. The failure to obtain more extensive hydrogenation may be due to the catalyst's inability to activate the more complex aromatics. Also, a portion of the substrate's unsaturation is undoubtedly very facile and a portion is very inert toward hydrogenation. As noted earlier, the Ni-Ziegler very rapidly hydrogenates mononuclear aromatics but is much more sluggish toward the polynuclear aromatics. For example, at 100 psig H_2, phenol is 92% converted to cyclohexanol in 0.2 hours at 150°C in the presence of 5 mole% Ni-Ziegler catalyst, while only 12.5% naphthalene is converted to decalin in 18 hours at 210°C in the presence of 2.5 mole% Ni-Ziegler (5). The more complex polynuclear aromatics, such as phenanthrene and pyrene that are contained in coal along with the heteropolynuclear aromatics such as indoles and dibenzothiphenes will undoubtedly be even more sluggish toward hydrogenation.

Another possible cause of low hydrogenation levels is the inaccessibility of catalyst to substrate. This is anticipated to be more of a problem for coal than for coal-derived products, due to their differences in physical properties, particularly pore size distributions and solubility in carrier solvent. The moderate to high solubility of the coal-derived materials will undoubtedly enhance catalytic hydrogenation over the relatively insoluble coal in a carrier solvent. The coal itself, having an extensive pore structure, will undoubtedly contain a significant number of micropores that are too small for access of the catalyst (16).

To obtain information on the reactivity of the solid hydrogenated product, thermogravimetric analyses were run and compared to those of the unhydrogenated coal. The results are shown in Figure 1 for hydrogenation products of Runs 25 and 36 as well as the parent coal. These thermogravimetric curves were all recorded with a duPont model 900 analyzer on 19-23 mg samples under a high-purity argon atmosphere at 20°C/min heating rate, from ambient to 800°C. Although the sample from Run 25 had a H/C of 0.148, greater than the parent coal, there is no qualitative difference in their thermogravimetric curves except for the small amount of moisture devolatilized from the parent coal at about 125°C. This type of thermogram is typical of many that have been reported for different ranks of coal (17). In sharp contrast, the sample from Run 36 which had a H/C increase of 0.291 over the parent coal reveals considerable structure in its derivative thermogram. In addition to the major peak at about 475°C which was also observed for the coal and sample from Run 25, two additional peaks were observed at considerably lower temperatures--one at 360°C and the other at 275°C. The volatile matter from this sample was 54% (ash-free basis) while that from Run 25 and the parent coal was 42% and 37% respectively (ash-free basis).

It is not surprising that the sample from Run 36 showed more structure in its thermogram and a larger amount of volatile matter production than the parent coal since it does have a significantly greater hydrogen content than the parent coal. What is somewhat surprising is that the sample from Run 25 acted very similar to the parent coal during thermolysis even though its H/C ratio lies between that of product from Run 36 and the parent coal. The hydrogenation in Run 36 is apparently extensive enough to saturate key unsaturated groups in coal leading to decreased thermal stability. It is generally recognized that cleavage of saturated groups is much more facile than unsaturated groups under thermolysis conditions.

To project the yield of liquid product from the thermogravimetric analysis we have assumed that a linear relationship exists between the percent volatile matter and liquid yield. This implies that if a coal is 25% volatilized and yields 10% oil then 35% devolatilization of this coal after pretreatment such as hydrogenation will yield 0.35/0.25 x 10% = 14% oil. Using this approximation one can readily estimate that prehydrogenation of the coal in Run 36 has increased yield of oil by 46% over the unhydrogenated coal upon pyrolysis. In practical terms this means that the prehydrogenation could be used to significantly increase the yield of liquids from coal by pyrolysis. For instance, the oil yield from the COED process (18) which uses multiple stage pyrolysis to produce gases, liquids and char would be increased from 1.5 bbl/ton to about 2.2 bbl/ton.

Hydrogenolysis

A hydrogenolysis experiment was conducted with the Hvab coal and its hydrogenated product under identical experimental conditions. The analysis of these -200 mesh feed materials was reported in Table I. Hydrogenolysis was carried out in the previously described 300 cc magnedrive autoclave. In the hydrogenolysis of Hvab, 10.0 g were charged to the autoclave along with 30.0 g (32 ml) tetralin and 0.5 g -100 mesh Co-Mo catalyst (Harshaw HT-400 containing 3% cobalt oxide and 15% molybdenum dioxide on alumina). In the hydrogenolysis of the prehydrogenated coal from Run 25 the same quantity of carbonaceous feed and tetralin were charged, but no catalyst was added. After materials were charged, the autoclave was purged twice with high-purity hydrogen and then pressurized with hydrogen to 1500 psig. Experimental hydrogenolysis conditions were 0.5 hour at 400°C, 2700 psig (1500 psig H_2 ambient) and 300 rpm. The autoclave was brought up to reaction temperature and cooled after the designated reaction time as rapidly as possible. It required about 30 minutes to reach 400°C from ambient and 10 minutes to cool from 400°C to 60°C. However, the time required to heat from 200°C to

400°C and cool from 400°C to 200°C was only about 2 minutes; i.e., the time spent at temperatures sufficiently high to potentially contribute to hydrogenolysis was significantly shorter than the overall heat-up and cool-down time.

Once the run was completed and the temperature had cooled to ambient the experiment was worked up according to the diagram in Figure 2. The gas volume was measured with a wet test meter and its composition determined by gas chromatography. The liquid and solid products were emptied into a tared extraction thimble whence the filtrate from the thimble was collected and saved. The autoclave was rinsed with benzene into the same extraction thimble. The solids were then extracted with the rinse benzene for 8 hours in a soxhlet apparatus whence the extraction thimble was vacuum dried and weighed. The benzene solution from the extraction was combined with the previously saved tetralin-laden filtrate. To this solution was added a three-fold excess of pentane, to precipitate asphaltenes which were then separated by filtration and vacuum dried. This filtrate was then vacuum distilled (≃25 mm Hg, 40-80°C) to remove the pentane and benzene from the tetralin soluble product (light oil). The yields of the various fractions obtained by following this procedure are all summarized on a maf charge basis in Table IV for the Hvab coal and the prehydrogenated coal.

The tabulated results (Table IV) for the hydrogenolysis of the coal and hydrogenated coal clearly indicate a significant difference in their product yields. The yield of gas and light oil was greater for the hydrogenated than the nonhydrogenated coal (30% and 49.0% versus 31% and 48.6%). Perhaps even more significant is the lower asphaltene yield observed for the hydrogenated sample (3.0% versus 10.3%). The overall conversion was essentially the same for both samples (90 versus 89%) and there was little difference in their product gas compositions.

In examining the differences between these two hydrogenolysis experiments one is reminded that while a Co-Mo catalyst was used with the coal, there was no addition of catalyst to the hydrogenated coal. However, the hydrogenated coal did contain nickel that was apparently deposited on the coal from the Ni-Ziegler catalyst during the homogeneous catalytic hydrogenation. This nickel undoubtedly participated as a catalyst in the hydrogenolysis of the sample. So in essence, we are comparing the hydrogenolysis of this hydrogenated sample containing nickel with the Hvab coal to which was added a conventional Co-Mo catalyst. An analysis[a] of the nickel content in the hydrogenated coal

---

[a] The analysis of the hydrogenated sample was performed by first ashing the sample at 850°C in a muffle furnace. This revealed 27.0% ash. The ash was then digested in hot aqua regia solution and filtered. The filtrate was then diluted with water and analyzed for nickel by atomic adsorption spectroscopy to give 1.2% Ni in the hydrogenated sample.

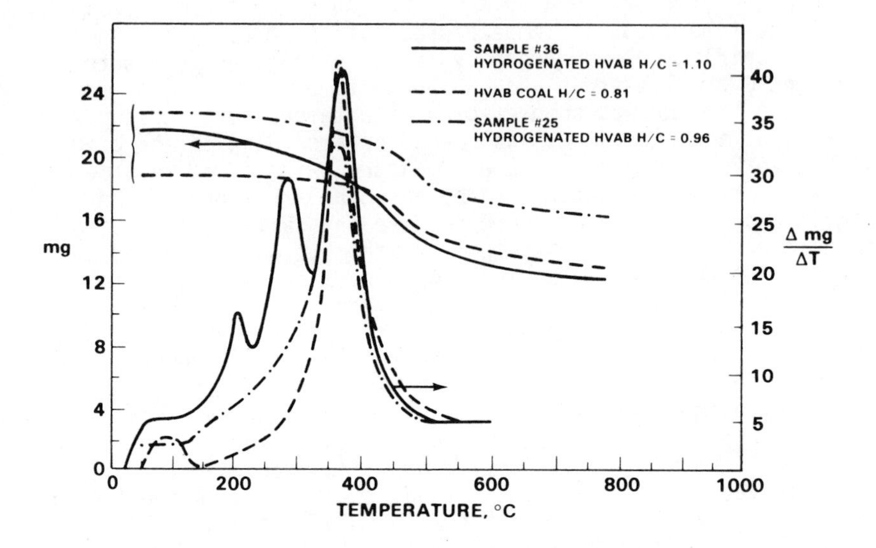

Figure 1. Thermograms and DTG of coal and hydrogenated coal

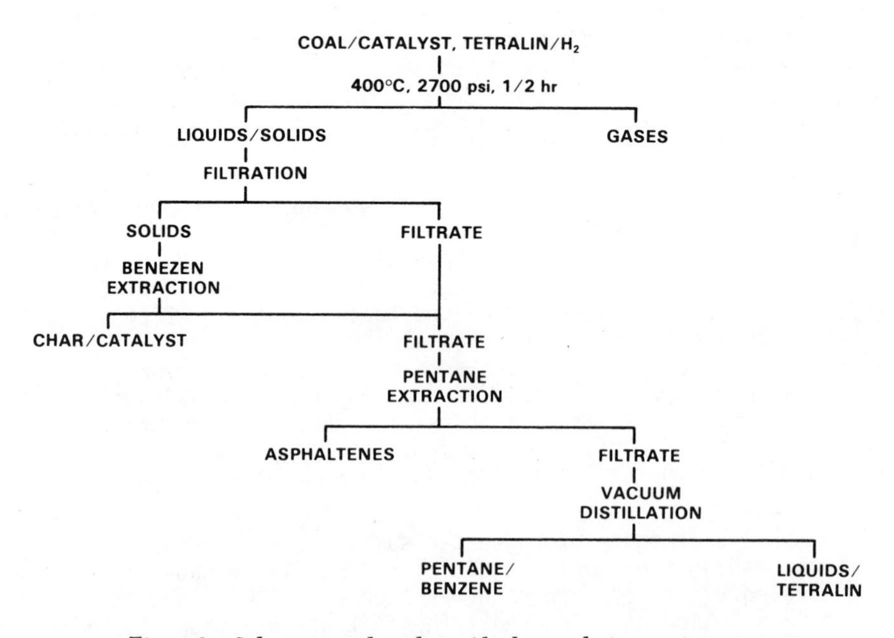

Figure 2. Schematic and workup of hydrogenolysis experiments

TABLE IV. Hydrogenolysis Experimental Results

	Hvab Coal	Hydrogenated Coal Run 25
Reaction Conditions		
feed, g	10.0	10.0
tetralin, g	30.0	30.0
catalyst, g	0.5	none
temperature, °C	400	400
pressure, psig	2660	2770
time, hr	1/2	1/2
Conversion, %	90.0	88.9
Product Yields, %		
gas	30.8	38.0
oil	58.8	52.1
light oil	48.6	49.0
asphaltenes	10.2	3.0
char	10.3	9.9
Gas Composition, mole %		
H_2	91.2	87.7
CO_2	0.5	0.4
C_2H_4	trace	trace
C_2H_6	2.4	2.6
O_2	trace	0.3
N_2	0.5	2.5
CH_4	4.1	3.9
CO	0.1	0.2
C_3H_8	1.3	2.3
C_4H_{10}	0.2	0.1

revealed 1.2% Ni which can be compared to 5% Co-Mo catalyst in the Hvab coal. Since the nickel catalyst was apparently decomposed on the hydrogenated coal sample it almost certainly had better catalyst-substrate contacting than the Co-Mo which was mixed with the Hvab coal. This in itself could account for the higher hydrogenolysis yields of gas and light oil and lower asphaltenes than with the Co-Mo catalyst. In addition, nickel catalysts generally show higher gas yields under hydrogenolysis than supported Co-Mo catalysts. Finally, the hydrogenated sample is expected to produce higher yields of gases and light oils and lower asphaltenes under hydrogenolysis simply because it has been hydrogenated.

In an attempt to better define the nature of the liquid product from the hydrogenolysis of these two samples their light oil fractions were subjected to gas chromatographic-mass spectral analysis. The more prominent components of the tetralin soluble fraction (light oil) from hydrogenolysis of the two samples were analyzed on a Hewlett-Packard (HP) 5980A quadruple mass spectrometer with a HP 5710A GC and HP 5934A data system. Chromatography was performed on a 6-foot SP 2250 column temperature programmed from 80 to 250°C at 8°C/min using He carrier at 40 ml/min. Mass spectrometry was conducted with electron impact at 60 electron volts. Additional chromatography of the light oil from each sample was performed on an HP 5830A GC equippped with an HP 18850A terminal at the previously described run conditions.

A typical chromatogram of these samples is shown in Figure 3 along with retention times while Table V contains the mass spectral identification of the more prominent components of both samples. Upon comparing the chromatograph data their similarity is particularly striking. When the solvent (tetralin) is subtracted from both samples as has been done in tabulating the area percent in Table VI we quickly realize that only a few of the more than 22 integrated peaks account for the majority of the sample. In fact, there are only five compounds, all with two or greater area percent, that account for greater than 90 area percent of the entire sample. These compounds are common to each sample and include the following in descending order of abundance: naphthalene (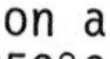) >methyl indan (—CH_3) >benzene ( ), C_4-benzene (—C_4H_9) >decalin (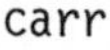). Of these five compounds the first three account for at least 80% of the light oil in each sample. The broadness of many of the chromatogram peaks, particularly those at the longer retention time, is a good indication that they consist of more than one component. Hence, the number of chemical compounds actually in the light oil sample is probably at least twice the number of integrated gas chromatograph peaks.

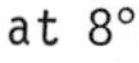

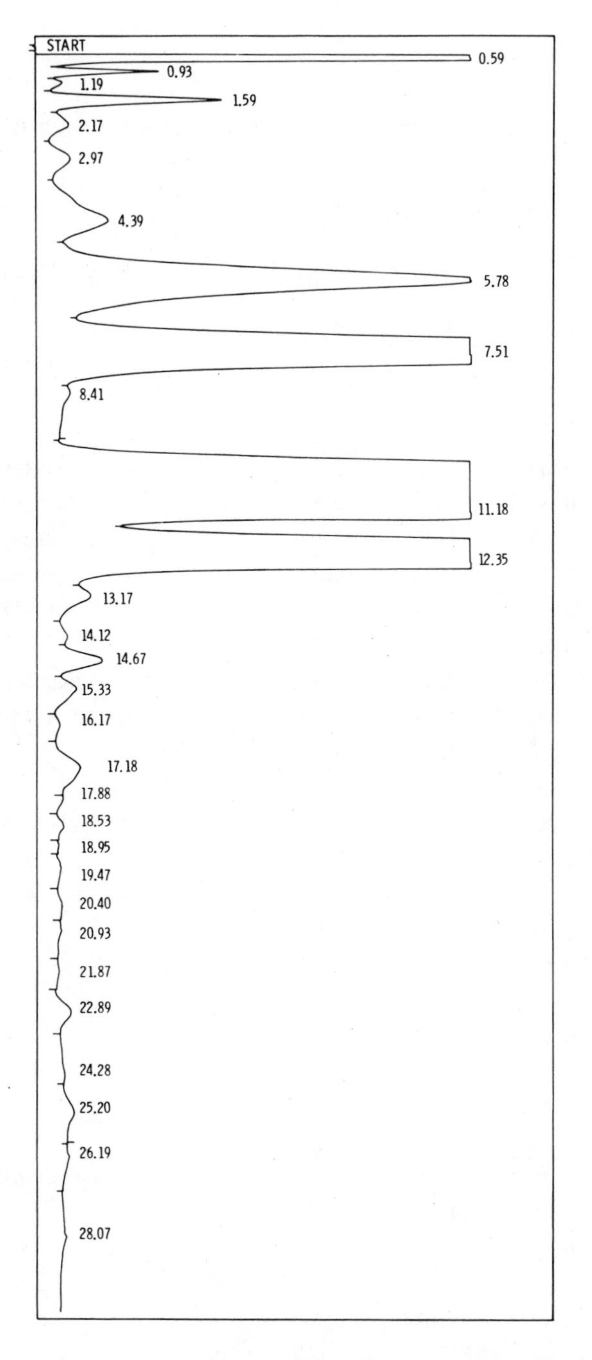

Figure 3. Gas chromatogram of light oil from hydrogenolysis of hydrogenated Hvab coal

TABLE V. G. C. Mass Spectral Analysis of Light Oil

G. C. Retention Time, Area %				
Hvab Coal		H_2-Hvab Coal		Mass Spec Assignments
0.59	18.59	0.58	16.02	benzene
0.93	0.44	0.93	0.52	toluene
1.19	0.15	1.19	0.20	
1.60	0.67	1.59	1.32	xylene
2.15	0.35	2.17	0.52	xylene
2.99	0.46	2.97	0.68	C_3-benzene
4.41	2.06	4.39	2.03	decalin
5.73	8.10	5.78	10.37	C_4-benzene
7.44	20.47	7.51	22.74	1-methylindane
8.38	0.93	8.41	1.04	
12.32	40.54	12.35	40.61	naphthalene
13.15	0.80	13.17	0.49	methyl-tetrahydronaphthalene
14.07	0.50	14.12	0.14	
14.65	0.82	14.67	0.60	
15.30	0.65	15.33	0.29	methyl-naphthalene
16.13	0.42	16.17	0.05	
17.38	1.17	17.18	0.54	C_2-tetralin
17.87	0.32	17.88	0.08	
18.51	0.68	18.53	0.10	
---	---	18.95	0.03	
19.54	0.31	19.47	0.10	
20.34	0.04	20.40	0.12	
20.90	0.04	20.93	0.12	
21.86	0.05	21.87	0.07	
22.87	0.29	22.89	0.28	methyl-biphenyl
24.25	0.12	24.28	0.27	
25.15	0.66	25.20	0.52	
26.12	0.00	26.19	0.02	
28.00	0.18	28.07	0.12	

Utility

Even though homogeneous catalytic hydrogenation has had considerable practical utility in the hydrogenation of specialized chemicals such as certain fats, oils and pharmaceuticals, its economic and technical utility in processing carbonaceous feedstocks such as coal, oil or their derived intermediates is very uncertain. We have seen that prehydrogenation of coal can significantly increase the amount of liquids obtained by pyrolysis compared to the unhydrogenated coal. It has also been shown that the hydrogenolysis of prehydrogenated coal produces less asphaltenes and more light oil and gas than the catalytic hydrogenolysis of the parent coal. Finally it was shown that homogeneous catalytic hydrogenation can effectively increase the atomic hydrogen to carbon ratio of the carbonaceous materials including coal and materials derived from coal by pyrolysis and solvent refining.

Although the amount of experimental data generated here in support of homogeneous catalytic hydrogenation is infinitesimal in regard to that needed for a sound technical judgment concerning its utility in the area of fuels processing, there is a clear indication that the approach has merit. The extrapolation of the experimental data indicates that homogeneous catalytic hydrogenation has considerable potential as a preliminary processing step for increasing the atomic H/C ratio over conventional methods for carbonaceous materials such as coal, and possibly oil shale and tar sands. The derived benefit from such a unit operation aside from the addition of hydrogen to these hydrogen-deficient materials is to increase the yield and quality of the products over that now obtainable by conventional processing techniques at such mild conditions. Yet another potential application of homogeneous catalytic hydrogenation is as an intermediate step in fuel processing or conversion schemes in which the H/C ratio is increased to produce a superior quality product. Finally, potential applications of homogeneous catalytic hydrogenation are foreseen in the area of basic research studies where it is used as an analytical tool or technique for investigating complex carbonaceous substrates.

Much of the impetus for the use of homogeneous catalysts is the prospect of reducing temperature and pressure required for conversion, increasing reaction specificity and obtaining the most efficient use possible of the active metal component. In some instances of homogeneous catalytic hydrogenation, all these prospects have been realized. It is reasonable to expect that homogeneous catalysts will eventually be developed that are capable and effective in hydrogenolysis reactions of carbon-carbon bonds. This indeed would be an extremely significant break-through with respect to coal liquefaction.

The major technical drawback to the use of homogeneous as well as heterogeneous catalysts is the difficulty of recovery from processing streams and poisoning. Since catalysts are extremely expensive processing materials, as a result of their manufacturing costs and cost of their component constituents, only small losses can be economically tolerated. In homogeneous catalytic hydrogenation Run 25 where 7 mmole of Ni-Ziegler catlyst were employed per 15 g coal, the cost of nickel alone at zero recovery (i.e., 54.8 lb/ton coal) would amount to about $110/ton coal hydrogenated. If the cost of the triethylaluminum and the carboxylate of nickel are taken into consideration the cost of catalyst materials alone is estimated to exceed $150/ton coal hydrogenated. Clearly, in order to contemplate such a use of these catalysts, they must either be recovered or used in a much reduced concentration or preferably both. Even though no attempts were made at catalyst recovery or use of reduced amounts in this study it was found that the hydrogenated coal in Run 25 contained 1.2% Ni. This figures to be 31% of the nickel used in the homogeneous catalytic hydrogenation. The remaining 69% of the nickel catalyst apparently remained in the carrier solvent, which in practice would be recycled. Obviously, even higher catalyst recovery is necessary to promulgate its economic viability.

That catalysts can be used effectively and economically in bulk chemical processes is amply demonstrated in the hydrocarbon processing industries. Although these catalysts have been for the most part heterogeneous, homogeneous catalysts have found a home in at least two notable areas. One is the use of Ziegler catalysts in coordination polymerization and the other is in the hydroformylation process. These two examples show that the technical and economic problems so often associated with the use of homogeneous catalysts in industrial processes can be overcome.

Acknowledgement

The authors wish to acknowledge the Division of Basic Energy Sciences, Department of Energy for financial support of this work.

Literature Cited

1. Kwiatek, J., Mador, I.L., and Seyler, J.K., Adv. in Chem. Ser. (1963) (37), 201-215.
2. Osborn, J.A., Jardine, F.H., Young, J.F., and Wilkerson, G., J. Chem. Soc., A, (1966), 1711-32.
3. Vaska, L., and Rhodes, R.E., J. Amer. Chem. Soc., (1965) 87, 4970.
4. Vaska, L., Inor. Nucl. Chem. Lett., (1965), 1, 89.
5. Lapporte, S.J., Ann. N.Y. Acad. Sci., (1969), 158, (2), 510.

6. Bressan, G., and Broggi, R., Chem. Abstr., (1969), 70, 37257k.
7. Friedman, S., Metlin, S., Svedi, A., and Wender, I., J. Org. Chem., (1959), 24, 1287-89.
8. Efimov, O.N., et al., J. Gen. Chem., USSR, (1968) 38, (12), 2581.
9. Efimov, O.N., et al., Izv. Akad. Nauk. SSSR, Ser. Khim. (1969), (4), 855-8.
10. Holy, N., Nalesnik, T., and McClanahan, s., Fuel, (1977), 56, (10) 272-77.
11. Muetterties, E.L., and Hirsekon, F.J., J. Amer. Chem. Soc. (1974), 96, (12), 4063-7.
12. Cotton, F.A., and Wilkinson, G., "Advanced Inorganic Chemistry", 3rd Edition, 772-6, Interscience Publishers, New York, 1972.
13. Coates, G.E., Green, M.L.H., Powell, P., and Wade, K., "Principles of Organometallic Chemistry", 197-8, Methuen and Co., Ltd., London, 1968.
14. Heredy, L.A., and Neuworth, M.B., Fuel (London), (1962), 41, 221.
15. Ouchi, K., Imuta, K., and Yamashita, Y., Fuel, (London), (1965), 44, 29.
16. Gan, H., Nandi, S.P., and Walker, P.L. Jr., Fuel (London) (1972), 51, (10), 272-77.
17. Howard, H.C., in "Chemistry of Coal Utilization, Supplemental Volume", H.H. Lowry, ed., 363-70, John Wiley & Sons, New York, 1963.
18. Jones, J.F., "Project COED: Clean Fuels from Coal Symposium" Institute of Gas Technology, Chicago, Ill., September 10-14, 1973.

RECEIVED February 10, 1978

15

Hydrotreatment of Coal with $AlCl_3$/HCl and Other Strong Acid Media

J. Y. LOW and D. S. ROSS

SRI International, 333 Ravenswood Avenue, Menlo Park, CA 94025

Most current processes for upgrading coal to cleaner fuels require stringent reaction conditions of high temperatures and pressure. Less severe reaction conditions are needed to make coal upgrading economically feasible. The objective of this work was to investigate catalyst systems for upgrading coal to clean fuels under moderated conditions. In this work, homogeneous acid catalysts are of particular interest because they allow intimate contact with the coal and are not liable to coal ash founding.

The most common homogeneous catalysts studied in coal upgrading belong to the general class of molten salt catalyst (1-5) and include halide salts of antimony, bismuth, aluminum, and many of the transition metals. Most often, these molten salts have been studied at high temperature and in massive excess (1-5). We have performed a systematic study of the use of some of these molten salts as homogeneous acid catalysts for upgrading of coal at relatively low temperatures and in moderate quantities.

In our initial work to establish relatively mild reaction conditions that would still give relatively good conversions, we conducted a series of experiments to determine the role of HCl, $AlCl_3$, and H_2 in coal hydrocracking. We examined the effects of temperature and residence time, studied catalyst/coal weight ratios of 1/1 to 3/1, then chose the standard reaction conditions for the screening of several acid catalysts. The effectiveness of the catalysts was judged by the solubility of the treated coal in THF and pyridine or both, and by the gas yields. In some cases where gasification was significant, gas yields were the only criteria used.

Experimental Studies

We used Illinois No. 6 coal pulverized by ball milling under nitrogen to -60 mesh and then usually dried in a vacuum oven at 115°C overnight. Pennsylvania State University supplied beneficiated coal samples (PSOC-26) as well as an unbeneficiated sample (PSOC-25) for use in some experiments. The beneficiated

0-8412-0427-6/78/47-071-204$05.00/0

coal has the following elemental analyses: C, 77.2%; H, 5.05% N, 1.69%; S, 2.08%; ash, 2.0%.

The experiments were carried out in either a rocking 500-ml autoclave fully lined with Teflon or in a 300 ml-Hastelloy C MagneDrive stirred autoclave from Autoclave Engineers. In general, the reactor was charged with coal and catalyst, evacuated, then filled with 0.7 mole of HCl (~ 500 psi) and 800 psi of hydrogen. The reaction mixture was heated to and kept at the reaction temperature for a given period. After the reaction, the reactor was cooled slowly to room temperature. The gases were analyzed by gas chromatography to determine hydrogen, methane, ethane, and propane contents. The reactor was then depressurized and the reaction mixture washed with water until the washings were neutral. The filtered coal products were then allowed to dry in a vacuum oven at 115°C overnight.

The treated product coal was usually characterized by elemental analyses (C, H, N), by molecular weight determinations, and by solubilities in THF and pyridine. THF and pyridine solubilities were determined by stirring a 0.50 g sample of the product coal in 50 ml THF or pyridine at room temperature for 1 hr, filtering the mixture in a medium porosity sintered glass filter, and then washing the residue with fresh solvent (~ 50 ml) until the washings were clear.

Results and Discussion

$AlCl_3/HCl$. In a series of runs in a rocking Teflon-lined autoclave, we first studied the role of HCl, $AlCl_3$, and H_2 in coal hydrocracking using 5 g each of $AlCl_3$ and coal, at 190°C (just above the melting point of $AlCl_3$), for 15 hr. As shown in Figure 1, one or more of the three components were absent in Runs 1 to 6 and in Run 9, and in each case, no increase in THF and pyridine solubilities was observed. In Run 10, where all three components were present, solubilities increased substantially, suggesting that the $AlCl_3/HCl$ system was active. The purpose of Runs 7 and 10 was to assess the importance of HCl in the system under these conditions; however, the results are not unequivocal. Here, the presence in the coal of proton sources, such as phenolic groups and traces of water, undoubtedly hydrolyzes some of the $AlCl_3$, producing HCl. These runs indicate that no added HCl is required for coal hydrocracking at these lower temperatures.

At higher reaction temperatures (210°C) and shorter reaction time (5 hr), the added HCl clearly increases the conversion (Runs 21 and 25), suggesting that the effective catalyst in the system must contain the elements of HCl and $AlCl_3$.

We also studied the effect of potential H-donor hydrocarbons and temperature. We based our work on the results of Siskin (6), who found that saturated, tertiary hydrocarbons serve as effective hydride donors in the strong acid-promoted hydrogenolysis of benzene. In our system, they proved ineffective (Runs 17, 22, 24,

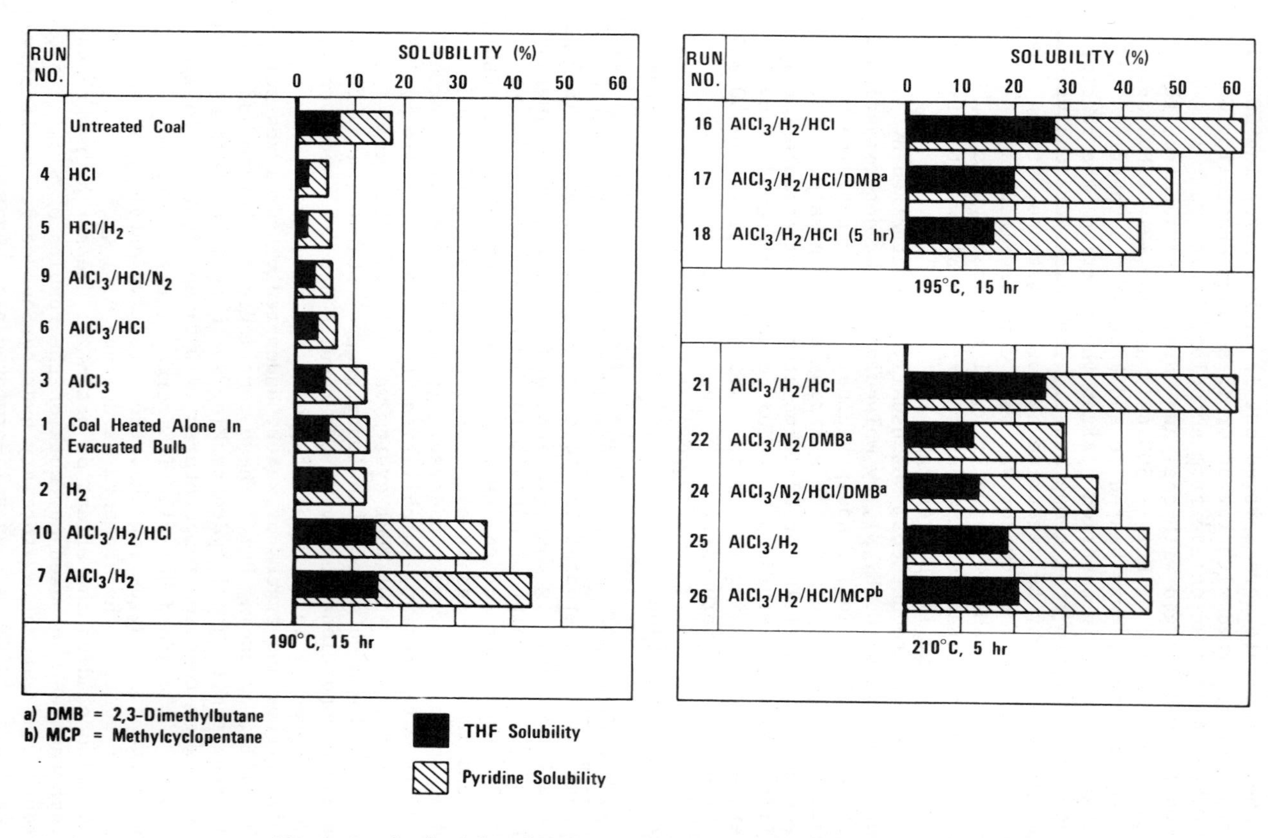

Figure 1. Acid-catalyzed hydrocracking of beneficiated Illinois No. 9 coal

and 26, Figure 1). Higher temperatures allowed shorter reaction times. The results for Run 17 (only 5 hr at 195°C) are comparable to those for Runs 7 and 10 (15 hr, 190°C). Runs at 195°C for 15 hr were far more effective, and the conversion for Run 15 at 210°C for 5 hr is about the same as that for Run 16 at 195°C for 15 hr.

Next, we studied the effects of the catalyst/coal weight ratio on product character and coal product yields in both the Teflon-lined and the Hastelloy C autoclaves. At a weight ratio of 1.0, the two systems yielded products with strikingly different pyridine solubilities: about 13% and 60% with the metal and Teflon equipment, respectively (Figure 2). Increasing the catalyst/coal ratio to 2.0 increased solubilities to above 90% for both systems, but a further ratio increase actually caused solubilities to decrease slightly. The solid product recovery also decreased with increasing catalyst/coal ratio.

As shown in Figure 2, at a 2.0 ratio only about half the coal was recovered as a solid product. The other half was converted to a mixture of methane and ethane. The softening point for the THF-soluble fraction was about 150°C; however, the pyridine-soluble fraction did not melt even at temperatures up to 280°C.

Figure 3 presents data on the H/C ratios for products from both systems. The coal products from the Teflon-lined reactor have consistently higher H/C ratio than those from the Hastelloy C reactor. Results in the Hastelloy C autoclave were unchanged when a loosely fitting Teflon liner was used. We have no detailed explanation for the effect of autoclave surface on the results, but passivation of the metal surface by some minimum quantity of catalyst is part of the answer. Whatever the mechanism, the Teflon surface is helpful.

The catalyst system gasifies some of the coal directly to methane and ethane. This result and the effects of temperature on coal conversion are shown in Table I. The table shows data from runs at 210°C for reaction times from 45 min to 5 hr, and at 300°C for a 90 min reaction time. The 210°C data are from an earlier phase of our work, where the gasification was not quantified, and the gasification was determined by difference. For the 300°C work, the quantities of gases and residue were determined independently, and thus, the mass balances for these runs are not exactly 100%.

The 300°C runs are all for 90 min, and Runs 83 and 85 show striking degrees of gasification. More than 90% of the carbon in the coal was converted to a 50/50 mixture of methane and ethane in these experiments. In the next three runs, no HCl was present, and we observe a cumulative effect of its absence. The degrees of gasification decline severely, and all the solid coal products recovered have declining pyridine solubilities and H/C ratios. Kawa et al. (2) observed a similar effect for HCl.

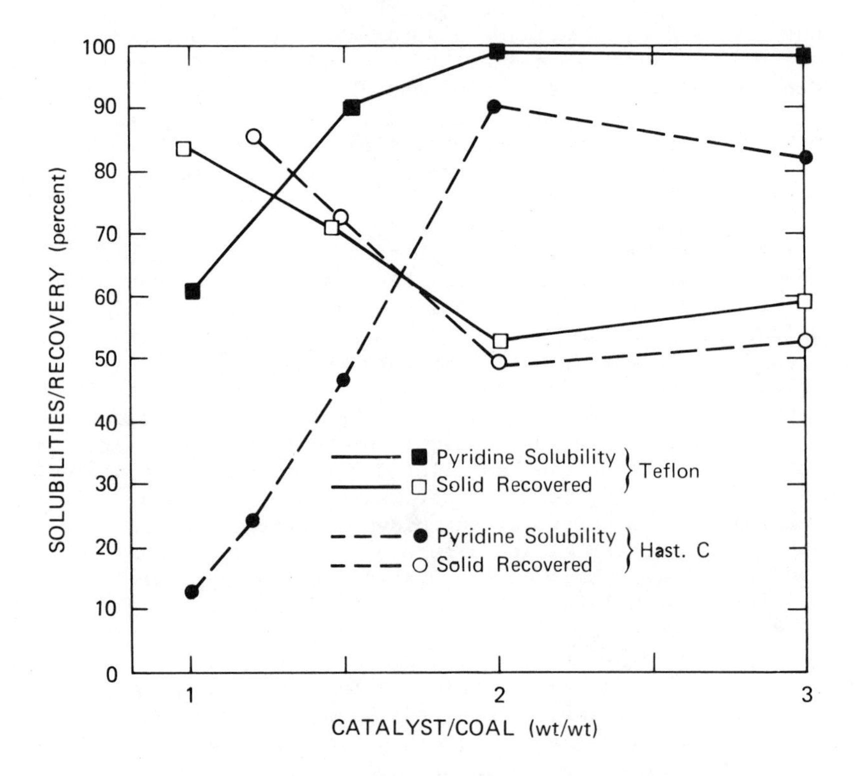

Figure 2. Effect of autoclave surface and catalyst/coal ratio on coal conversion.

The $AlCl_3/HCl$ system was used with reaction temperature of 210°C, for 5 hr in a stirring Hastelloy C autoclave or rocking Teflon-lined autoclave.

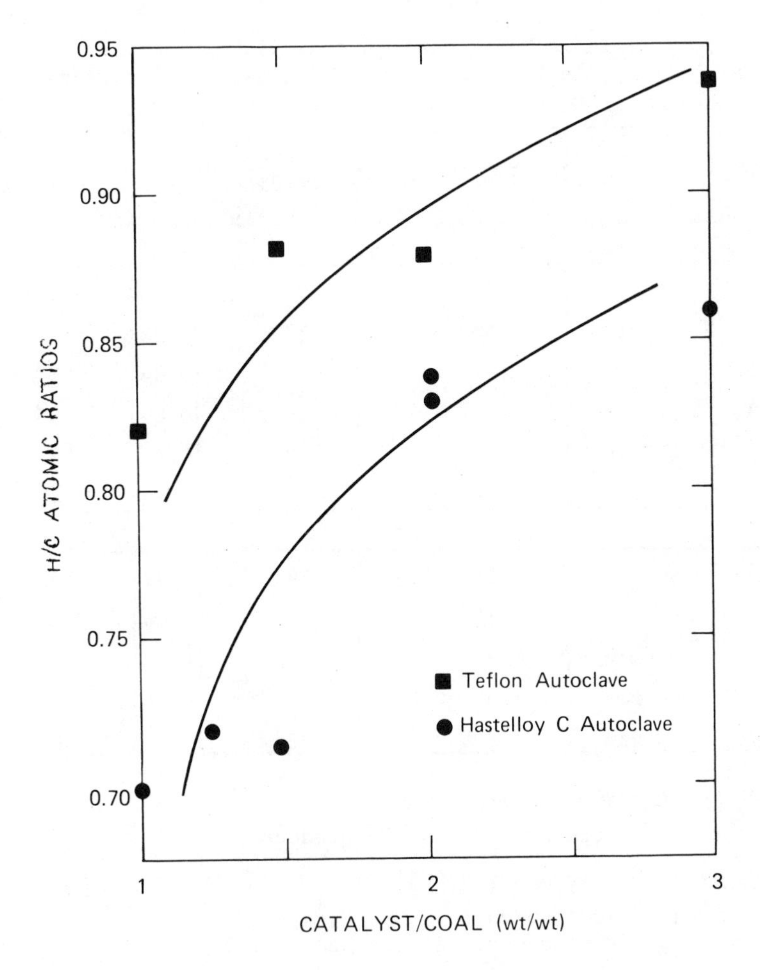

Figure 3. Comparison of H/C atomic ratio of treated product coal vs. effects by different reactors and catalyst/coal ratios. For runs at 210°C and 5 hr.

Table I

THE EFFECT OF RESIDENCE TIMES ON COAL GASIFICATION

(4 g Illinois No. 6 coal, 8 g $AlCl_3$, 500 psi HCl, 800 psi H_2, in 2 300 ml stirred Hastelloy C autoclave)

Run	Residence Time	Coal Residue			% Coal Gasified[b]
		%Recovered[a]	%Pyridine Solubility	H/C	
210°C					
67	5 hr	39	97	0.83	61[b]
46[c]	5 hr	48	87	0.82	52[b]
41	4 hr	49	91	0.84	51[b]
66	90 min	47	96	0.85	53[b]
69[c]	45 min	72	93	0.82	28[b]
71	45 min	66	97	0.82	34[b]
300°C					
83	90 min	18	78	0.72	96[d]
85	90 min	18	83	0.74	90[d]
86[e]	90 min	30	60	0.75	72[d]
87[e]	90 min	49	31	0.68	56[d]
88[e]	90 min	68	28	0.64	36[d]

[a] Based on 4 g of coal.

[b] Based on unaccounted for solid product.

[c] Run with 3 g coal and 6 g $AlCl_3$.

[d] Determined independently. $CH_4/C_2H_6 \simeq 1/1$. Traces of propane were seen in Runs 86, 87, and 88.

[e] Run without HCl

These data can be explained by the following scheme:

$$\underset{H/C = 0.79}{\text{Coal}} \xrightarrow[H_2/\text{catalyst}]{1} \text{Coal product} \xrightarrow[H_2/\text{catalyst}]{2} C_1 + C_2$$

$$\text{Coal} \xrightarrow[H_2/\text{no catalyst}]{3} \underbrace{H/C = 0.75 \rightarrow 0.64}_{\text{Char}}$$

Steps 1 and 2 are important in the presence of an effective catalyst and step 3 becomes competitive with no effective catalyst present. At lower temperatures, $k_1 > k_2$. Thus, the hydrogen-rich, pyridine-soluble coal product accumulates and can be isolated. At higher temperatures, the relative rates of steps 1 and 2 are reversed, $k_2 > k_1$, and gasification is the major effect.

Clearly, both $AlCl_3$ and HCl are necessary for effective catalysis. When HCl is eliminated, catalyst effectiveness is reduced, steps 1 and 2 are suppressed, and step 3 becomes dominant. With the lessening degrees of gasification, the coal residue appear to be increasingly cross-linked and depleted in hydrogen, possibly a result of chemistry at the autoclave surface. HCl alone (Figure 1) not only is ineffective, but also promoted char formation.

Other Lewis/Bronsted Combinations. Several acid catalysts were screened in two series of tests at two catalyst concentration levels, 210°C, 5 hr, 800 psi H_2 (cold), and 0.7 mole HX (X = Br, Cl, or F). We found catalyst activity to vary considerably from one series to the next (Figure 2). Catalyst/coal weight ratio for the first series was 1/1. All catalysts studied at this ratio, except $AlBr_3$ and $AlCl_3$, were ineffective, reducing THF and pyridine solubilities significantly, perhaps because of internal condensation in the starting coal. The coal products in these runs are probably highly cross-linked. $AlCl_3$ was considerably more effective than $AlBr_3$, and HBr alone (Run 30) was, not surprisingly, ineffective. These results indicate a catalyst effectiveness of $AlCl_3 > AlBr_3 \gg SbCl_3 \simeq SbF_3 \simeq ZnCl_2 \simeq TaF_5 \simeq NiSO_4 \simeq CoSO_4 \simeq HBr$. In the second series, a constant molar quantity of catalyst was used: 0.045 moles catalyst /4 g coal, equivalent to 6 g $AlCl_3$/4 g coal. The ordering here is $SbBr_3 \simeq SbCl_3 > AlBr_3 > AlCl_3 > Ni(AA)_2 > TaF_5 \gg SbF_5 \simeq MoCl_5 \simeq WCl_6$ (AA = acetylacetonate).

Thus TaF_5, which Siskin (6) found effectively hydrocracked benzene to mixed hexanes, is not at all effective under our conditions. Similarly, $ZnCl_2$, the well known coal conversion catalyst, is not effective under these conditions, perhaps because under our relatively mild conditions, $ZnCl_2$ is not molten (mp 283°C). Finally, the favorable antimony bromide and chloride results are similar to those reported by Shell (1).

In Run 35 (Table II), with $AlCl_3/HCl$, we used unbeneficiated coal. Here, the THF solubility of the product coal increased by almost a factor of 2, to 40%. The pyridine solubility increased slightly, to 66%. Since pyridine is generally a better solvent for coal liquids than is THF, the considerable increase in THF solubility suggests that more lower molecular weight products are obtained when unbeneficiated coal is used. Also, the mineral matter present in the unbeneficiated coal clearly aids the hydrocracking process, suggesting that the mineral matter in the coal is an effective catalyst under acid conditions.

Summary

In studies with mixtures of Lewis and Bronsted (proton) acids as catalysts, we have found that with hydrogen some mixtures convert Illinois No. 6 coal to fully pyridine-soluble products and 50/50 mixtures of methane and ethane. The $AlCl_3/HCl$ system was studied most extensively, and it was found that at 210°C (410°F) and 45 min residence time, about 70% of the coal was converted to a hydrogen-enriched, fully pyridine-soluble product and the remainder, to methane and ethane. A 90 min or longer residence time yielded 50% gases. At 300°C (572°F), more than 90% of the carbon in the starting coal was converted to methane and ethane. Neither HCl nor $AlCl_3$ alone was effective; when used along, either system yielded a solid product relatively low in pyridine solubility and depleted in hydrogen relative to the starting coal. Reaction runs in an autoclave fully lined with Teflon gave far better results than those in a Hastelloy C autoclave. A screening study of several acids showed that at a 1/1 catalyst/coal weight ratio, the order of effectiveness is $AlCl_3 > AlBr_3 \gg SbCl_3 \simeq SbF_3 \simeq ZnCl_2 \simeq TaF_5 \simeq NiSO_4 \simeq CoSO_4$. With a constant ratio of a given molar quantity of catalyst to mass of coal, the order is $SbBr_3 \simeq SbCl_3 > AlBr_3 > AlCl_3 > Ni(AA)_2 > TaF_5 \gg SbF_5 \simeq MoCl_5 \simeq WCl_6$. (AA = acetonylacetonate.)

Acknowledgment

The financial support for this work from DOE under Contract No. EF-76-C-01-2202 is gratefully acknowledged.

Table II

TREATMENT OF ILLINOIS NO. 6 COAL
WITH H_2/STRONG ACID SYSTEMS

Run No.	Catalyst System	Pressure (psi)		Solubilities (%)[b]	
		H_2	HX	THF	Pyridine
	(a) Constant Weight[a]				
28	$AlCl_3/HCl/H_2$	800	500	23	58
35	$AlCl_3/HCl/H_2$[c]	800	500	40	66
29	$AlBr_3/HB_2/H_2$	980	33g	9	27
27	$AlBr_3/HBr/H_2$	820	50g	11	32
30	HBr/H_2	1000	31g	2	6
31	$SbCl_3/HCl/H_2$	800	500	-	1
32	$TaF_5/HF/H_2$	1100	22g	-	13
33	$SbF_3/HF/H_2$	1100	16g	< 1	4
44	$ZnCl_2/HCl/H_2$	800	500	-	9
56	$CoSO_4/H_2SO_4/H_2$	1300	68.6g	< 1	< 1
57	$NiSO_4/H_2SO_4/H_2$	1300	68.6g	< 1	< 1
	(b) Constant Molar Quantity[d]				
45	$AlCl_3/HCl/H_2$	800	500	25	47
48	$AlBr_3/HBr/H_2$	1100	70g	30	59
49	$SbCl_3/HCl/H_2$	800	500	-	95^+
50	$TaF_5/HF/H_2$	900	14g	11	20
52	$SbF_5/HF/H_2$	1150	14g	< 1	< 1
54	$MoCl_5/H_2$	1300	-	8	16
55	WCl_6/H_2	1300	-	6	12
61	$SbBr_3/HBr$	850	59g	43	95^+
62	$Ni(AA)_2/HCl/H_2$	800	500	ND[e]	38
70	$MoCl_5/HCl/H_2$	800	500	< 1	4

[a] In this series of experiments, 5 g of coal was treated at 210°C for 5 hr in a rocking Teflon-lined autoclave.

[b] Moisture-ash-free basis. [c] Unbeneficiated coal was used.

[d] These experiments used 0.045 moles of catalyst per 4 g coal. A stirred Hastelloy C autoclave was used.

ND = Not determined.

Literature Cited

1. Wald, M., U.S. Patent 3,543,665 (November 24, 1970).
2. Kawa, W., S. Freidman, L. V. Frank, and R. W. Hiteshue, Amer. Chem. Soc., Div. Fuel Chem. (1968), Preprint 12 (3), 43-7.
3. Qader, S., R. Haddadin, L. Anderson, and G. Hill, Hydrocarbon Process (1969), 48 (9), 147.
4. Kiovsky, T. E., U.S. Patent 3,764,515 (October 9, 1973).
5. Zielke, C. W., R. T. Struck, J. M. Evans, C. P. Costanza, and E. Gorin, I & E C Process Design and Development (1969), 5, (2), 151, and references therein.
6. Siskin, M., J. Amer. Chem. Soc. (1974), 95, 3641.

RECEIVED February 10, 1978

16

Characterization of Coal Products by Mass Spectrometry

H. E. LUMPKIN and THOMAS ACZEL

Exxon Research and Engineering Co., P.O. Box 4255, Baytown, TX 77520

I. Introduction and Background

In studies on the organic chemistry of coal, the researcher, unfortunately, is unable to examine a complete coal molecule. He must instead be content to analyze bits and pieces of coal molecules produced by solvent refining, liquefaction, pyrolysis, or extraction. Knowledge of the composition of these pieces helps in understanding the organic chemistry of coal and is vital for the development of coal liquefaction processes and the further upgrading of the liquefaction products.

Mass spectrometry is the prime technique used in our laboratories (1,2) and in other laboratories (3) to determine the composition of the very complex mixtures derived from coal. Petroleum fractions have been analyzed by mass spectrometry (MS) for over 30 years. As heated inlet systems evolved (4,5) and instrumental resolving power increased, MS was applied to higher boiling ranges and more complex mixtures. When research in coal liquefaction began in our laboratories about 10 years ago, we had well-developed instrumentation, data handling procedures, and quantitative analyses for petroleum (6,7). Extension of these techniques to coal products required only minor changes and extensions (8,9). In this paper we describe some of the MS procedures we use and give some typical examples of analyses.

II. Equipment and Data Handling Procedures

A very repeatable low resolution instrument is used for streams in which the major components have been previously identified. The unseparated naptha boiling range, separated saturate fractions, and mid-boiling range samples (when detailed knowledge of the hetero-atom components is not required) fall in this category. For mid-boiling and high-boiling fractions requiring more complete breakdown of aromatic, hydroaromatic, and aromatic hetero-compounds, spectra are obtained on a high-resolution double focusing instrument.

0-8412-0427-6/78/47-071-215$05.00/0

Both of the instruments are automated. A digital readout system senses peaks and converts analog signals to digital signals, records digital data on printed paper tape and on magnetic tape, and a larger computer reads the data from the magnetic tape and further processes it employing proprietary computer programs. A list of the equipment is shown below:

Item	Manufacturer	Model
Low Resolution MS	Cons. Electrodynamics Corp.	21-103C
High Resolution MS	Assoc. Elec. Industries, Ltd.	MS50
MS Readout System	Columbia Scientific Ind.	CSI-260
Printer	Mohawk Data Systems	2016
Computer	International Bus. Machines	370

III. Methods and Results

A. Naphtha Boiling Range. High ionizing voltage, low resolution spectra are adequate to determine paraffins, naphthenes, 2-ring naphthenes, C_6-C_{11} benzenes, C_9-C_{10} indanes and tetralins, C_9-C_{10} indenes, C_{10}-C_{11} naphthalenes, and C_{10}-C_{12} phenols in the C_5 to 450°F boiling range. The calibration data were derived primarily from scans of pure compounds and assembled in a 20 component matrix. A summary analysis is given in Table I.

Table I

Component	Wt.%
Total Saturates	76.6
Total Benzenes	13.9
Indanes/Tetralins	2.3
Indenes	0.1
Naphthalenes	0.0
Total Phenols	7.1
Total	100.0
Wt.% Carbon	85.80
Wt.% Hydrogen	13.09
Wt.% Oxygen	1.11

With some assumptions regarding the molecular weight distributions of the paraffins and naphthenes, a useful elemental analysis can be readily calculated.

Occasionally more detailed data for the saturated components is desirable. A 42 component combined MS and gas chromatographic procedure determines the aromatic and phenolic components listed previously and further breaks down the paraffins into iso- and normal types by carbon number and the naphthenes into cyclohexanes and cyclopentanes by carbon number. Calibration data were obtained from pure compounds and from concentrates separated by molecular sieve and gas chromatography.

B. Higher Boiling Range Saturate Fraction. Fractions boiling above the naphtha range can be separated into saturate, aromatic and polar fractions employing a modified version of the clay-gel adsorption chromatographic method, ASTM D-2007. The saturate fraction is analyzed by the high ionizing voltage MS method, ASTM D-2786. A typical analysis of a 430-950°F saturate fraction from a Synthoil product (10) is given in Table II.

Table II

Compound Type	Wt.%
Paraffins	34.2
1-Ring Naphthenes	18.4
2-Ring Naphthenes	16.3
3-Ring Naphthenes	14.6
4-Ring Naphthenes	10.3
5-Ring Naphthenes	3.0
6-Ring Naphthenes	1.1
Monoaromatics	2.1

Normal paraffins generally comprise the major part, 80-90%, of the total paraffins in coal liquefaction products and a lesser part in coal extracts. When a split between iso- and normal-paraffins is desired, gas chromatography rather than mass spectrometry is normally the method of preference in higher boiling fractions. Iso-paraffins are not usually identified. However, we have recently identified the isoprenoid paraffins pristane, 2, 6, 10, 14-tetramethyl pentadecane, and phytane, 2, 6, 10, 14-tetramethyl hexadecane in coal extracts and liquefaction products. These components were separated and identified by gas chromatography, corroborated by MS (10).

Isoprenoid paraffins are used in organic geochemistry to group oils into "families" (11). It is possible that coals from different seams or deposits might be differentiated by the relative ratios of these isoprenoids or by the isoprenoid to n-paraffin ratio. Pristane is thought to be derived from the diterpenic alcohol phytol (12), that comprises about 30% of the chlorophyll molecule, and its presence and concentration might be related to the environment of the marsh in which a particular coal bed or seam was formed.

C. Higher Boiling Range Aromatic and Polar Fractions. If data on the saturate portion are not required, the aromatic and polar aromatic components are determined on the unseparated sample. This is done with a high resolution instrument operated in the low ionizing voltage mode (13). The same procedure can be applied to separated aromatic and polar fractions, and this is preferred if there is a significant concentration of polar components.

With low ionizing voltage electrons only those components containing double bonds, such as aromatics and olefins, are ionized and only the molecular ion is produced. Thus, the spectra, the spectral interpretation, and the calibration data are simplified, as there is no interference between components.

A full discussion of high resolution mass spectrometry is beyond the scope of this paper, so the technique will be described here only briefly. Different combinations to form molecules of the atomic species found in coal products will have different molecular weights. For example, from the atomic weights of the most abundant species given below one calculates the molecular weight of methyl

Atomic Species	Atomic Weight
Carbon	12.000
Hydrogen	1.0078
Oxygen	15.9949
Nitrogen	14.0031
Sulfur	31.9721

acenaphthene, $C_{13}H_{12}$, to be 168.0939, and the molecular weight of dibenzofuran, $C_{12}H_8O$, to be 168.0575. The high resolution MS resolves these two peaks having the same nominal molecular weight and the resolving power required is 4615 ($Mass/\Delta Mass = 168/0.0364 = 4615$). Other molecules require even greater resolving power to separate, particularly those containing nitrogen or sulfur (14). The MS and its auxiliary apparatus must also provide data from which precise mass measurements can be calculated. By measuring the time at which each peak occurs in a repeatable logarithmic scan of the spectra and by introducing compounds having peaks at known masses, the masses of the sample peaks can be determined very precisely, and the mass determines the molecular formula.

The scheme we use from mass spectrometer to final quantitative analysis is given in the data flow scheme below. The MS, MS readout system, printer, computer, computer programs, and people are required.

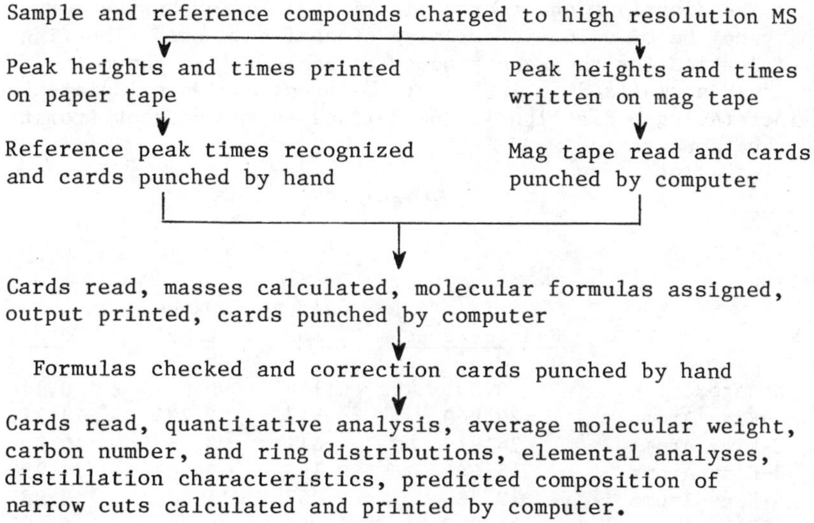

Sample and reference compounds charged to high resolution MS

Peak heights and times printed on paper tape

Reference peak times recognized and cards punched by hand

Peak heights and times written on mag tape

Mag tape read and cards punched by computer

Cards read, masses calculated, molecular formulas assigned, output printed, cards punched by computer

Formulas checked and correction cards punched by hand

Cards read, quantitative analysis, average molecular weight, carbon number, and ring distributions, elemental analyses, distillation characteristics, predicted composition of narrow cuts calculated and printed by computer.

The most detailed information calculated from the high resolution spectra is the quantitative amount of each compound type at each carbon number. This tabulation is printed on 6 pages, 50 rows and 12 columns per page. This is more data than most engineers care to examine; therefore, summary tables, distributions and other items are calculated from these detailed data. Excerpts from the compound type summary of a Synthoil product (10) are given in Table III. This is the initial summary made from the detailed data.

Table III

Compound Type	Wt.%	Average Mol.Wt.	Average Carbon No.	C Atoms in Sidechains
Alkyl Benzenes	1.74	160.8	11.9	5.9
Naphthalenes	11.02	176.4	13.5	3.5
Dibenzothiophenes	0.33	210.9	13.9	1.9
Fluorenothiophenes	0.16	241.2	16.4	2.4
Benzofurans	0.56	210.9	14.6	6.6
Dibenzofurans	2.55	255.0	16.1	4.1

The distribution of aromatic rings is a further summary which may be of value in refining of coal products. The ring distribution for the same product of Table III is shown in Table IV normalized to 100%, but the program also calculates and prints the same distribution normalized to the percent aromatics in the sample.

Table IV

	Hydrocarbons	Sulfur Comp.	Oxygen Comp.	Totals
Nonaromatics	0.0	0.144	0.0	0.144
1-Ring Aroms	20.960	0.517	2.235	23.712
2-Ring Aroms	36.919	0.611	3.465	40.996
3-Ring Aroms	15.644	0.153	2.202	17.999
4-Ring Aroms	12.388	0.067	1.470	13.925
5-Ring Aroms	1.997	0.020	0.208	2.225
6-Ring Aroms	0.856		0.084	0.940
7+Ring Aroms	0.068			0.068
Totals	88.832	1.512	9.664	100.008

Additional calculated items, such as distillation characteristics (15), can be of great value to a researcher. If there is insufficient sample available for actual distillation, say from a bench-scale experiment, a few milligrams will suffice for a high resolution MS run. The calculated MS values, GC distillation, and 15/5 distillation are in good agreement.

The same high resolution scheme can also be applied to the polar fractions from the clay-gel separation. But the analysis of polars can become very tedious--the composition is much more complex as the polars contain many of the same hydrocarbon species as the aromatic fractions in addition to the polar heteroaromatic oxygen, nitrogen, and sulfur compounds. In addition, our computer programs for some of these classes of components are not yet fully integrated into the final quantitative analysis program, and separate programs must be run and the results meshed. An indication of the complexity of the polar components in coal products is provided by a very small portion of spectra of the polar fraction of a Synthoil product given in Table V, in which some typical multiplets resolved by the high resolution MS are shown. The data were obtained at a resolving power of about 40,000.

Table V

Mass	Formula	Intensity	General Formula	Possible Structure
254.0764	$C_{16}H_{14}SO$	230	$C_nH_{2n-18}SO$	C_2-Hydroxythio-phenoacenaphthene
254.1306	$C_{17}H_{18}O_2$	822	$C_nH_{2n-16}O_2$	C_4-Dihydroxyfluorene
254.1671	$C_{18}H_{22}O$	511	$C_nH_{2n-14}O$	C_6-Hydroxyacenaph-thene
381.1517	$C_{29}H_{19}N$	337	$C_nH_{2n-39}N$	C_2-Dibenzoperylenide
381.1729	$C_{26}H_{23}NO_2$	219	$C_nH_{2n-29}NO_2$	C_5-Dihydroxybenzo-chrysenide
381.2092	$C_{27}H_{27}NO$	363	$C_nH_{2n-27}NO$	C_7-Hydroxydibenz-carbazole
381.2456	$C_{28}H_{31}N$	267	$C_nH_{2n-25}N$	C_9-Chloranthridine
394.1357	$C_{30}H_{18}O$	110	$C_nH_{2n-42}O$	C_2-Hydroxybenzo-coronene
394.1569	$C_{27}H_{22}O_3$	225	$C_nH_{2n-32}O_3$	C_5-Trihydroxybenzo-perylene
394.1933	$C_{28}H_{26}O_2$	275	$C_nH_{2n-30}O_2$	C_6-Dihydroxybenzo-chrysene
394.2295	$C_{29}H_{30}O$	507	$C_nH_{2n-28}O$	C_9-Hydroxybenzo-pyrene

III. Conclusions

The national need to develop liquid fuels from coal to augment diminishing petroleum fuels is a challenge to the coal chemist. Analytical characterization of these coal liquids is a challenge to the analytical chemist. We believe that mass spectrometry, both low resolution and high resolution, plays an important role in responding to this challenge, and have given examples of the application of the technique to various coal product samples.

More detailed information on the use of high resolution mass spectrometry to analyze hetero-compounds in coal extracts and liquefaction products is given in our paper in the "Symposium on Refining of Coal and Shale Liquids," Division of Petroleum Chemistry, National ACS meeting, Chicago, 1977.

Literature Cited

(1) Aczel, T., Foster, J. Q., and Karchmer, J. H. Paper presented at the 157th National Meeting of the American Chemical Society, Minneapolis, Minnesota, April 1969.

(2) Aczel, T., Reviews of Analytical Chemistry, 1, 226 (1971).

(3) Sharkey, A. G., Schultz, Janet, Friedel, R. A., Fuel, 38, 315 (1959).

(4) Lumpkin, H. E. and Johnson, B. H., Anal. Chem., 26, 1719 (1954).

(5) O'Neal, M. J., Jr. and Weir, T. P., Jr., Anal. Chem., 23, 830 (1951).

(6) Aczel, T., Allan, D. E., Harding, J. H., and Knipp, E. A., Anal. Chem., 42, 341 (1970).

(7) Johnson, B. H. and Aczel, T., Anal. Chem., 39, 682 (1967).

(8) Aczel, T. and Lumpkin, H. E., "MS Analysis of Coal Liquefaction Products," presented at 23rd Annual Conference on Mass Spectrometry and Allied Topics, Houston, Texas, May 25, 1975.

(9) Aczel, T. and Lumpkin, H. E., "Mass Spectral Characterization of Heavy Coal Liquefaction Products," presented at 24th Annual Conference on Mass Spectrometry and Allied Topics, San Diego, California, May 9, 1976.

(10) Aczel, T., Williams, R. B., Pancirov, R. J., and Karchmer, J. H., "Chemical Properties of Synthoil Products and Feeds," Report prepared for U.S. Energy Research and Development Administration, FE8007, 1977.

(11) Barbat, W. N., American Association Petroleum Geologists Bulletin, 51, 1255 (1967).

(12) Bendoraitis, T. G., Brown, B. L., and Hepner, L. S., Anal. Chem., 34, 49 (1962).

(13) Lumpkin, H. E., Anal. Chem., 36, 2399 (1964).

(14) Lumpkin, H. E., Wolstenholme, W. A., Elliott, R. M., Evans, S., and Hazelby, D., "The Application of Ultra High Resolution Dynamic Scanning to the Analysis of Sulfur Containing Petrochemicals," presented at 23rd Annual Conference on Mass Spectrometry and Allied Topics, Houston, Texas, May 25, 1975.

(15) Aczel, T. and Lumpkin, H. E., "Simulated Distillation by High Resolution-Low Voltage Mass Spectrometry," presented at 18th Annual Conference on Mass Spectrometry and Allied Topics, San Francisco, California, June 14, 1970.

RECEIVED February 10, 1978

17

Field Ionization and Field Desorption Mass Spectrometry Applied to Coal Research

G. A. ST. JOHN, S. E. BUTTRILL, JR., and M. ANBAR

Stanford Research Institute, Mass Spectrometry Research Center, Menlo Park, CA 94025

Mass spectrometry offers a unique way to characterize coal liquefaction products. Molecular weight profiles of such complex mixtures of organic materials may be considered as the first step in the understanding of their nature in molecular terms. Molecular weight profiles may be produced by nonfragmenting mass spectrometry, which almost exclusively yields molecular ions. Field ionization produces molecular ions from most organic compounds (1). When a complex mixture is analyzed by this mass spectrometric technique, we obtain a single peak for each constituent or for a groups of constituents that shares the same nominal molecular weight. By repeated multiscanning, we can obtain a quantitative molecular weight profile of complex mixtures (2).

The detailed molecular weight profiles attainable by field ionization mass spectrometry are much more informative than molecular weight profiles obtained by gel permeation chromatography (GPC). Moreover, GPC is subject to artifacts caused by associations of solutes or by solute-solvent complex formations. The average molecular weight profiles obtained by vapor phase osmosis (VPO) contain minimal chemical information and are useful mainly in conjunction with prior chromatographic separation. The cost per mass spectrometric analysis is higher than by these two techniques, but the information obtained on each individual constituent or group of constituents would cost much more if obtained separately by other techniques. The advantage of the mass spectrometric technique is its universality; for the same sample, the same molecular weight profile will be obtained by different investigators using different mass spectrometers.

Low energy electron impact ionization, which induces relatively little fragmentation, has been proposed as an appropriate alternative technique for the analysis of complex organic mixtures (3,4) including fuels (5). However, an excellent systematic study by Scheppele et al (6), has shown that field ionization is by far superior for this purpose. This

0-8412-0427-6/78/47-071-223$05.00/0

recent study has shown that the relative ionization efficiencies by field ionization of many different classes of organic compounds are very similar, ranging only over a factor of two. This can be compared with a range of over an order of magnitude for low energy electron impact on the same substrates. Moreover, it has shown (6) that once corrected for the small differences in ionization efficiencies, which can be programmed and calculated for known homologous series, the analysis of complex mixtures, like fuels, can produce quantitative results with a significantly lower variance than obtainable with low energy electron impact, following an identical computational correction. Without such a correction, when dealing with unknown constituents, field ionization is by far a superior technique for obtaining semiquantitative information on the composition of highly complex mixture.

Field ionization is facilitated by the high field gradient that can be produced with very high curvatures. A cathode with a radius of curvature of about 0.1 μ requires less than 1000 V to produce field ionization. Such a configuration is readily attainable in a reproducible manner by the appropriate technology. At SRI, we have developed a novel field ionization source, the preactivated foil slit type source (7). This source is superior to the now classical SRI multipoint source (2,8) because of its lower sensitivity to deactivation in the presence of oxygen-, sulfur-, or halogen-carbonaceous dendrites are deposited from pyridine vapor at high temperature under a high electrostatic field (7).

The mass analyzer used by us for multicomponent analysis is a 60° sector, 25 cm instrument and has been described before (9). The temperature of the sample can be controlled independently of the temperature of the source. The temperature of the source is maintained constant and higher than the maximum temperature the sample is subjected to; this prevents memory effects and results in more controlled ionization conditions. This ionization source, which may be operated up to 400°C, may thus handle thermally stable compounds of very low volatility (9).

The ionization efficiency of our sources is 5×10^{-4} or higher, (10) which is comparable to that of advanced electron impact sources. However, owing to the relatively large area of our ionization source, the high energy of the ions produced, and their divergence, less than 10^{-4} of the ions produced are detected after mass separation. The overall efficiency of the present generation of field ionization mass spectrometers is about 2×10^{-8} ions/molecule for instruments with a magnetic sector analyzer and a resolution of 700.

Since most organic compounds have similar field ionization efficiencies (6), the molecular weight profile obtained by integrating all the spectra while evaporating the sample to completion truly represents the composition of the mixture. A number of examples of molecular weight profiles of different coal liquefaction products have been presented elsewhere (9). These spectra were obtained on our mass spectrometric system before it was interfaced with a PDP-11 computer. In this mode of operation, the mass range of interest was scanned repeatedly and synchronized with a 4096-multichannel analyzer operating in the multiscaler mode (8). The instrument integrates the spectra produced in such scan into a composite mass spectrum. The integration over time is necessary because the sample is evaporated slowly and the composition of the vapor phase changes because of the wide range of volatility of the different components. This field ionization mass spectrometric system thus facilitates the quantitative analysis of molecular weight profiles of mixtures that may contain constituents varying in their vapor pressure by many orders of magnitude (estimated range, 10^{10}) over a mass range up to 2000 amu with a resolution of $M/\Delta M = 800$.

Computer Controlled FI Multicomponent Analysis

The published field ionization spectra of coal liquefaction products (9) are just the first step in the full utilization of field ionization mass spectrometric multicomponent analysis. These are chart recordings from a 4096-channel analyzer and, although the information in each channel is digitized, these spectra give us only a means of visual inspection of the gross feature of the spectra. Accurate mass assignment and the integrated ion counts under each peak are lacking, and these are necessary for any detailed quantitative interpretation of these complex spectra.

We would also like to know the "history" of each peak--the actual rate of accumulation of the ions of a given nominal molecular weight as a function of time and temperature of the analyzed sample. This information is necessary for estimating the number of materials of the same nominal molecular weight that contributed to a given peak. Moreover, from the temperature profile of a given peak, it may be possible to deduce whether some of the contributing ions originate from a chemical process (e.g., pyrolysis, dehydrogenation) that took place in the sample while the sample probe was being heated. Obtaining, for instance, a material with a molecular weight of only 150 when the probe temperature reaches 300°C suggests that it may be a secondary pyrolytic decomposition product. The temperature profile analysis may help us to distinguish between these two possibilities and even determine the activation energy for the appearance of the given species.

To achieve these goals, we have interfaced our mass spectrometric system with a PDP 11/10 dedicated computer (Fig. 1). The computer controls the magnet scan of the mass spectrometer by means of the 12-bit digital-to-analog converter (DAC). The data acquisition program increments the input to the DAC at precisely controlled time intervals so that each channel is receiving ions counted for exactly the same amount of time. At the end of each time interval, the computer causes the ion counts accumulated by the 10-MHz counter to be transferred to the 12-bit buffer register. The counter is cleared and restarted in less than one microsecond, so the interface has a negligible dead-time and no ion counts are missed. The DAC input is incremented by one unit, and the ion count in the buffer is transferred to the computer and added to the previous ion counts for that channel. The time spent at each channel, or in other words, the scan rate of the mass spectrometer, is variable and is controlled by the operator through the data acquisition software. Actual time intervals are measured within the computer by a programmable clock based on a very stable quartz crystal oscillator.

The output from the DAC is a linear voltage ramp, since each of the 4096 possible channels is active for an equal amount of time. Since the mass of the ions focused on the mass spectrometer's detector varies as the square root of the magnet current, the magnet scan control unit is used to convert the linear voltage ramp into a signal that drives the magnet power supply to produce a linear mass scan.

The 12-bit analog-to-digital converter (ADC) is connected to a temperature programmer for the solids probe. At the end of each mass spectrometer scan, the temperature of the probe is recorded for later use in the printed reports or data analysis. Since the temperature programmer is digitally driven, a given temperature program can be very precisely reproduced to allow meaningful comparisons between samples. Our experience indicates that most pure compounds are volatilized over a narrow temperature range of 10-30°C. Thus, separate peaks will be observed in the temperature profile of a single mass if there is more than one component of the sample with that particular molecular weight. It is possible to distinguish between genuine low molecular weight components of a sample and those resulting from the thermal decomposition of much larger molecules because these two different types of species appear at very different temperatures. Our experiments show no indication of significant pyrolysis of coal liquefaction products or of crude oils.

The report program produces reports listing the masses and total ion counts for each peak in the spectrum. Two different formats are available: one is a simple table of the peak mass and intensity, and the other is the same information arranged in a fixed format with 14 masses in each row. The advantage of this second format is that homologs are all listed in the same column, making it easy to pick out groups of peaks that may have similar chemical structures.

The plotting program produces simple bar graphs of the mass spectra on the X-Y recorder. Full-scale intensity is arbitrarily chosen as 50 and the entire spectrum is automatically

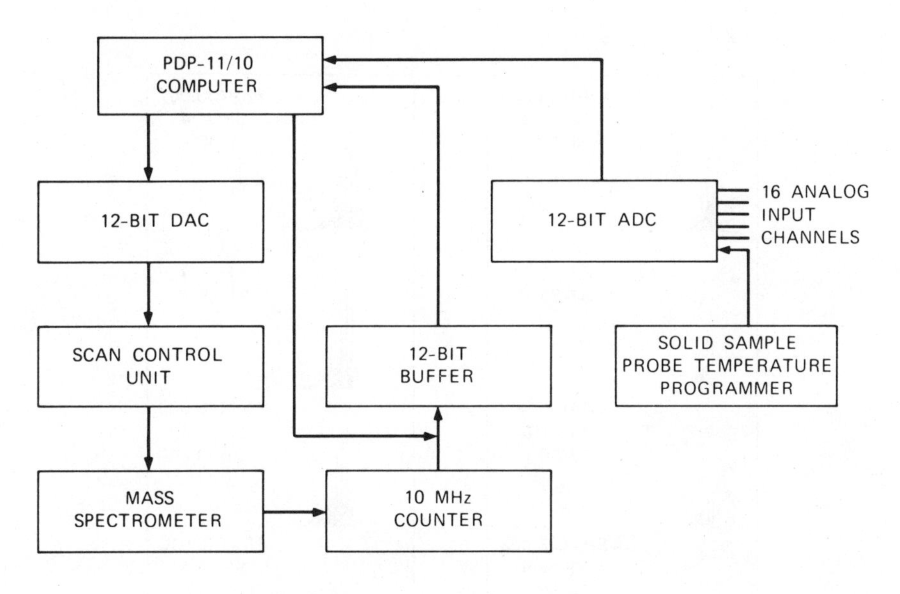

Figure 1. Mass spectrometer–computer interface

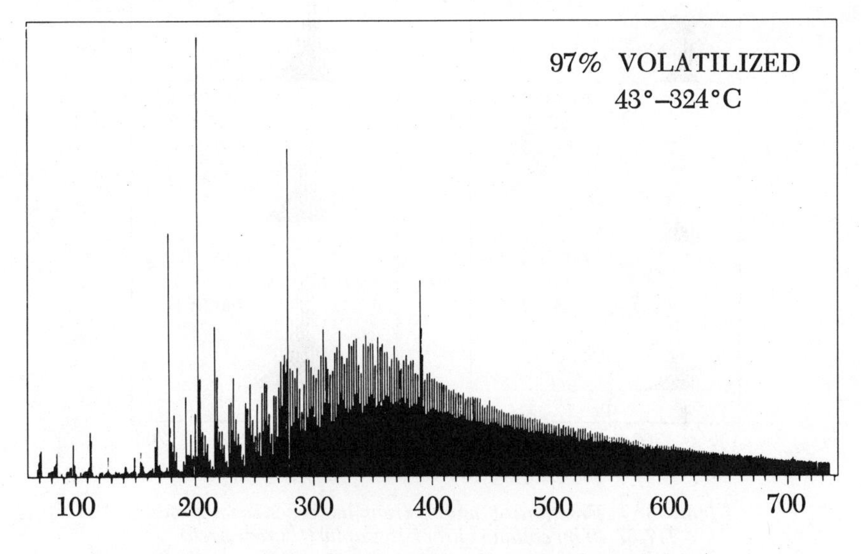

Figure 2. Kentucky 9/14 SRC, oils fraction (source: Arco)

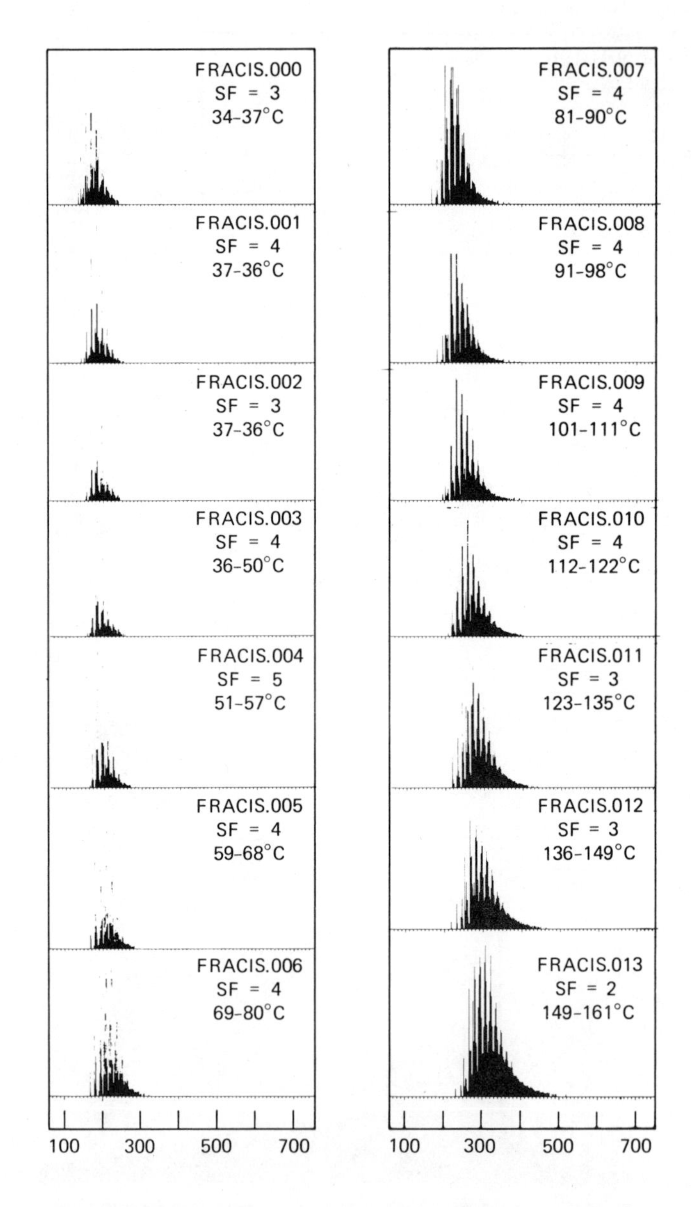

Figure 3A. Fractions 1 and 2 from hydrogen–coal product 177-57-49 by column chromatography (source: Arco)

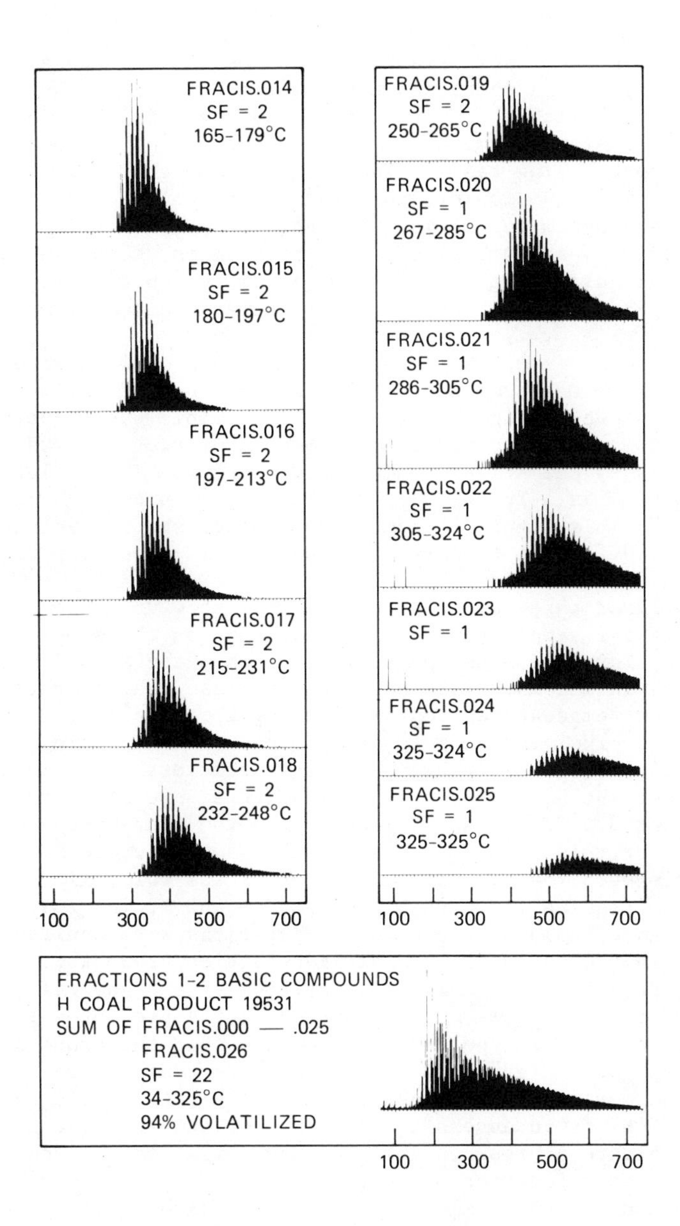

Figure 3B. Fractions 1 and 2 from hydrogen–coal product 177-57-49 by column chromatography (source: Arco)

scaled, if necessary, by dividing all peak intensities by an integer. Figure 2 is an example of a spectrum plot obtained in this manner. These spectra are evidently superior in quality to those obtained with the multichennel analyzers (9).

Additional programs are available for performing simple but useful data handling tasks. These include programs for listing on the terminal the ion counts in each channel of a raw data file and a program for summing the data in several files into a composition spectrum. The latter program is useful for obtaining the molecular weight profile of a complex multicomponent mixture by adding together all the spectra obtained from a sample.

Additional examples of the types of information currently available from the combination of FIMS and the PDP 11/10 computer are shown in Figs. 3a and 3b. The sample was fractions 1 and 2 of basic compounds from an H-coal product and was provided to us by the Atlantic Richfield Company. The evolution of this spectrum as a function of temperature is presented in Figs. 3a and 3b. This figure presents the plotted spectra integrated within the different temperature ranges during the evaporation of a single sample.

The first three spectra in Figure 3a show volatile components of the sample that came off as soon as the sample was introduced into the mass spectrometer. When the signal produced by these volatile materials began to decrease, the operator started the temperature program, heating the sample at about 2°C per minute. The heating rate was increased twice during the run to maintain a reasonably high signal as the less volatile components were being analyzed. Finally, three spectra were recorded at 325°C, (Figure 3b) which was the final probe temperature for this sample. Weighing the sample before and after analysis showed that 94% of this material was volatilized.

These results illustrate the vast amount of information to be obtained by combining chemical separations with nonfragmenting FIMS in the analysis of coal liquefaction products.

Figure 4 is an example of a spectrum of a crude oil analyzed by computer in the same manner as the samples presented in Figures 2 and 3. The only difference was that the crude oil sample was "weathered" in the probe at room temperature to remove the most volatile constituents. The same sample was analyzed 5 times to assess the variance of the analytical procedure. Figure 5 presents the standard deviation of each of the mass peaks as a function of molecular weight. One can see here that the constituents below 250 amu have a high variance due to irreproducible preevaporation ("weathering") but in the mass range 250 to 550, the standard deviation is in the range of 3 to 6%, which is very satisfactory for such a complex analysis. It should be noted that even the most abundant constituents in our complex mixture amount to just about 0.6% of the total. At higher molecular weights, there is an increase in the variance

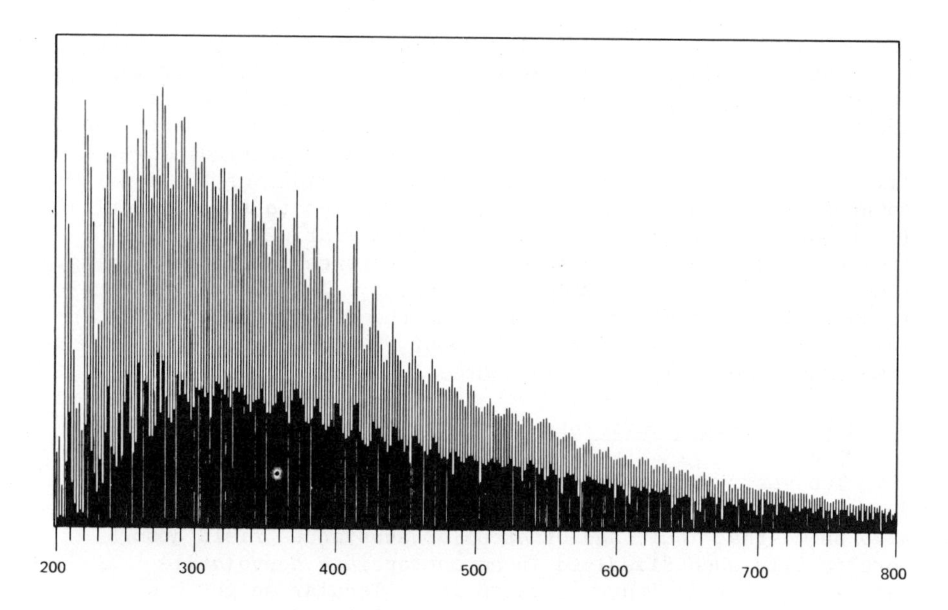

Figure 4. Venezuelan oil weathered in mass spectrometer

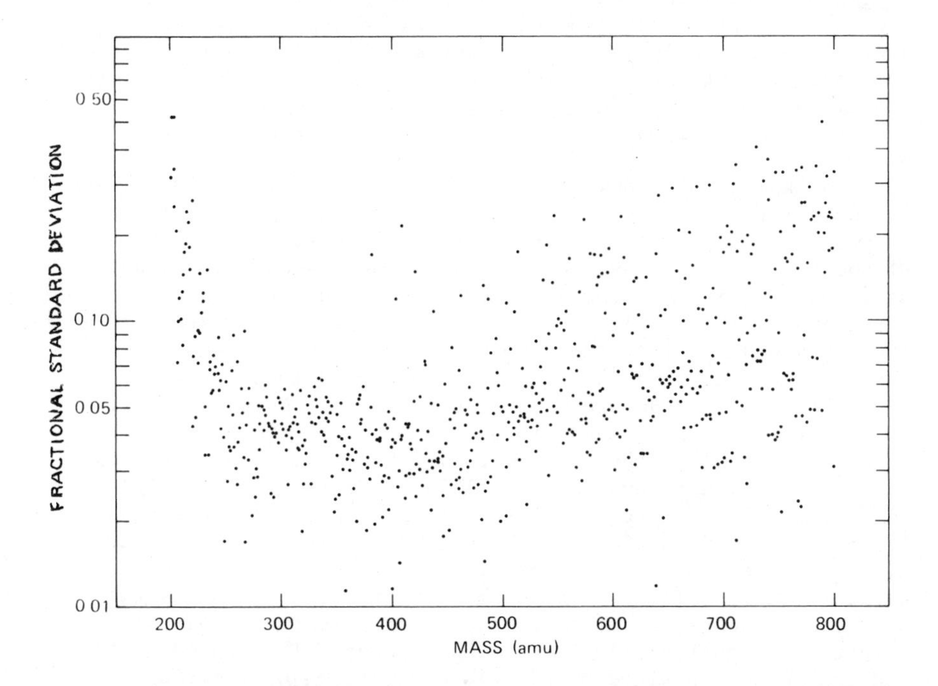

Figure 5. Standard deviation of each of the mass peaks as function of molecular weight

predominantly because of the lower abundance of these constituents and possibly also because of some irreproducible pyrolysis or polymerization of such minor components. The effect of abundance on the variability of the individual constituents can be seen in Figure 6 which also presents the theoretical lower limit of variance due to statistical fluctuations of the ions counted. The two lines "200" and "800" designate the theoretical limit for these two extreme cases of molecular weights. The limits are different because of the difference in the monitoring time per amu during the magnetic scanning. The actual variance is about 2 to 3 times higher than the theoretical lower limit. This is fairly satisfactory in view of the complexity of the sample and the analytical procedure.

Field Desorption Experiments

Two years ago, we developed at SRI a novel type of field desorbing source that used a broken metal tip (11). We have also shown that ionic and other polar substances field desorb more readily when dissolved in an appropriate nonvolatile matrix. Our preliminary tests on low molecular weight hydrocarbon polymers were highly encouraging (9). Recently, we have extended our experiments, using broken graphite rods and bundles of graphite rods and bundles of graphite fibers as field desorbing sources, with even greater success. We applied these sources to the analysis of asphaltenes. Figure 7 shows a field desorption spectrum (obtained on a multichannel analyzer) of the asphaltenes of SRC produced from Illinois No. 6 coal. The spectrum was obtained from a graphite fiber field desorbing bundle at 205°C.

An example of a computer handled FD spectrum of ashphaltenes (produced from Kentucky coal) is presented in Fig. 8. In order to provide a direct comparison of the characteristics of FD and FI spectra currently available, Figures 8-12 show the raw data as it was acquired by the computer, rather than the normal bar-graphs. The resolution of the FD spectrum is inferior to that obtained by FI of the same sample (Fig. 9), probably due to the wider energy spread of the field desorbed ions, as well as to the fluctuating nature of the FD ion beam. Note, however, the significant mass peaks at about 395 amu in the FD spectrum which are absent in the FI spectrum. This feature appears again at somewhat higher temperatures (157-158°C) as shown in Figure 10. Figure 11 a-d present FD spectra in the temperature range 160 to 187°C. Comparison of these spectra with the FI spectrum over the same temperature range (Fig. 9) and over a higher temperature range (Fig. 12) shows that under FD, one ionizes the same constituents, or other compounds of comparable molecular weight, at significantly lower temperatures. The reproducibility and resolution obtained by FD are, however, major limiting factors in the application of this technique as a quantitative

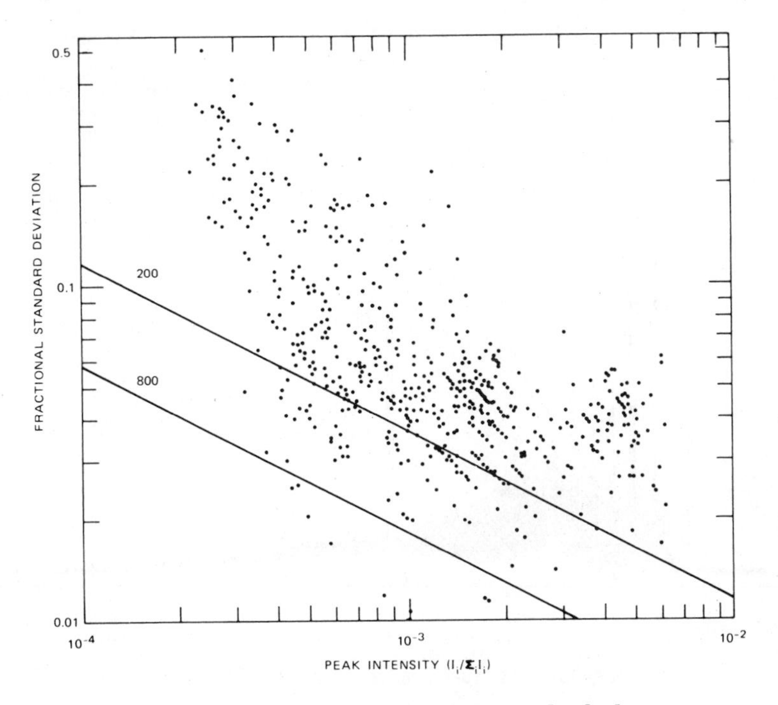

Figure 6. Effect of abundance on variability of individual constituents

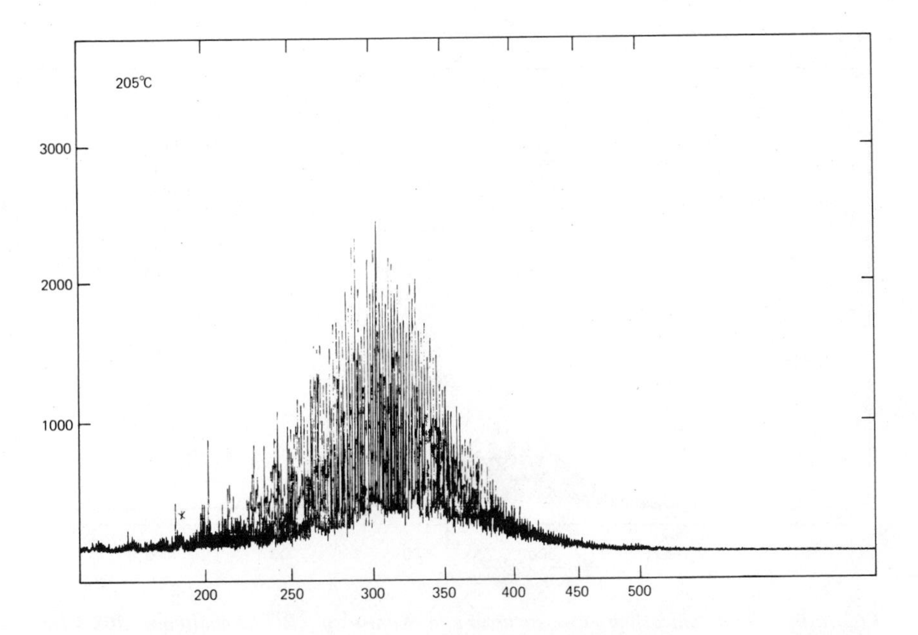

Figure 7. Asphaltenes from Illinois No. 6 coal, SRC product, Arco sample No. 9634, graphite fibers source—FDMS

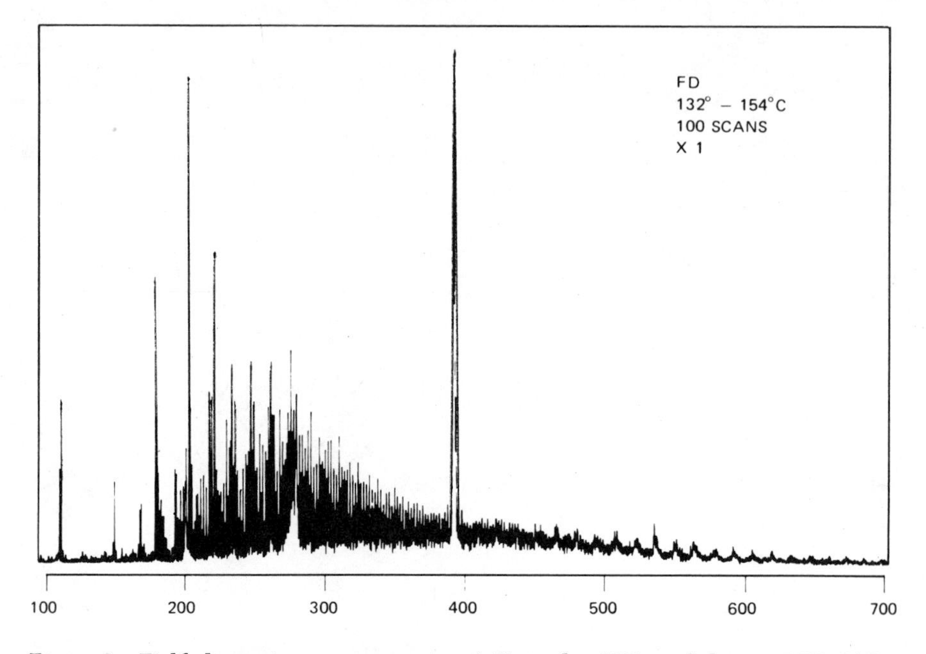

Figure 8. Field desorption mass spectrum of Kentucky SRC asphaltenes, 132°–154°C

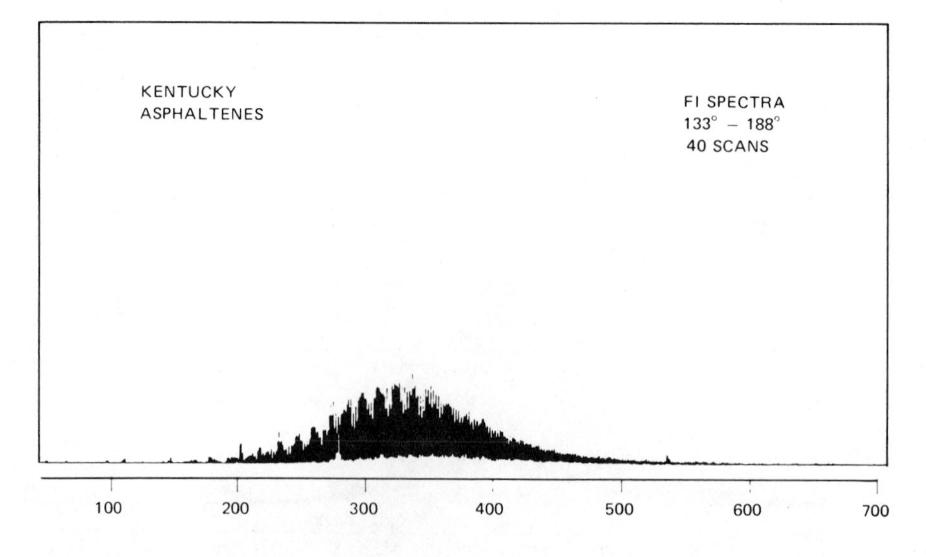

Figure 9. Field ionization mass spectrum of Kentucky SRC asphaltenes, 133°–188°C

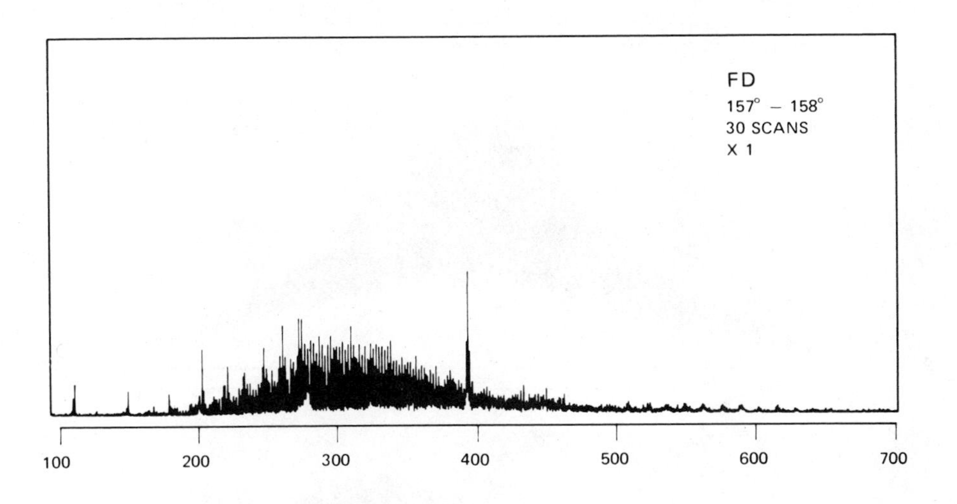

Figure 10. Field desorption of mass spectrum of Kentucky SRC asphaltenes, 157°–158°C

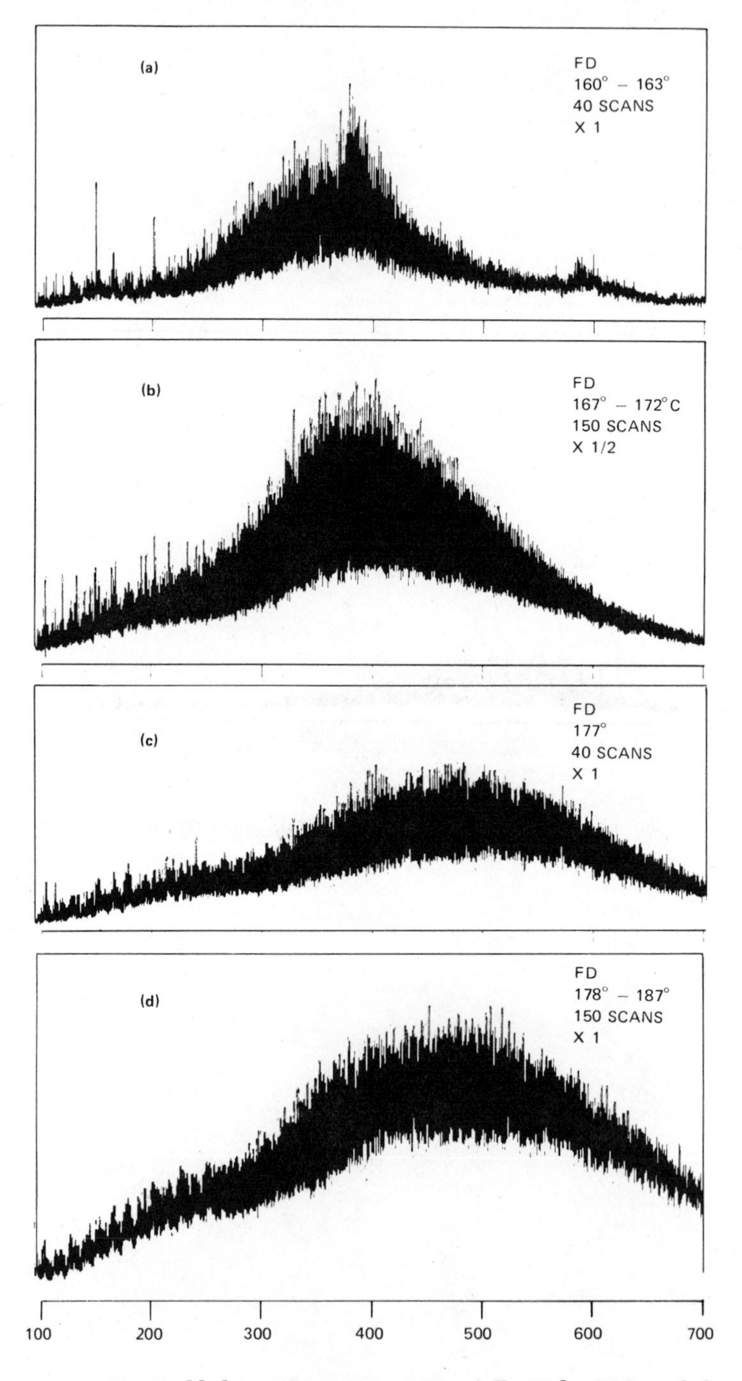

Figure 11. Field desorption mass spectra of Kentucky SRC asphaltenes, 160°–187°C

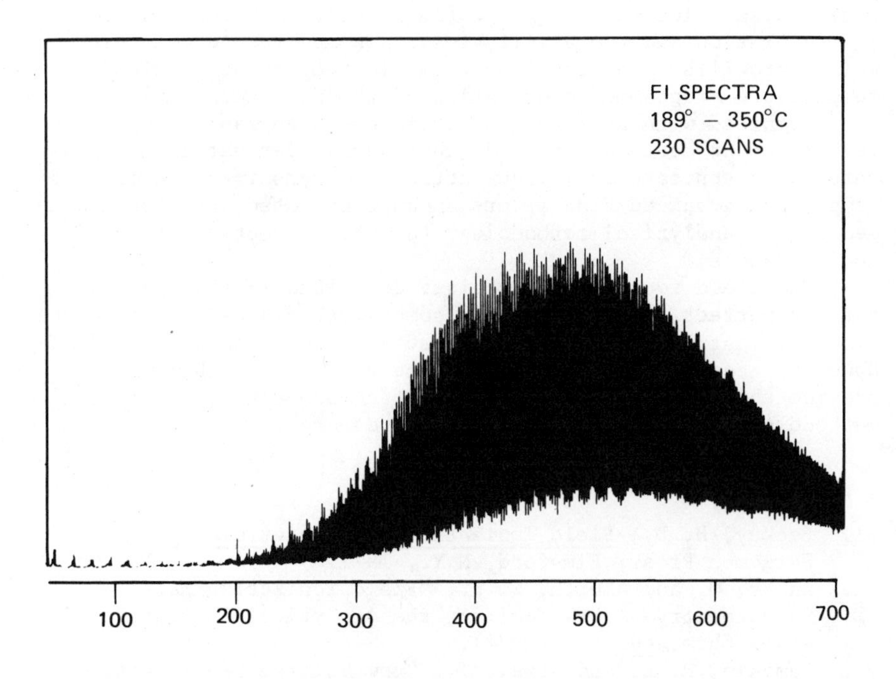

Figure 12. Field ionization mass spectrum of Kentucky SRC asphaltenes, 189°–350°C

way to characterize coal products. Some of these limitations could be overcome by focal plane simultaneous ion collection techniques (e.g., photoplate mass spectrography).

Summary

We have shown that mass spectrometric multicomponent analysis techniques are ideal for the characterization of coal liquefaction and fuel products. These include the capability of determining molecular weight profiles up to 1000 amu with unit amu resolution and the ability to obtain and record molecular weight profiles as a function of sample temperature during a temperature programmed evaporation of the analyzed sample.

Combined with appropriate liquid chromatographic separation techniques or certain fast and quantitative derivatization procedures to separate coal liquefaction products into families of compounds, advanced FIMS offers an unprecedented, precise and meaningful analytical methodology for the characterization of coal products.

The field ionization technique described in this paper is not yet perfect, and it requires some further development in the areas of instrumentation, sample pretreatment, and data handling. However, there is sufficient evidence that this technique can provide the basis for one of the most comprehensive analytical methodologies ever available to coal research.

Literature Cited

1. Beckey, H. D., Field Ionization Mass Spectrometry, Pergamon Press, Elmsford, N.Y., 1971.
2. Anbar, M. and Aberth, W. H., "Field Ionization Mass Spectrometry--A New Tool for the Analytical Chemist," Anal. Chem. 46, 59A (1974).
3. Lumpkin, H. E. and Aczel, T., "Low Voltage Sensitivities of Aromatic Hydrocarbons," Anal. Chem. 36, 181 (1964).
4. Schultz, J. L., Sharkey, Jr., A. G. and Brown, R. A., "Determination of Mass Spectrometric Sensitivity Data for Hydroaromatic Compounds," Anal. Chem. 44, 1486 (1972).
5. Aczel, T. and Lumpkin, H. E., "MS Analysis of Coal Liquefaction Products," 23rd Annual Conference on Mass Spectrometry and Allied Topics, Houston, Texas (1975), p. 228.
6. Scheppele, S. E., Grizzle, P. L., Greenwood, G. J., Marriott, T. D. and Perreira, N. B., "Determination of Field Ionization Relative Sensitivities for the Analysis of Coal-Derived Liquids and Their Correlation with Low Voltage Electron Impact Relative Sensitivities," Anal. Chem. 48, 2105 (1976).

7. Cross, R. H., Brown, H. L. and Anbar, M., "Preactivated Highly Efficient Linear Field Ionization Source," Rev. Scien. Instrum. 47, 1270 (1976).
8. Scolnick, M. E., Aberth, W. H. and Anbar, M., "An Integrating Multiscanning Field Ionization Mass Spectrometer," Int. J. Mass Spectrom. Ion Phys. 17, 139 (1975).
9. Anbar, M. and St. John, G. A., "Characterization of Coal Liquefaction Products by Molecular Weight Profiles Produced by Field Ionization Mass Spectrometry," Fuel, in press (1977).
10. Brown, H. L., Cross, R. H. and Anbar, M., "Characterization of Multipoint Field Ionization Sources," Int. J. Mass Spectrom. Ion Phys. 23, 63 (1977).
11. Anbar, M. and St. John, G. A., "FI-FD Source for Nonfragmenting Mass Spectrometry," Anal. Chem. 48, 198 (1976).

RECEIVED February 10, 1978

18

Heteroatom Species in Coal Liquefaction Products

F. K. SCHWEIGHARDT, C. M. WHITE, S. FRIEDMAN, and J. L. SHULTZ

U.S. Department of Energy, Pittsburgh Energy Research Center, 4800 Forbes Avenue, Pittsburgh, PA 15213

An assessment of the nitrogen and oxygen heteroatom species in coal-derived products is a complex yet important analytical problem in fuel chemistry. Principally, this is because the system is a multifarious molecular mixture that does not easily lend itself to direct analysis of any one component or functional group. Albeit this problem is not new, the characterization of these heteroatoms is of immediate importance to further processing of these fuels. Methods and techniques used to rapidly isolate and/or characterize both nitrogen and oxygen heteromolecular species are described. Utilization is made of solvent separations, functional group type separation, chemical derivatization, HCl salt formation and the use of chromatographic and spectrometric analytical methods to quantitate results. Specifically, the kind and distribution of nitrogen and oxygen heteromolecules in a coal liquefaction product and in a recycle solvent used in solvent refined coal (SRC) processing were determined. The coal liquefaction product was first solvent separated into oils, asphaltenes, preasphaltenes and ash, while low boiling oils (light oils) trapped from knock-out tanks and the SRC recycle solvent were treated directly. Nitrogen bases were complexed as HCl adducts or separated on ion-exchange resins. Hydroxyl-containing species from the separated fractions were quantitated by infrared spectroscopy or by formation of a trimethylsilyl ether (TMS) and subsequent analysis by ^{1}H NMR and mass spectrometry. Hydroxyl species were also isolated on ion-exchange resins or by selective gradient elution from silica gel.

EXPERIMENTAL

Solvent Separation

Coal liquefaction products were solvent separated (1) by

0-8412-0427-6/78/47-071-240$05.00/0

first freezing the coal liquids in liquid nitrogen and grinding them to fine particles. This frozen oil can be easily transferred to a stainless steel centrifuge tube. Pesticide grade solvents were then used to solubilize specific fractions--oils (pentane), asphaltenes (benzene), preasphaltenes (2) (tetrahydrofuran) and coal-derived ash (insoluble in all solvents used). By starting with a 3-4 gram sample, one (1) liter of each solvent in four or five 200 ml portions was usually sufficient to extract the solubles. Insolubles were removed by centrifugation at 10^4 rpm at 6°C for 10 minutes. Solvents were removed by nitrogen flush on a Rotovap using a water bath (65-85°C). Asphaltenes were treated differently at the final solvent removal step; a 20 ml solution of benzene/asphaltenes was swirled in a flask and flash frozen in liquid nitrogen, and the solvent was sublimed at 10^{-1}-10^{-2} torr for 2-3 hours.

HCl Treatment

The objective of this procedure was to separate and/or concentrate both nitrogen heteromolecules and hydroxyl-containing species from coal-derived material (3). Gaseous HCl was bubbled through a benzene or pentane solution of the coal product to form an insoluble HCl adduct with molecules containing a basic nitrogen atom. The adduct, after being washed free of other components, was back titrated with dilute NaOH solution to free the base nitrogen into an organic phase, usually diethyl ether, methylene chloride or benzene. The two fractions recovered contain acid/neutral and nitrogen base material, respectively.

Hydroxyl Silylation

Oils, asphaltenes and preasphaltenes were treated with hexamethlydisilazane (HMDS) to form a trimethylsilyl ether (TMS) of active hydroxyl groups (4,5). A 50 mg sample of coal-derived product was dissolved in 25 ml of benzene containing 50 µl of pyridine-d_5. To this solution 500 µl each of HMDS and N-trimethylsilyldimethylamine were added. This mixture was maintained as a closed system except for a small Bunsen valve and mildly refluxed for one hour with occasional swirling of the flask. After the reaction was completed, solvents and unreacted reagents were removed under nitrogen flush on a Rotovap and finally freeze dried from 5 ml of benzene for 30 minutes. A portion of the final product was checked by infrared spectroscopy (IR) for disappearance of the OH band at 3590 cm^{-1} (6). The remaining sample was dissolved in benzene-d_6 and its proton NMR spectrum taken and integrated. From the relative areas of the peaks in the proton NMR spectrum, a percent H as OH was calculated (Equation 1) (7).

$$\frac{(\frac{\text{TMS Area}}{9})}{(\frac{\text{TMS Area}}{9}) + (\text{Remaining Proton Area})} \times 10^2 = \%\ \text{H as OH} \quad (1)$$

From an elemental analysis of the original sample, one can calculate the weight percent oxygen as OH on a moisture and ash free basis (MAF).

Combined Gas Chromatography-Mass Spectrometry (GCMS)

The combined GCMS analyses were performed using a Dupont 490 mass spectrometer* interfaced to a Varian 1700 Series gas chromatograph, equipped with an 80:20 glass splitter and a flame ionization detector. The spectrometer was also coupled to a Hewlett-Packard 2100A computer used for spectrometric data storage and reduction. The mass spectrometer was operated at a resolution of 600 and an ionizing voltage of 70 eV. The ion source, jet separator and glass line from the chromatograph to the mass spectrometer were held at 275°C. The chromatographic effluent was continuously scanned at a rate of four seconds per decade by the mass spectrometer.

The gas chromatographic separations were effected using a variety of conditions. The nitrogen bases and acid fractions from the coal liquefaction product were chromatographed on a 10' x 1/4" OD glass column packed with 100-120 mesh Supelcoport coated with 3% OV-17. Bases from the SRC product were chromatographed on a 10' by 1/8" OD glass column packed with 100-120 mesh Chromasorb-G coated with 2% OV-17. Gas chromatographic separation of bases from the light oil was achieved using a 10' x 1/8" OD glass column containing acid washed and silyl treated 100-120 mesh Supelcoport coated with 3% Carbowax 20M. In each case the He flow rate was 30 cc/min and the analyses were performed using appropriate temperature programming conditions.

Column Chromatographic Separation

Coal-derived liquids, soluble in pentane, were separated into five fractions: acids, bases, neutral nitrogen, saturate hydrocarbons and aromatic hydrocarbons. Acids were isolated using anion-exchange resins, bases with cation-exchange resins, and neutral nitrogen by complexation with ferric chloride adsorbed on Attapulgus clay. Those pentane soluble hydrocarbons remaining were separated on silica gel to give the non-adsorbed saturates and the moderately retained aromatics. This method is commonly referred to as the SARA technique (8).

RESULTS AND DISCUSSION

The centrifuged coal liquid product (CLP) was produced

using Ireland Mine, Pittsburgh seam, West Virginia coal in the 1/2 ton per day SYNTHOIL Process Development Unit (PDU) (9). Operating conditions for this experiment were 4000 psi hydrogen pressure, 450°C and no added catalyst. The light oils were derived from a catalytic experiment (Harshaw 0402T) using Homestead Mine, Kentucky coal, under 4000 psi pressure and 450°C.

The coal liquefaction product was solvent separated by the method previously described to yield the distribution of fractions given in Table I. Figure 1 gives the atom weight percent distribution of nitrogen and oxygen in the solvent separated fractions listed in Table I. The pentane soluble oils were subsequently separated into five fractions using the SARA chromatographic scheme. Table II lists the weight percents of the individual fractions. The asphaltenes were treated with HCl to form acid/neutral and base subfractions, 63 and 37 weight percent, respectively.

Table I. Solvent Separation Weight Percent Distribution.

Solvent Fraction	Wt. % of CLP
Oil	68.0
Asphaltenes	26
Preasphaltenes	4
Ash	2

Table II. SARA Chromatography Fractions, Weight Percent Distribution.

Chromatography Fraction	Wt. % of CLP
Saturates	4.1
Aromatics	37.3
Acids	6.3
Bases	10.2
Neutral Nitrogen	3.9
Loss	7.4
Total Oil	68.8

The acid and base fractions from the SARA separation of the oils were subjected to analysis by combined GCMS and low voltage low resolution mass spectrometry (LVLR). Figures 2 and 3 reproduce the gas chromatograms of the base and acid fractions, respectively. The oxygen containing species shown in Figure 3 have been classified as alkylated phenols, indanols/tetralinols, phenylphenols, and cyclohexylphenols. Table III lists the carbon number range and tentative compound type assignments for the nitrogen heteromolecules in the acid and base fractions as determined by LVLR mass spectrometry (10). Table IV lists the

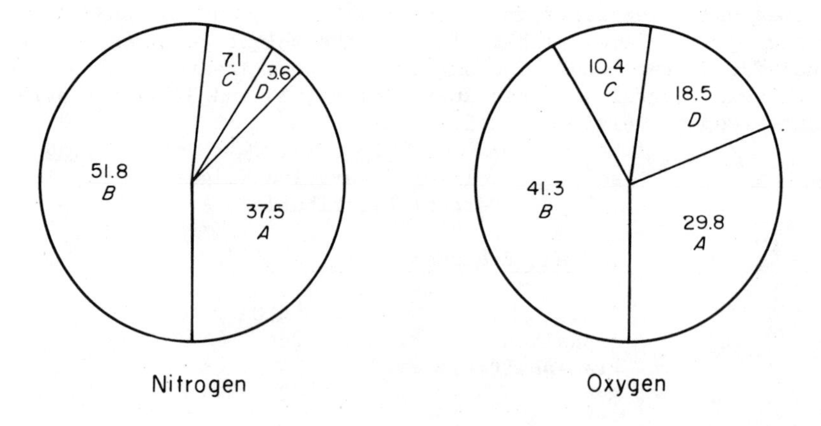

Figure 1. Atom percent distribution of heteroatoms: (A) oils, (B) asphaltenes, (C) preasphaltenes, and (D) residue.

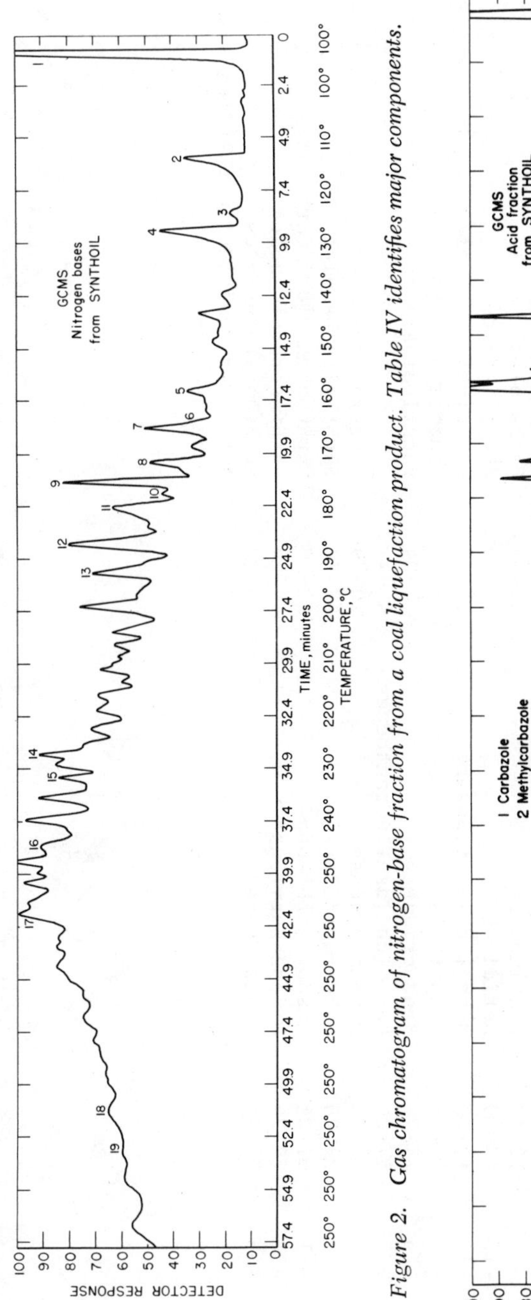

Figure 2. Gas chromatogram of nitrogen-base fraction from a coal liquefaction product. Table IV identifies major components.

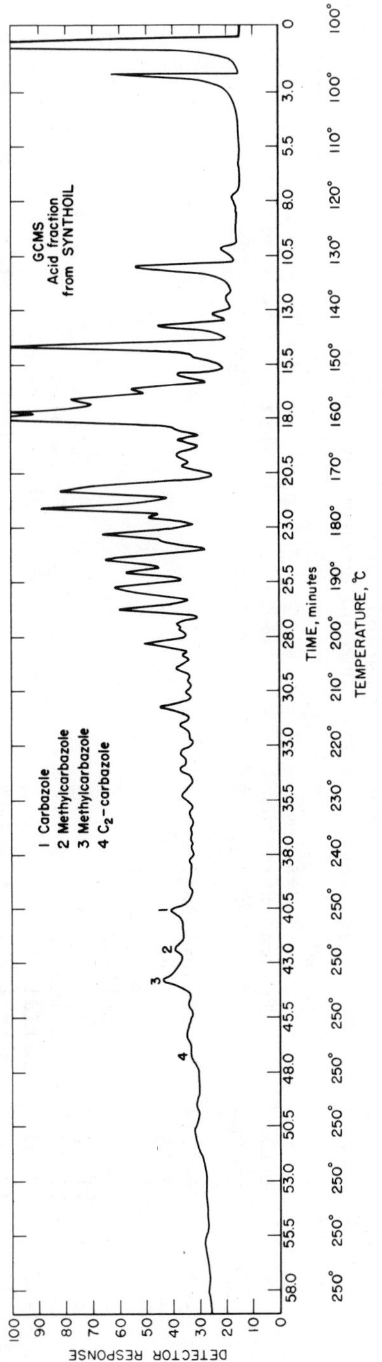

Figure 3. Gas chromatogram of acid fraction from a coal liquefaction product

Table III. Carbon Number Distribution for the Acid and Base Fractions From SARA Chromatography Separation.

	Bases		Acids	
Z#	LRLV C# Range	Typical Structural Nitrogen Analogs	LRLV C# Range	Typical Structural Nitrogen Analogs
- 5	6–12	Pyridines, Anilines	--	--
- 7	8–13	Azaindans	--	--
- 9	9–15	Dihydroquinolines, Indols	--	--
-11	9–15	Quinolines	--	--
-13	11–17	Phenylpyridines	--	--
-15	12–19	Azafluorenes	12–20	Carbazoles
-17	13–19	Acridines	14–20	Phenylindoles
-19	14–21	Benzo[ghi]azafluorenes	14–22	Benzo[def]carbazoles
-21	15–21	Azapyrenes	16–24	Benzocarbazoles
-23	17–26	Benzacridines	18–24	Phenylcarbazoles
-25	17–26	Benzo[ghi]azafluoranthenes	--	--
-27	19–26	Benzazapyrenes	20–29	Dibenzocarbazoles
-29	21–26	Dibenzacridines	22–29	Naphthylcarbazoles
-31	21–26	Azaanthanthrenes	22–29	Naphthobenzo[def]carbazoles
-33	23–26	Dibenzazapyrenes	24–29	Anthracenocarbazoles
-35	23–27	Azacoronenes	26–29	Anthranylcarbazoles
-37	--	--	28–29	Anthracenobenzo[def]carbazoles
-39	--	--	28–30	Dinaphthocarbazoles

Table IV. Compounds Found in the Nitrogen Base Fraction of Figure 2.

Peak Number	Compound
1	Solvent (Carbon Disulfide)
2	Phenol
3	Cresol
4	Cresol
5	Tetrahydroquinoline
6	Quinoline
7	6-Methyltetrahydroquinoline
8	Methylquinoline
9	Isoquinoline
10	Methylquinoline
11	Methylquinoline
12	C_2-Quinoline
13	C_2-Quinoline
14	Octahydro-N-3-Ring (Octahydroacridine)
15	Methyloctahydro-N-3-Ring (Methyloctahydroacridine)
16	Phenanthridine + Acridine
17	Methylacridine
18	Azapyrene
19	Methylazapyrene

compound types assigned to the base fraction of the oils by GCMS.

Table V lists the carbon number range data from the high resolution mass spectrometry (HRMS) analysis of the asphaltenes and their acid/neutral and base subfractions. It must be noted that at the operating conditions of the solids inlet, 300°C, 10^{-6} torr, less than 50% of the these materials could be volatilized. These preliminary studies have also indicated the presence of a limited number of diaza-species from Z = -8 to -18, where Z is the hydrogen deficiency in the general formula, C_nH_{2n-z}.

The SYNTHOIL PDU contains several knock-out traps that condense low boiling components, light oils (11). Nitrogen bases in the light oils were isolated by their precipitation with gaseous HCl and back titrated with NaOH into diethyl ether. These nitrogen bases constituted 3% by weight of the light oils. The gas chromatographic profile of these bases is given in Figure 4. An earlier study of these light oils characterized the saturates, aromatics and acidic components separated by Fluorescence Indicator Analysis (FIA) (12). The present investigation has resulted in the first quantitative analysis of pyridines and anilines in an oil produced by the hydrogenation of coal.

Table VI summarizes the quantitative results from the chromatogram of Figure 4. It is of interest to point out that during this investigation, though numerous substituted pyridines were quantitated, no evidence for the parent was found. Because the techniques employed recovered components with boiling points near that of pyridine it is suggested that this observation may be significant. If free pyridine was trapped within the coal macromolecular structure it surely would have been found in either the light oils or the pentane soluble oils. If, on the other hand, pyridine was attached exo-, via a single C-C bond, to a more complex molecular network, the hydrogenation process should have freed it intact. But if the nitrogen heteroatom was an integral part of the original coal macromolecule, then hydrogenation would have cleaved a number of Cα-Cβ bonds to produce a wide distribution of methylpyridines. Table VI shows this methyl substitution trend. Quantitative results indicate that 2,3,6-trimethylpyridine is seven times more abundant than 2,3-dimethylpyridine and approximately twice as abundant as any other methylpyridine.

The source of anilines and, in particular, the observation of both the parent and the methyl substituted anilines are of interest. Anilines can arise from hydrogenation of the hetero-ring in a fused ring system followed by breaking of the bond between nitrogen and an aliphatic carbon (13). Therefore quinoline and its alkyl derivative could be a source of the anilines. Table IV lists seven quinolines found in the oils that could be the precursors. The presence of the parent ani-

Table V. Carbon Number Range Data for Nitrogen Heteromolecules in Asphaltenes.

Z#	Asph	Asph A	Asph B	Possible Structural Types
	Carbon Number Range			
- 5	5-7	6,8,9	6	Pyridines
- 7	8-11	9-12	8-12	Azaindans
- 9	8-14	9-15	8-14	Dihydroquinolines; Indoles
-11	9-14	9-15	9-14	Quinolines
-13	11-16	11,12,14-17	11-16	Phenylpyridines
-15	11-19	11,12,14-19	11-17	Azafluorenes; Carbazoles
-17	13-20	14-20	13-18	Acridines
-19	14-20	14-21	14-19	Azabenzo[ghi]fluorenes
-21	15-21	15-22	15-20	Azapyrenes; Benzocarbazoles
-23	17-21	18-23	17-21	Benzacridines
-25	17-21	19-22	17-21	Azabenzo[ghi]fluoranthenes
-27	19-23	20-23	19-22	Azabenzopyrenes; Dibenzocarbazoles
-29	21-23	21-23	21	Dibenzacridines
-31	21-23	21,22	21,22	Azabenzoperylenes
-33	23,24			Azadibenzopyrenes

Formulas derived by HRMS represent isomers and fragment ions as well as molecular ions.

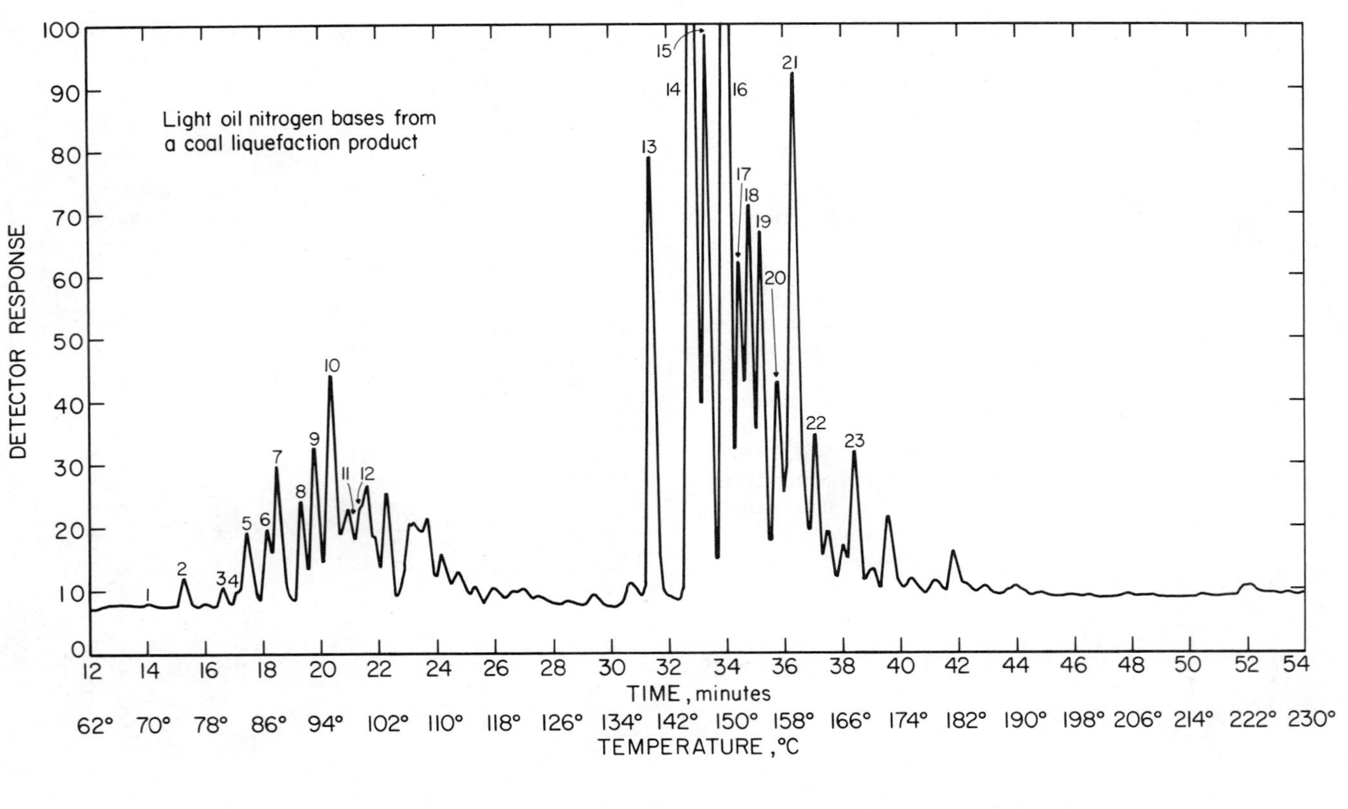

Figure 4. Gas chromatogram of nitrogen bases from a coal liquefaction light oil

line, that should readily be hydrodenitrogenated (13), is significant. If dealkylation was the same for all species, why didn't we see any of the parent pyridine? It could mean that the potential aniline moieties are located near the periphery of the coal macromolecule in contrast to the pyridines. The quantitative results indicate that methylanilines (toluidines) are in abundance in the order meta > ortho >> para > parent aniline, and all are greater than the dimethylanilines. These interpretations are based upon the more detailed analysis of the light oils. To date direct evidence for the presence of significant amounts of alkylated anilines and pyridines in the pentane soluble oils from the CLP have not been reported.

To complete this initial investigation of nitrogen species we chose to look at the nitrogen compounds present in the recycle solvent used for SRC processing and compare them with those found in coal liquefaction product oils. After extracting the gross benzene solubles, they were treated with HCl to isolate nitrogen bases. This particular sample had a slight residue that was benzene insoluble. Figure 5 gives the gas chromatographic profile of these nitrogen bases and summarizes the prominent structural isomers. The base fraction from the SRC solvent was less complex than the nitrogen bases found in the liquefaction oils, but the same principal molecular species were found in both samples.

The presence of hydroxyl groups in coal-derived materials has long been established. Our present interest is to define quantitatively the OH as a percentage of the total oxygen. The separation methods described concentrate a high percentage of the hydroxyl groups by anion exchange resin chromatography (acids) or the HCl treatment (acid/neutral). Once the separation/concentration has been made the sample is treated with a derivatizing reagent to form a trimethylsilyl ether, $Ar\text{-}O\text{-}Si(CH_3)_3$.

It has been shown that all of the hydroxyl groups contributing to the 3590 cm^{-1} infrared band can be quantitatively removed with the derivatizing reagent (6). The TMS ethers are next examined by proton NMR. The signals near 0 ppm represent the trimethylsilyl $(CH_3)_3$ protons from each of the hydroxyl derivatives. By integrating the area under the total proton spectrum and allowing for the 9-fold intensity enhancement for the TMS area, the percent H as OH can be calculated. Table VII lists some representative determinations of hydroxyl content from oils and asphaltenes. The silyl derivatization quantitation of hydroxyls in asphaltenes has been compared to the infrared spectroscopic method of standard additions (14). Our results agreed to within 10%. Infrared data, and those from others working on similar fractions (15) indicates that there is little if any carbonyl oxygen (C=O) present in coal liquefaction products produced in the SYNTHOIL PDU. Therefore, we conclude that substantially all of the oxygen exists as either hydroxyl (phenolic or benzylic) or in an ether linkage (e.g. furan).

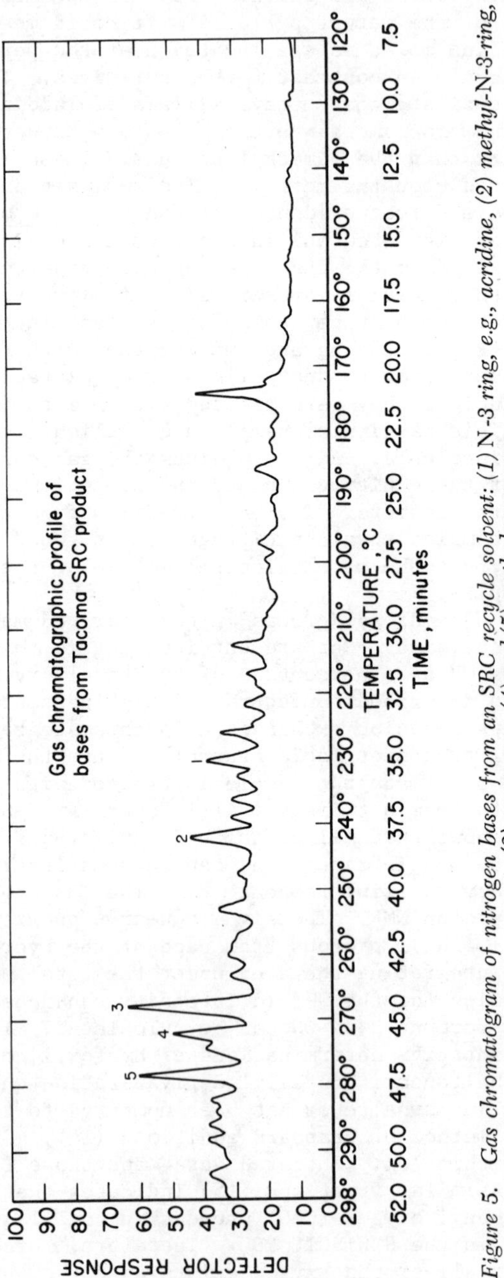

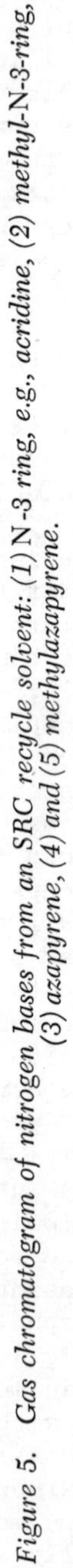

Figure 5. Gas chromatogram of nitrogen bases from an SRC recycle solvent: (1) N-3 *ring, e.g., acridine, (2) methyl-*N-3*-ring, (3) azapyrene, (4) and (5) methylazapyrene.*

Table VI. GCMS Results of the Analysis of the Nitrogen Base Components (16).

Peak #	Compound	Wt. % of Base Fraction*
1	2-Methylpyridine	0.1
2	2,6-Dimethylpyridine	0.5
3	2-Ethylpyridine	0.4
4	3-Methylpyridine and 4-Methylpyridine	0.3
5	2-Methyl-6-ethylpyridine	1.2
6	2,5-Dimethylpyridine	1.2
7	2,4-Dimethylpyridine	2.2
8	2,3-Dimethylpyridine	1.7
9	2,4,6-Trimethylpyridine	2.4
10	2,3,6-Trimethylpyridine	3.5
11	3,5-Dimethylpyridine	1.5
12	2-Methyl-5-Ethylpyridine	1.5
13	Aniline	6.4
14	o-Methylaniline	12.8
15	p-Methylaniline	8.1
16	m-Methylaniline	15.1
17	2,6-Dimethylaniline	4.9
18	2,4-Dimethylaniline	5.7
19	2,5-Dimethylaniline	5.3
20	2,3-Dimethylaniline	3.2
21	3,5-Dimethylaniline and quinoline	7.6
22	C_3-Aniline and Isoquinoline	2.4
23	1,2,3,4-Tetrahydroquinoline	2.2
		90.2

*These values are based on the assumption that all pyridines have the same response factor as 2,4-dimethylpyridine and all anilines have the same response factor as aniline. The assumption may introduce a small error into the quantitative data.

Table VII. Hydroxyl Distribution in Solvent Separated Fractions Determined By TMS Derivatization.

Fraction	% H as OH $\pm 5\%$	% O as OH $\pm 10\%$
Oils	1.2	67
Asphaltenes	2.0	43
Asphaltenes Acid/Neutral	2.0	44
Asphaltenes Base	1.5	49
Preasphaltenes	3.0	35

Table VIII. TMS Derivatized Hydroxyl Species in Acid Components From SARA Separation.

Molecular Weights		Molecular Weights	
Phenol Series	TMS Ether	Indanol Series	TMS Ether
94	166	134	206
108	180	148	220
122	194	162	234
136	208	176	248
150	222	190	262
164	236		276

Molecular Weights		Molecular Weights	
Acenaphthenol Series	TMS Ether	Fluorenol Series	TMS Ether
170	242	196	268
184	256	210	282
198	270	224	296
212	284	238	310
226	298	252	324
240	312	266	338
254	326	280	352

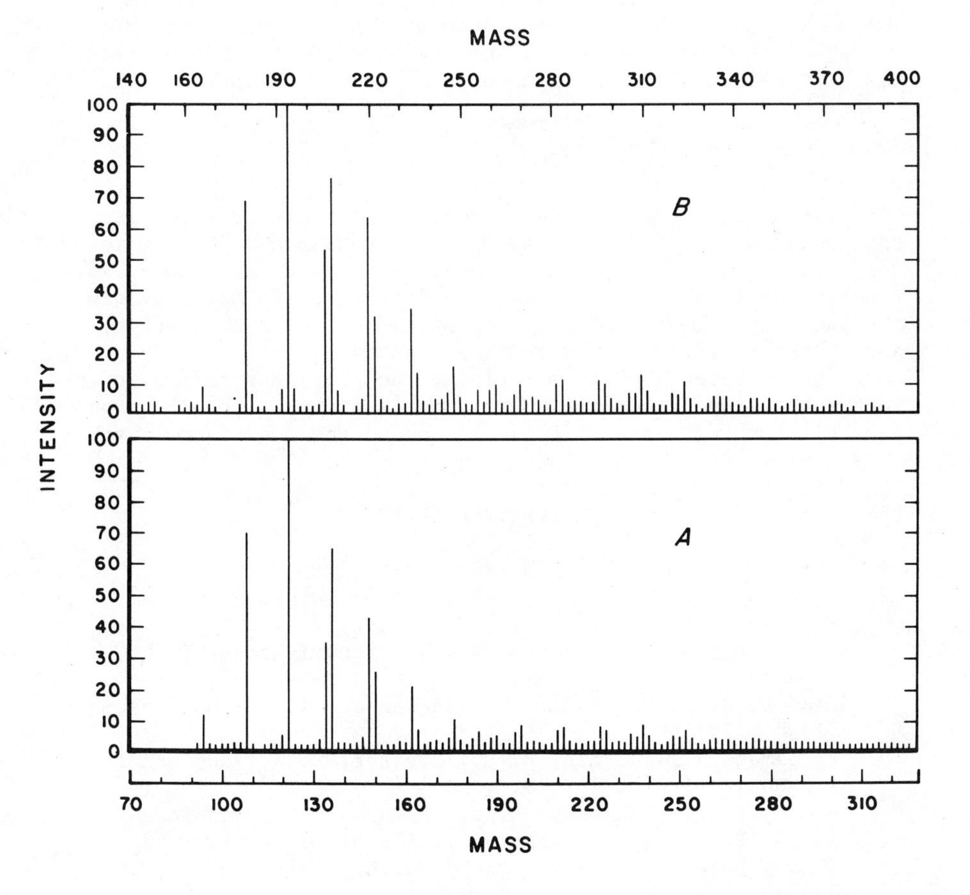

Figure 6. Mass spectra of acids before (A) and after (B) TMS derivatization. Note change in mass scale.

A useful corroboration of the NMR data and of characterizing the acid fraction of the oils is its mass spectrum before and after TMS derivatization. Figure 6 A and B shows the acid components from the pentane soluble oils before and after TMS derivatization, respectively. Note that the mass peaks are shifted 72 amu to give a nearly identical mass distribution. Table VIII lists those hydroxyl containing compounds that definitely formed a TMS ether. From the mass spectral data there was also evidence for trace amounts of indenol, naphthol and phenanthrol derivatives.

ACKNOWLEDGEMENT

The authors acknowledge the cooperation of the following PERC personnel: Sayeed Akhtar and Nestor Mazzocco for providing the coal liquefaction products and associated data, Dennis Finseth for taking the infrared spectra, Y. C. Fu for providing the SRC samples, Joseph Malli for providing the mass spectra, and Thomas Link for taking the NMR spectra.

LITERATURE CITED

1. Schweighardt, F. K., ERDA-PERC/RI-77/3.
2. Sternberg, H. W., Raymond, R., and Schweighardt, F. K., Preprints, Div. Fuel Chem., (1976), 21(7), 1.
3. Sternberg, H. W., Raymond, R., and Schweighardt, F. K., Science, (1975), 188, 49.
4. Langer, S. H., Connell, S., and Wender, I., J. Org. Chem., (1958), 23, 50.
5. Friedman, S., Kaufman, M. L., Steiner, W. A., and Wender, I., Fuel, (1961), 40, 33.
6. Brown, F. R., Makovsky, L. E., Friedman, S., and Schweighardt, F. K., Appl. Spectrosc., (1977), 31, 241.
7. Schweighardt, F. K., Retcofsky, H. L., Friedman, S., and Hough, M., Anal. Chem., in press.
8. Jewell, D. M., Weber, J. H., Bunger, J. W., Plancher, H., and Latham, D. R., Anal. Chem., (1972), 44, 1391.
9. Yavorsky, P. M., Akhtar, S., and Friedman, S., Chem. Eng. Prog., (1973), 69, 51.
10. Shultz, J. L., White, C. M., and Schweighardt, F. K., ERDA-PERC/RI-77/7.
11. Schultz, H., Gibbon, G. A., Hattman, E. A., Booher, H. B., and Adkins, J. W., ERDA-PERC/RI-77/2.
12. White C. M. and Newman, J. O. H., ERDA-PERC/RI-76/3.
13. Poulson, R. E., Preprints, Div. Fuel Chem., (1975), 20(2), 183.

14. Finseth, D. H., private communication.
15. Brown, F. R., private communication.
16. White, C. M., Schweighardt, F. K., and Shultz, J. L., Fuel Processing Technology, in press.

RECEIVED February 10, 1978

19

Characterization of Liquids and Gases Obtained by Hydrogenating Lumps of Texas Lignite

C. V. PHILIP and RAYFORD G. ANTHONY

Department of Chemical Engineering, Texas A&M University, College Station, TX 77843

The increased cost of natural gas and the high cost of fuel oil have increased the importance of Texas lignite as a potential source of chemicals and as a fuel for generating electricity. The lignite resource in Texas has been estimated (1) as 10 billion tons at depths less than 200 feet and 100 billion tons at depths of 200 to 5000 feet. The shallow basin lignite can be recovered by strip mining but the deep basin lignite will probably have to be recovered by *in situ* mining methods. *In situ* liquefaction and comminution have considerable potential for recovering deep basin lignite (2,3,4).

In order to evaluate the potential of underground liquefaction, autoclave experiments have been conducted at pressures of 500 to 5000 psi and temperatures of 650 to 800°F. The charge to the autoclave has been cylinderical cores, 1 1/2 inch in diameter and 3 to 5 inches long, hydrogen, helium and hydrogen donor solvents. In order to gain insight into the reaction mechanisms underlying the conversion process, the liquid and gas products have been analyzed by use of several methods. Solvents such as pentane, cyclohexane, benzene, and pyridine have been used to obtain fractions of coal-derived liquids enriched with aliphatic, aromatic, polar and asphaltenic species. These solvent separations are time consuming and it is difficult to reproduce the composition of each fraction. The complete removal of the solvent always poses an additional complication prior to characterization of the components in a given fraction. In a few cases absorption column chromatography has been used to separate the chemical species in a coal-drived liquid. We, therefore, have developed gas chromatographic methods for the analysis of coal-derived products with boiling points less than 500°C. GC-MS is used to characterize most of the GC peaks. Both proton and carbon-13 magnetic resonance techniques are used to provide information on structural properties, molecular dynamics, and chemical composition of the liquid samples.

0-8412-0427-6/78/47-071-258$05.00/0

Experimental

Three Gow-Mac gas chromatographs, Model 69-550, with thermal conductivity detectors were used for simultaneous analysis of gases and liquids. The oven temperatures were manually programmed. Commercially available helium is used as carrier gas. Helium is purified by passing it through molecular seive 5A (3 ft. X 1 1/2" O.D. stainless steel column) and through a high capacity purifier (Supelco Carrier Gas Purifier) to remove the traces of oxygen and water. The gas chromatographs accept only 1/4" columns with a maximum length of 10 ft. Samples were injected directly into the column to avoid the recovery loss in the injection port. The products from lignite liquefaction experiments, which were analysed by gas chromatography can be classified as follows: (a.) gases; (b.) low boiling point liquids (boils below 100°C); (c.) high boiling point liquids (boils above 100°C).

Porapak N was used for identifying methane, carbon dioxide, ethylene, ethane, hydrogen sulfide, propane, water, isobutane and *n*-butane. After an initial 1 min. hold at 25°C, a shotgun temperature program - 25°C to 140°C at a rate of 15 to 20°C per min. - was used to get excellent separations. Molecular Sieve 5A could separate hydrogen, carbon monoxide, oxygen and nitrogen from the gas samples at room temperature. Porapak S also gave a separation similar to Porapak N but did not separate propane and water under identical conditions. The thermal stability of Porapak S (max. temp. 250°C vs. 190°C for Porapak N) favored it as a choice for a few samples.

The low boiling liquids were separated on two Durapak columns. Durapak *n*-octane on poracil C is good for separating aliphatic components while Durapak OPN on poracil C separates the aromatic compounds. The sample is run on both columns simultaneously under identical conditions. A shotgun temperature program from 25 to 150°C gave fairly good separation of the components.

The high boiling point liquids contain nonvolatile components as well as lignite fragments which may deposit on the columns. The samples were cleaned to get the desired boiling point range by using a fractional sublimator. The sublimator consists of two concentric glass tubes. The outer tube holds the sample and the inner tube contains a coolant, like liquid nitrogen or dry ice-acetone mixture. The space between the tubes is evacuated while the outer tube is heated by a jacket type furnace. The temperature of the furnace is controlled by a Thermolyne proportional temperature controller. The sample temperature is recorded on a strip chart recorder. The sample evaporates and deposits on the outside of the inner tube (liquid nitrogen cold finger). The sample temperature and the degree of vacuum controls the volatility of the fractions deposited.

The clean samples from the sublimator were analysed using five different 8 ft. columns with packings that can withstand column temperatures well above 300°C without appreciable bleeding. The column temperature was programmed from 80°C to 280°C at a rate of 1.5 to 2°C per minute. The same sample was analysed with different columns under identical conditions. A hydrocarbon standard of *n*-alkanes ranging from C_{10} to C_{36} along with pristane and phytane was used to qualitatively identify the boiling point range of the components separated on different columns.

Detailed analysis of the lignite derived products were done on GC-MS. The apparatus mainly consists of a Hewlett-Packard 5710A Gas Chromatograph. A 5980A Mass Spectrometer, a 5947A Multi Ion Detector and a 5933A Data System. The gas chromatograph is able to accept packed columns as well as glass capillary columns. A 30 ft. X 1/8" stainless steel column packed with 3% OV 101 on 80/100 mesh Chromosorb W-HP and a 30 M glass capillary coated with OV 101 were used for most of the GC-MS studies. OV 101 is a methylsilicone polymer similar to the SP 2100 used in the Gow-Mac gas chromatographs. The hydrocarbon standard was used to determine the boiling point range of the components as well as the fragmentation pattern of the *n*-alkane series.

The proton nmr spectra of the samples dissolved in CCl_4 were taken on a Varian T-60 nmr spectrometer. JEOL PS-100-PFT was used for scanning ^{13}C nmr spectra of samples in $CDCl_3$. Samples used for these studies were not sublimated. The samples, therefore, contained high molecular weight species as well as minute suspended particles.

Results and Discussion

The gaseous products from different lignite liquefaction experiments were composed of the same components but the composition varied depending on the experimental conditions and the lignite sample cores used. The gaseous components were identified using known standards and simple chemical tests. Figure 1 is a typical gas chromatogram for the gas sample obtained during the hydrogenation of wet Texas lignite. Carbon dioxide is the major component. Hydrogen sulfide is present in an appreciable concentration. Once both carbon dioxide and hydrogen sulfide were removed from the gaseous mixture, the product has a composition comparable to commercial natural gas containing a series of low molecular weight hydrocarbons with methane in large proportion.

Texas lignite is a low grade coal (8000 BTU per pounds) with a high oxygen content (up to 30% of dry weight) and about a medium level of sulfur residues (nearly 2% of dry weight). Most of the carbon dioxide represents a major portion of the chemically bound oxygen in lignite which may exist as carboxylic groups. Hydrogen sulfide could be liberated from the sulfhydryl groups (thiols, sulfides, disulfides and chelated sulfur residues) and elemental sulfur (at least a small fraction) in the lignite.

The lignite-derived liquid obtained in this work is less

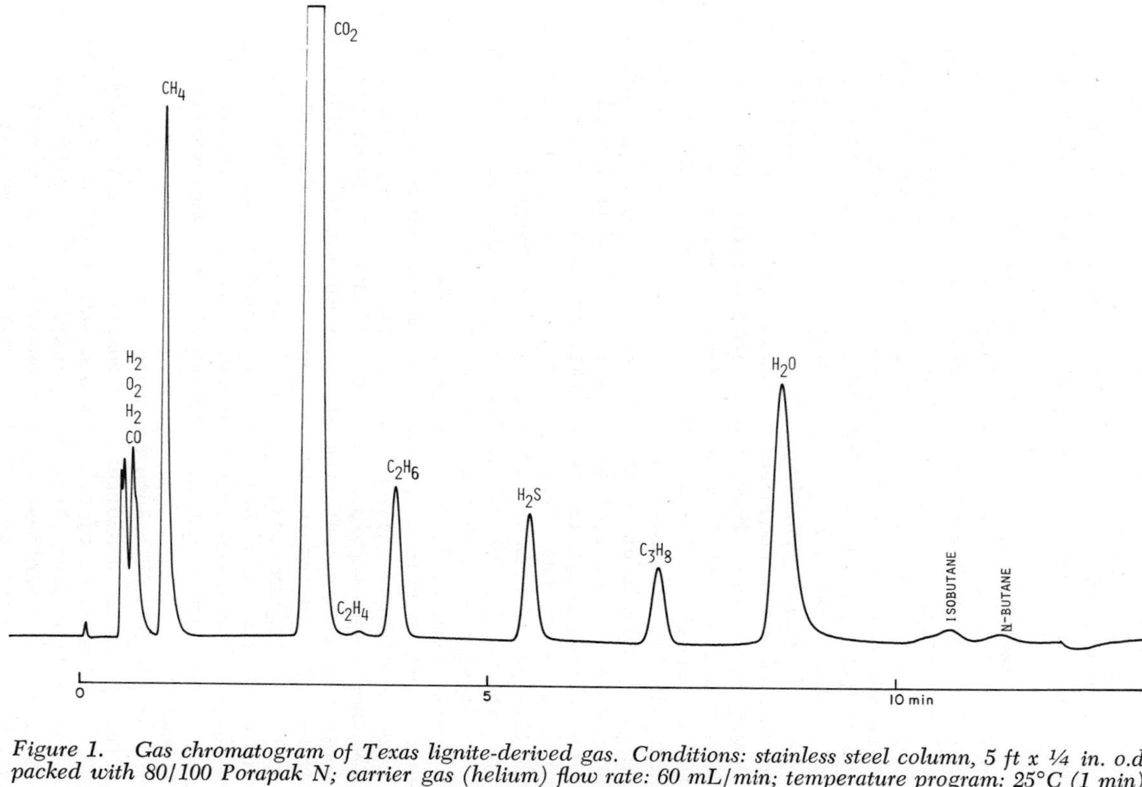

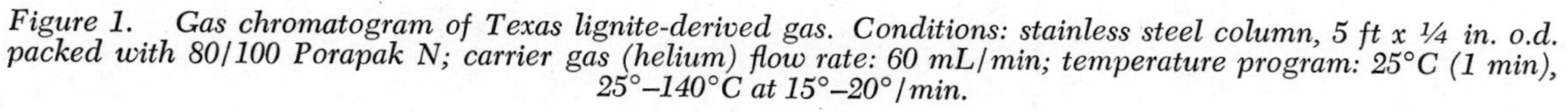

Figure 1. Gas chromatogram of Texas lignite-derived gas. Conditions: stainless steel column, 5 ft x ¼ in. o.d. packed with 80/100 Porapak N; carrier gas (helium) flow rate: 60 mL/min; temperature program: 25°C (1 min), 25°–140°C at 15°–20°/min.

complex than the bituminous coal-derived liquid. The lignite-derived liquid was divided into low boiling liquid and high boiling liquid in order to use two Durapak columns which have an upper temperature limit of about 150°C for the separation of aromatic and aliphatic compounds. As a matter of fact both the low and high boiling point liquids could be separated on any of the five columns used for high boiling liquid but the Durapak columns gave a much better resolution for the low boiling point liquid.

The low-boiling liquid is a clear colorless liquid which turns dark and cloudy on exposure to air at room temperature for a few hours. Figure 2 shows the total ion monitor chromatogram of the liquid using a 30 ft. 1/8" column packed with 3% OV 101 on Chromosorb W-HP. Table I summarizes the identification of major components and gives an overview of the general nature of the most common chemical species present in the low-boiling liquid. Aliphatic hydrocarbons, alkylated aromatics, furans and small amounts of thiophenes constitute the bulk of the sample. The mass spectra of these components clearly indicate the substance type, however, in cases where two or more hydrogen atoms have been substituted by alkyl groups, a large number of different patterns is possible. The mass spectra of some of these isomers are quite similar and so the identification has been done by using known standards or using individual boiling point range. All the possible isomers of some alkylated species are identified.

High boiling-point liquid was cleaned using a fractional sublimator prior to gas chromatographic analysis. The residue from sublimation was about 20 to 40% of the charge to the sublimator. The sample was sublimated to limit the boiling-point range of the sample so that the column temperature could be set for an upper limit of 280°C. Column bleeding was the major problem in GC-MS studies. Figure 3 shows the chromatogram of a sublimated sample and Table II list all the components identified. The same sample was separated on a Dexsil 300 GC column (Figure 4). Comparing the chromatogram of the same sample on five different column helps to resolve some components which may not separate on a particular column under identical conditions. For the GC-MS analysis a 30 ft. x 1/8" stainless steel column packed with 3% OV 101 on 80/100 mesh Chromosorb W-HP gave a better analysis than a 30M glass capillary column coated with OV 101 under similar conditions.

The Dexsil 300 GC column separated components into sharper symmetrical peaks in a shorter time compared to other columns. SP 2250 is the slowest of all. When a sample containing naphthalene and tetrahydronaphthalene is used on five different columns, SP 2250 gave the best separation while SP 2100 gave no separation. The efficiency of separation in decreasing order can be listed as follows: SP 2250, Dexsil 410 GC, Dexsil 400 GC, Dexsil 300 GC and SP 2100. The bulk of the high-boiling point liquid sample consists of saturated hydrocarbons mostly *n*-alkanes

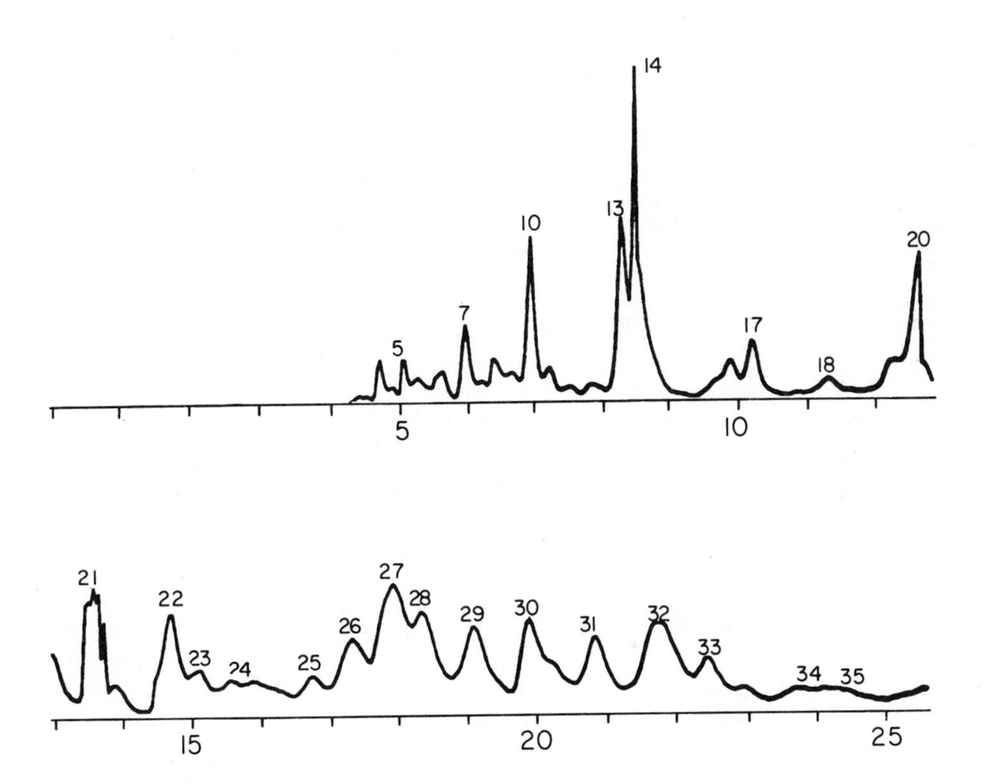

Figure 2. Total ion current monitor chromatogram of low-boiling liquid. Conditions: stainless steel column, 30 ft x 1/8 in. o.d. packed with 3% OV 101 on 80/100 Chromosorb W-HP; carries gas (helium) flow rate: 20 mL/min; temperature program: 75°–200°C at 2°/min. For identification of peaks see Table I.

Table I. Identification of Major Components in the Low Boiling Liquids

Peak No.	Compound
1	Acetone
2	1,1-Dimethylcyclopropane
3	C_6H_{12}
4	Ethylmethylketone
5	Hexane
6	Methylcyclopentane
7	1-Methylcyclopentene
8	Pentane-2-one
9	Heptene
10	2,4-Dimethylpentadiene
11	Dimethylcyclopentene
12	Heptene
13	C_7H_{12}
14	Toluene
15	2-Isopropylfuran
16	C_8H_{12}
17	Isopropylfuran
18	1-Ethylcyclohexene
19	Ethylbenzene
20	Xylenes
21	2,3-Dimethylthiophene
22	C_9H_{20}
23	2-Methyl 5-propylfuran
24	Cumene
25	$C_{10}H_{22}$
26	*t*-Butylcyclohexanone
27	*p*-Ethyltoluene + trimethylthiophene
28	Trimethylthiophene
29	Trimethylthiophene (iso.)
30	C_3-Alkylbenzene
31	*n*-decane
32	Tetramethylbenzene
33	Cyclopropylbenzene
34	Tetramethylthiophene
35	C_4-Alkylthiophene
36	$C_{11}H_{24}$

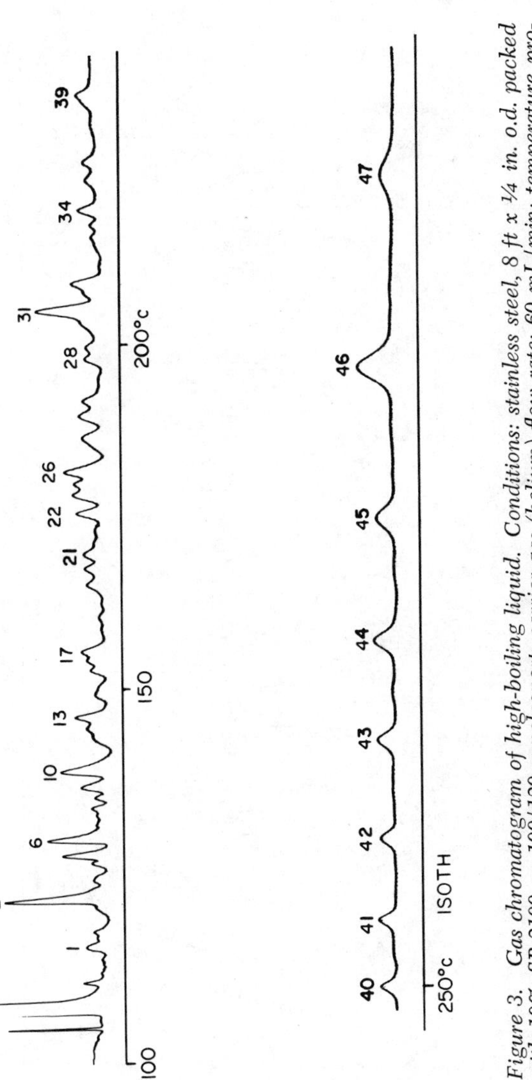

Figure 3. Gas chromatogram of high-boiling liquid. Conditions: stainless steel, 8 ft x ¼ in. o.d. packed with 10% SP 2100 on 100/120 supelcoport; carrier gas (helium) flow rate: 60 mL/min; temperature program: 100°–250°C at 2°/min. For identification of peaks see Table II.

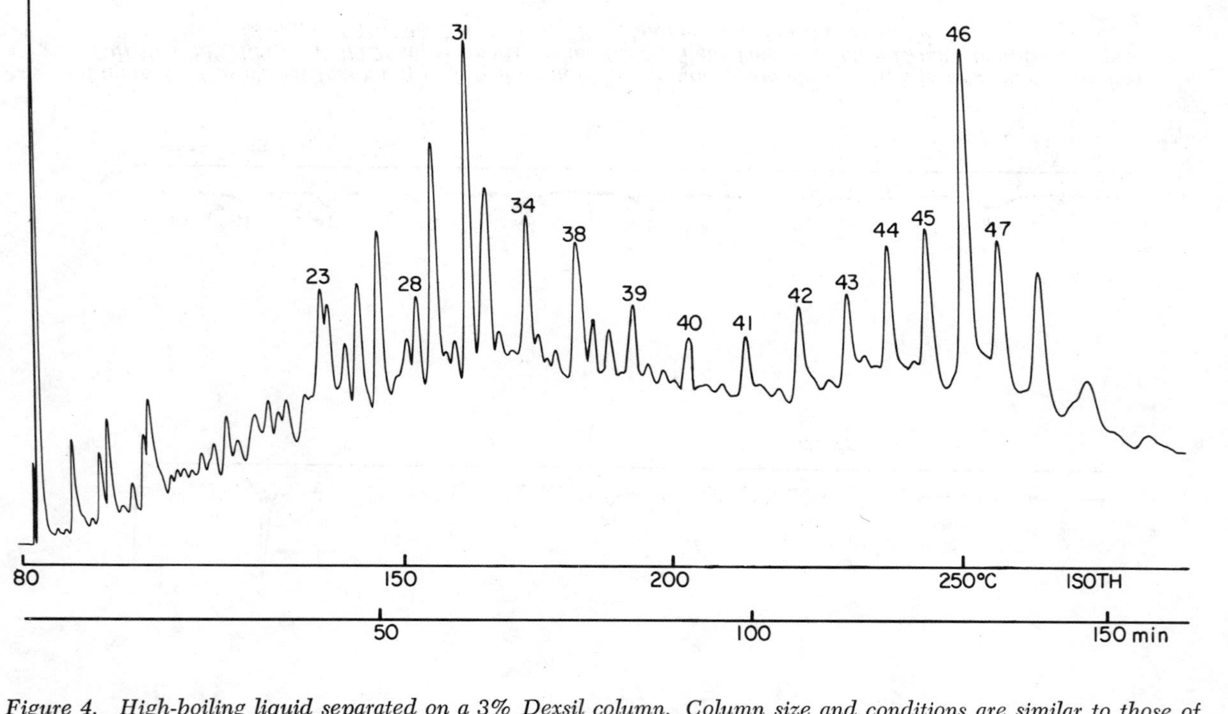

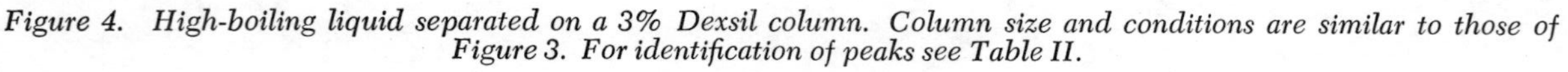

Figure 4. High-boiling liquid separated on a 3% Dexsil column. Column size and conditions are similar to those of Figure 3. For identification of peaks see Table II.

ranging from C_{10} to C_{36} distributed over the entire boiling point range. The aromatic species were predominantly alkylated phenols, benzenes, indenes hydrogenated indenes and naphthalenes. Aromatic hydrocarbons containing three or more rings were not detected in the sublimated sample. The *n*-alkanes are not distributed proportionately throughout the series, though not a single member is missing. Unusually large enrichment occures at C_{17} and C_{27}. Mass spectral data of these higher members is not good enough to distinguish between a *n*-alkane and a slightly branched alkane of a higher molecular weight. The peak assigned to *n*-C_{17} alkane may also be assigned to branched alkanes with more than 17 carbon atoms, namely pristane, a branched C_{19} alkane. Pristane is derived from the phytol residues of chlorophyll (7). The hydrogenation products of other diterpene residues in lignite may also contribute to peaks in the range of C_{17} through C_{19}. The branched C_{30} alkanes obtained by the hydrogenation of triterpene type residues may be responsible for the *n*-C_{27} alkane peak enhancement.

The high boiling liquid is composed of species with a very wide range of boiling points. Starting with phenol (181°C) at the low end and *n*-$C_{36}H_{74}$ (497°C) at the upper end. A careful examination of Table II reveals that fractional distillation or sublimation can be effectively used to separate the high boiling liquid into separate fractions enriched with phenols (180-230°C), aromatic hydrocarbons (230-300°C) and alkanes (300-500°C). Similarly the low boiling liquids can also be fractionated into enriched samples. The minor components of the high and low boiling liquids are concentrated in these fractions and can be identified by use of GC and GC-MS.

The proton nmr spectra show the distribution of chemically bound hydrogen among the aromatic rings, aliphatic chains and other carbon atoms with varying chemical shifts due to different functional groups. The spectra give only a very qualitative picture about the chemical nature of the numerous components present in the lignite derived products. An approximate estimation of the aromatic and the aliphatic moieties in the sample could be attempted with reasonable success. Figure 5 shows the proton nmr spectra of four different samples. Most of the components of the sample in Figure 5b and 5c are listed in Table I and II. Fluids derived from hydrogenated West Virginia subbituminous coal are composed of more alkylated aromatics compared to Texas lignite derived products. Hydrogenation of Texas lignite cleaves the lattice structure releasing the aromatic and aliphatic constituents while simple benzene extraction of lignite releases only a small amount of alkanes (Figure 5c and 5d). Since the comparative nature of products derived from one hydrogenation experiment to another does not change much, proton nmr can be used to see the products distribution and the extent of the reaction. Compared to GC and GC-MS, proton nmr requires a short time and the sample containing nonvolatiles can be used.

The proton nmr spectra of complex mixtures such as coal

Table II. Identification of Major Components in the High Boiling Liquids

Peak No.	Compound
1	Phenol
2	1-Ethyl-3-methylbenzene plus Decane
3	*o*-Cresol
4	*p*-Cresol
5	*n*-Undecane plus methylcresol
6	*o*-Ethylphenol
7	2,6-Dimethylphenol
8	*p*-Ethylphenol
9	*p*-Cymene
10	$C_{12}H_{26}$ plus 1,3-Dimethylindan
11	*n*-Dodecane plus 2-Methyl-6-ethylphenol
12	3-Methyl-6-ethylphenol
13	$C_{12}H_{16}$
14	3-Methyl-6-ethylphenol
15	$C_{13}H_{28}$ plus 1,6-Dimethylindan
16	1,2-Dimethylindan
17	*n*-Tridecane
18	$C_{11}H_{16}$(Methylated benzene)plus $C_{14}H_{30}$
19	*n*-Tetradecane
20	Dimethylnaphthalene
21	2,3-Dimethylnaphthalene
22	$C_{15}H_{32}$
23	*n*-Pentadecane
24	Pentamethylindan
25	C_6-Alkylindan
26	Trimethylnaphthalene (iso.)
27	$C_{16}H_{34}$ plus Trimethylnahthalene(iso.)
28	*n*-Hexadecane
29	Diethyl methylnaphthalene
30	$C_{17}H_{36}$ plus Tetramethylnaphthalene
31	*n*-Heptadecane
32	Alkylated naphthalene
33	$C_{18}H_{38}$
34	*n*-Octadecane
35	$C_{19}H_{40}$
36	$C_{19}H_{40}$
37	$C_{19}H_{40}$
38	*n*-Nonadecane
39	*n*-Eicosane
40	*n*-Heneicosane
41	*n*-Docosane
42	*n*-Tricosane
43	*n*-Tetracosane

(cont'd) Table II. Identification of Major Components in the High Boiling Liquids

Peak No.	Compound
44	*n*-Pentacosane
45	*n*-Hexacosane
46	*n*-Hepacosane
47	*n*-Octacosane

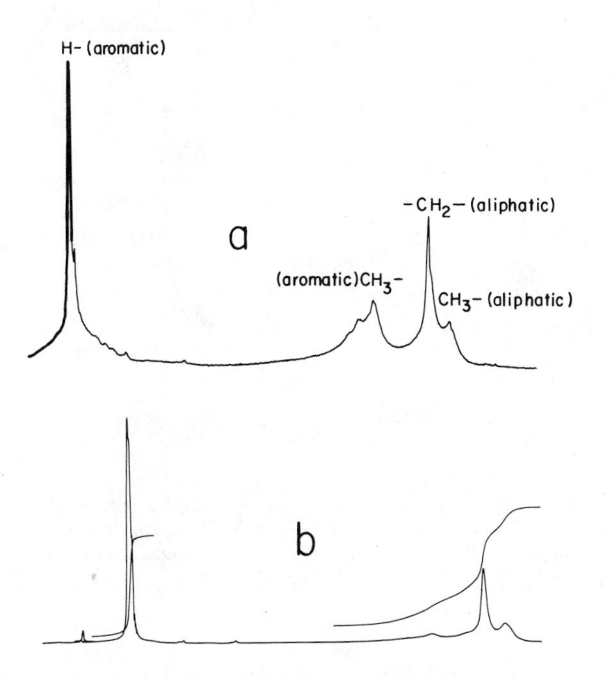

Figure 5. Proton NMR spectra of samples dissolved in carbon tetrachloride. (a) Fluids derived from hydrogenated West Virginia subbituminous coal. (b) Low boiling liquids from hydrogenated Texas lignite.

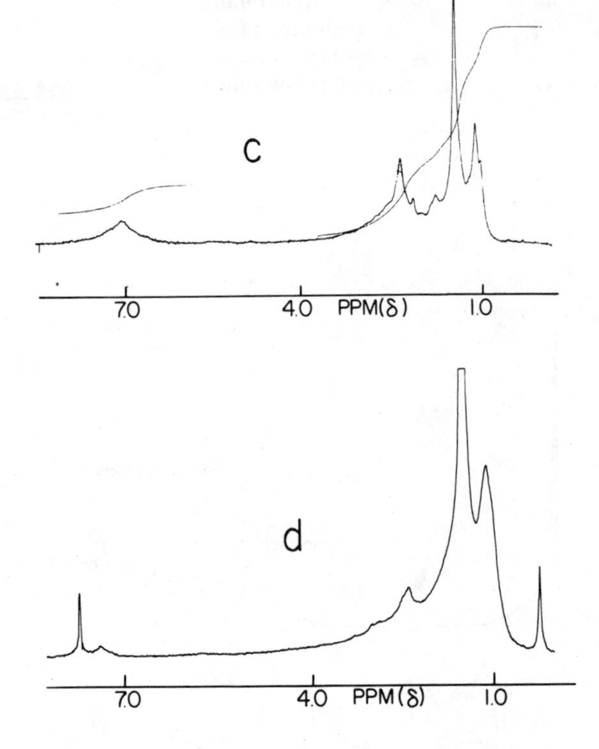

Figure 5. Proton NMR spectra of samples dissolved in carbon tetrachloride. (c) High boiling liquids from hydrogenated Texas lignite. (d) Benzene extract of Texas lignite.

derived fluids give a very broad absorption pattern due to complex proton splitting patterns and a relatively shorter range of proton chemical shifts. The increased range (compare chemical shift values in Table III and in Figure 5) of ^{13}C chemical shifts compared to proton values is mainly responsible for the unique power of ^{13}C magnetic resonance in structural identification. Since Carbon-13 accounts for about 1.1% of the natural carbon isotopes and carbon-13 (98.9) is nonmagnetic, a simplified spectral response is provided as direct coupling between adjacent carbon atoms is not observed. A high resolution ^{13}C nmr spectra devoid of complex splitting patterns can be obtained by eliminating proton splitting by means of broad band proton decoupling. Due to the high resolution power of the ^{13}C nmr spectroscopy a cumulative estimation of a sample containing many components may result in the disappearance of a large number of ^{13}C absorptions in the noise. The sample spectra will show few nmr absorption peaks representing carbon centers with nearly identical chemical shifts. Each narrow peak is formed by summing up the nmr absorptions of ^{13}C carbons from similar but different compounds. The ^{13}C nmr spectrum of a nonsublimated high-boiling liquid derived from hydrogenated lignite has about ten major absorption peaks. These peaks are listed in Table III. Three peaks are in the aromatic region (110-130 ppm) and seven in the aliphatic region (10-40 ppm). The peak at 14.19 ppm could be due to terminal methyl groups of saturated long chain hydrocarbons. The intense absorption at 29.91 ppm is due to methylene groups in the middle regions of one or more long chain saturated hydrocarbon compounds. Relatively very large area of the peak at 29.91 ppm suggests that *n*-alkanes are the bulk species in the lignite derived fluids. The ^{13}C nmr studies also show that most of the GC peaks of alkanes are due to straight chain rather than branched hydrocarbons. GC-MS studies clearly shows that saturated hydrocarbons are the major constituents of high boiling liquids derived from Texas lignite. The fragmentation pattern also shows that most of these hydrocarbons are straight chain alkanes. If we assume that straight chain alkanes with $nC<10$ is the predominant species in the high boiling liquids, the peaks at 14.19, 22.84, 32.1, 29.53 and 29.91 ppm belong respectively to α, β, γ, δ and ε carbons of aliphatic chains. Unlike proton nmr, the intensity of ^{13}C peaks varies depending upon the nature of ^{13}C centers. It is evident from a number ^{13}C nmr data (8) already reported for a number of straight chain aliphatic compounds that the intensity increases in the order α, β, δ, γ and ε carbon centers. By comparing the peak intensity at 32.1 and 29.91 the average chain length of alkanes in a nonsublimated sample obtained as number of carbons is equal to approximately 35. The GC-MS analysis is limited due to upper boiling point limitations and the loss on the column due to decomposition and irreversible adsorptions. ^{13}C nmr is a very powerfull tool in determining the structural identification of nonvolatiles as it can easily identify between condensed ring aromatics and straight chain aliphatics.

Conclusion

We have shown that GC, GC-MS, and nmr can be effectively used to characterize the products obtained by hydrogenating lignite. The liquid samples were found to consist primarily of saturated hydrocarbon mostly *n*-alkanes ranging from C_6 though C_{40+}. The aromatic species were predominantly alkylated phenols, benzenes, indenes, hydrogenated indenes and naphthalenes. Alkylated furans and a very small concentration of thiophenes were the only hetrocyclic compounds detected. ^{13}C nmr spectroscopic analysis of the samples complimented the GC-MS identification of the series of long chain *n*-alkanes. Aromatic hydrocarbons containing three or more rings were not detected.

Acknowledgements

The financial support of the Texas Engineering Experiment Station, the Texas A&M University Center for Energy and Mineral Resources, and Dow Chemical Co., Freeport, Texas is very much appreciated. The lignite was furnished by Alcoa at Rockdale, Texas and by Dow Chemical Co. The cooperation and assistance of Dr. Robert L. Spraggins of The Center for Trace Characterization, in analyzing the samples by GC-MS is appreciated.

Table III. ^{13}C NMR Data of Products from Hydrogenated Lignite

Peak No.	Chemical Shift ppm	Peak Height(rel)	Peak Area(rel)
1	14.19	7.16	22.17
2	19.77	6.62	27.30
3	22.84	11.07	38.04
4	29.53	20.94	65.52
5	29.91	122.68	603.50
6	32.10	11.77	42.25
7	37.64	5.72	25.24
8	115.54	6.22	10.52
9	128.37	8.95	57.72
10	129.57	7.67	22.30

References

1. Kaiser, W. R., "Texas Lignite: Near-Surface and Deep-Basin Resources" Report No. 79, Bureau of Economic Geology, The University of Texas at Austin, Austin, Texas (1974).
2. Anthony, R. G., Texas Engineering Experiment Bulletin No. 76-3, 4-13, (July, 1976).
3. Skidmore, Dr. R. and Konya, C. J., Preprints Div. of Fuel Chem., ACS, Dallas (1973).

4. Aldrich, R. G., Keller, Jr., D. V. and Sawyer, R. G., U.S. Patent Nos. 3815,82 (July 11, 1974), 3850,477 (November 26, 1974), and 3870,237 (March 11, 1975).
5. Shciller, J. E., Hydrocarbon Processing, p. 147 (January, 1977).
6. Bertsch, W., Anderson, E. and Holzer, G., J. Chromatography, 126 213-224 (1976).
7. Gilbert, J. M., DeAndrade Bruning, I.M.R., Nooner, D. W. and Oro, J., Chemical Geology 15, 209-215, (1975).
8. "Selected ^{13}C NMR Spectral Data," API Project 44, Thermodynamics Research Center, Texas A&M University, Volume I, (Loose-leaf Data Sheets) (1975-76).

RECEIVED February 10, 1978

20

Preparative GPC Study of Solvent-Refined Coal and Its Acid-Neutral-Base Components

D. J. WELSH, J. W. HELLGETH, T. E. GLASS, H. C. DORN, and L. T. TAYLOR

Department of Chemistry, Virginia Polytechnic Institute and State University, Blacksburg, VA 24061

One of the major obstacles encountered in the characterization of donor solvent coal liquefaction products is the separation of material into fractions which contain relatively few compounds of a specific type. This achievement appears to be a prerequisite if one is to obtain a fundamental understanding of (1) the chemical structure of coal liquids and (2) the chemical modifications taking place when coal is liquefied. Various solvent extraction procedures have been employed, and while less complicated fractions are obtainable by this method, these fractions are composed of probably more than fifty compounds. One of the more common of these procedures is the fractionation (1) of liquid product into pyridine solubles and insolubles and pentane solubles and insolubles. Equally complexed mixtures are produced by high performance liquid chromatographic (HPLC) procedures; however, more progress seems to have been made recently using this technique. Separations according to polarity or effective molecular size (i. e. gel permeation chromatography) have been the methods of choice. A combination of solvent extraction--chromatography has been applied in certain cases. Highly efficient separations which enable detailed absolute structural information (i. e. aromatic vs. heteroaromatic, phenolic vs. alcoholic, long chain aliphatic vs. short chain aliphatic, etc.) to be obtained have not come forward. Such information requirements demand both a preparative-size and very selective separation. As close to 100% of the coal liquid as possible should also be involved and near quantitative recovery of the chromatographic sample should be possible.

Analytical GPC separations have been reported on an unidentified coal extract using μ-styragel (2) and on tetrahydrofuran (THF) soluble Pittsburgh solvent refined coal (SRC) using Bio Beads SX-4 (3). The feasibility of employing affinity HPLC to characterize creosote oil, a coal derived start-up solvent, has been reported (4) on an analytical size column. A limited attempt has been made to separate acid/neutral and base components of asphaltenes from a Kentucky liquefied coal via silica

0-8412-0427-6/78/47-071-274$05.00/0

gel chromatography by elution with benzene and diethyl ether. Detailed structural information on these fractions was not reported (5). More recently (6) acid/neutral and basic components derived from asphaltenes of a Pittsburgh coal liquid have been separated via GPC on an analytical size μ-styragel column. Analysis of the separated fractions was not carried out. The separation via silica gel of the centrifuged liquid product derived from a Synthoil hydrodesulfurization coal liquefaction product has been published (7). Fractionation into four general molecular types by successively eluting with hexane, hexane/benzene and THF followed by n.m.r. analysis revealed the hydrogen aromatic/aliphatic distribution. The heavy coal product from a laboratory processing unit which utilized the $ZnCl_2$ hydrogenation of western bituminous coal has been separated (8) into saturates, monoaromatic, diaromatic and polyaromatic polar fractions utilizing gradient elution through dual-packed silica gel-alumina gel columns. Carbon-13 magnetic resonance spectra suggests a significant amount of material in the form of normal paraffinic material. Farcasiu (9) has recently reviewed the various experimental problems involved in the separation (especially SARA (10)) of whole coal liquids. Sequential elution with specific solvents on a preparative silica gel column of both benzene soluble and pyridine soluble Western Kentucky SRC has been carried out. Nine separate chemically different fractions were obtained. Sub-fractionation of each fraction was predicted to be necessary prior to any structural studies.

We have undertaken a study (11) of the preparative GPC separation of various THF soluble (>90%) solvent refined coals using Bio Beads, a cross-linked divinylbenzene polystyrene copolymer. One gram quantities of coal were separated into four fractions based on effective molecular size. Nuclear magnetic resonance (^1H, ^{13}C) spectra (12) indicated significant differences in the aromatic/aliphatic H and C ratio in going from the larger to the smaller sized material. These fractions were, nevertheless, too complex for detailed structural analysis. We, therefore, wish to report here our further efforts to achieve a better coal liquid product separation via GPC and a nuclear magnetic resonance analysis of selected fractions taken therefrom.

Results and Discussion

Two solvent refined coal (SRC) samples have been examined in this study (Wilsonville, Alabama). An eastern derived SRC, Pittsburgh #8, and a western derived SRC, Amax, were choosen. Pertinent reaction parameters in effect when these SRC samples were drawn are respectively listed for Pittsburgh and Amax: Temperature = 855°F, 853°F; Conversion = 90%, 79%; Reaction Time = 26.4; 20.5 minutes; H_2 Pressure = 1700, 2400 psi and H_2 Consumption = 2.7%, 3.3%.

To date Bio Beads (Bio Rad Laboratories, Richlands, California) has proven to be the most effective material for achieving sized separations of coal derived liquids on a preparative as well as analytical scale. Ease of column packing, high plate count, wide solvent compatibility, relative economic cost and low pressure/fast flow rate capability are just a few features which make Bio Beads especially attractive. Preparative studies on six inch columns have been completed whereby trace metal, n.m.r. and mass spectral analyses were performed on various fractions. In an effort to expand and improve our separations we have conducted studies employing 36 inch preparative size tappered columns. Specific samples for this study were tetrahydrofuran and chloroform soluble SRC and acid-base-neutral fractions of Pittsburgh and Amax SRC. A comparison of Figures 1 and 2 reveal the enhanced separation obtainable in going from a 6 to a 36 inch column using the same coal and packing materials. While greater resolution between large and small size molecules has been obtained, the resolution within each group has not been significantly improved. The net effect seems to have been a "spreading-out" of the material eluted from approximately 70 ml on the 6 inch column to 180 ml on the 36 inch column. This is supported by the fact that the weight profile reveals significant material eluting from the column at all volumes. Proton FT n.m.r. spectra of various 1 ml fractions throughout the elution range suggested at best minor improvements toward simplifying the fraction mixtures when compared with the six inch column.

However, an important characteristic of these fractions is the progression from high aliphatic hydrogen to aromatic hydrogen content as the average molecular weight of the fractions decreases as monitored by ^{1}H FT n.m.r. If the proposed progression to higher aromatic content with the lower molecular weight fractions is confirmed via ^{13}C FT n.m.r., this would appear to be consistent with the progression to higher aromatic carbon content during solvent refining as reported by the Mobil group (13) utilizing solid-state ^{13}C n.m.r. techniques. This would also be consistent with the relatively high aliphatic carbon content recently reported (14, 15) for many raw coals utilizing solid-state ^{13}C n.m.r. Obviously, another interpretation of our ^{1}H FT n.m.r. results is simply progression from large polycondensed aromatic ring systems, (i. e., low (H/C) aromatic ratios) to smaller aromatic compounds with decreasing molecular size of the fractions.

GPC separation via THF elution coupled with weight profile analysis on Amax $CHCl_3$ soluble SRC was next attempted. Since $CHCl_3$ dissolves only 80% of SRC compared to 93% for THF, a simpler starting coal sample was envisioned. In addition it was believed that the higher molecular weight-less soluble coal liquid would be excluded from the chromatography sample. Figure 3 shows that this assumption has been partially verified. The

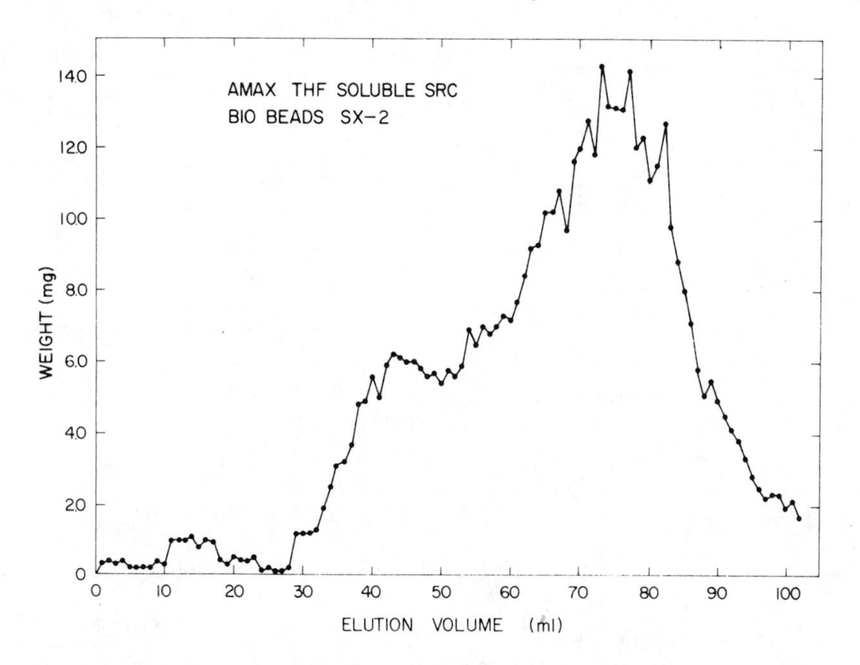

Figure 1. Chromatographic fraction weight profile for Amax THF soluble SRC employing Bio Beads SX-2 (200–400 mesh). Six-inch glass column (see Ref. 11 for dimensions; THF flow rate = 2.8 mL/min; pressure = 30 psi; refractive index change detection, sample = 0.44 gms dissolved in 15 mL of THF.

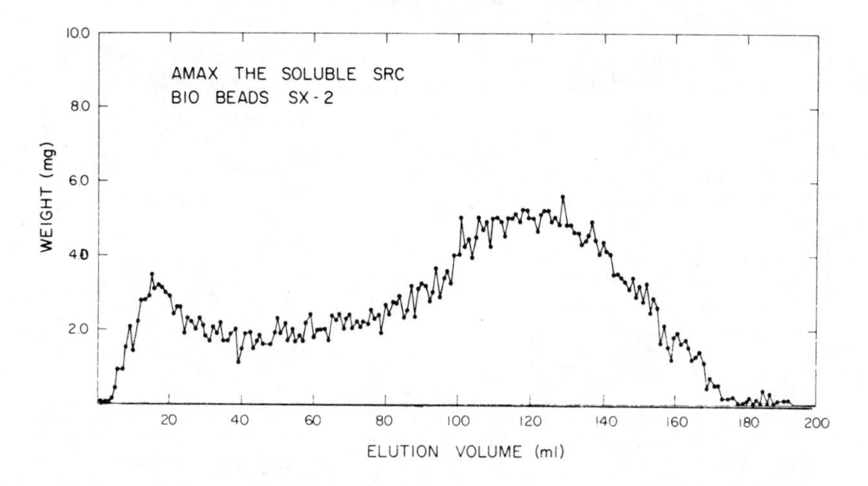

Figure 2. Chromatographic fraction weight profile for Amax THF soluble SRC employing Bio Beads SX-2 (200–400 mesh). Glass column (36 in. x 1¼ in.); THF flow rate = 1.6 mL/min; pressure = 40 psi; refractive index change detection, sample 0.5 gms dissolved in 5 mL of THF.

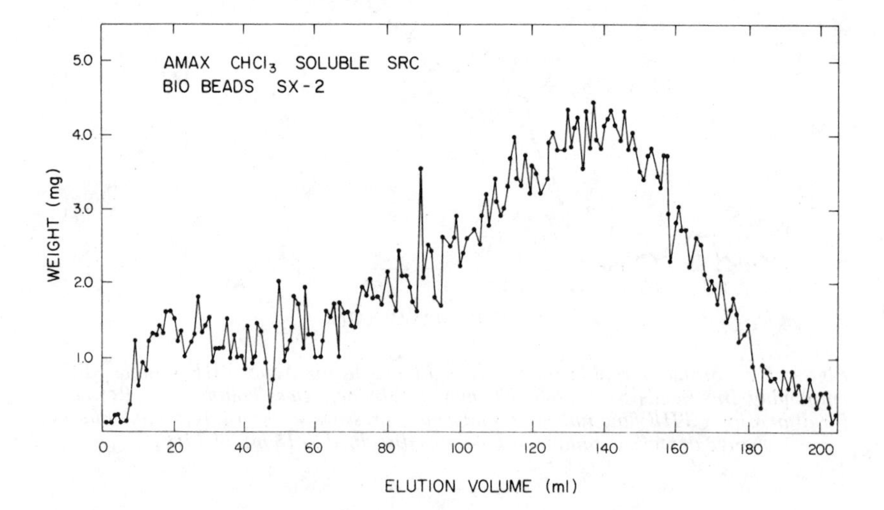

Figure 3. Chromatographic fraction weight profile for Amax $CHCl_3$ soluble SRC employing Bio Beads SX-2. See Figure 2 for experimental parameters.

fraction of large size molecules has decreased relative to the smaller molecules; however, the range of "sized" molecules coming from the column for THF and $CHCl_3$ soluble Amax appears to not have changed based on a comparison of sample eluted volumes. In other words $CHCl_3$ extraction does not totally exclude only large size molecules. Again ^1H FT n.m.r. analysis of 1.0 ml fractions indicated very complexed fractions and an increase in aromatic content in going to the smaller molecular-sized materials. While a more selective solvent extraction may result in a better overall separation, one naturally begins to examine a smaller and smaller percentage of the whole coal liquid which is not in keeping with our original premise.

An acid-neutral-base separation of the Pittsburgh and Amax THF soluble SRC has been developed in hopes of ultimately achieving the optimum chromatographic separation. The method is in principle very much like Sternberg et al. (16); however, it differs considerably in practice. For example Sternberg's procedure involved only asphaltene toluene soluble coal liquids whereas our procedure deals with approximately 95% of the original coal liquid dissolved in THF. The initial step involves precipitation of the basic materials from THF with HCl(g) at atmospheric pressure. In order to avoid "salting-out" a minimum amount of neutral materials and to be as consistent as possible from one run to the next, 0.2 liter of HCl(g) per gram of dissolved SRC was reacted. After several hours the insoluble basic hydrochloride salts were filtered from the soluble acids and neutrals. Schemes I and II reveal the methods whereby HCl-free acid-neutral-base materials were obtained. The basic precipitate (Scheme I) was suspended in basic (pH > 13) salt water (3 NaOH: 10 NaCl: 50 H_2O) and continuously extracted with THF to yield a black THF layer, a colorless H_2O layer and a dark-colored precipitate. A portion of this precipitate could be dissolved by re-suspension in H_2O (pH = 6) followed by THF extraction. This behavior is suggestive of amphoteric materials. The original hot THF layer was filtered then evaporated to dryness. The dry precipitate was stirred with warm water for 12 hours, in order to remove extraneous chloride ion, followed by filtration of the precipitate. This precipitate was washed with water until a negative test for chloride ion was obtained. The SRC bases were then re-constituted in THF, filtered, solvent removed and dried in vacuo at 75°C.

Scheme II outlines the isolation of SRC neutrals and acids. The THF filtrate was initially extracted with basic (pH > 13) salt water to yield a black THF layer, a black H_2O layer and insoluble material which was removed by filtration. The THF layer which contained SRC neutrals was evaporated to dryness, washed to remove residual chloride ion and reconstituted in the same manner as the SRC bases. The insoluble material and H_2O layer were combined, acidified and extracted with THF. The black THF layer which resulted was evaporated, washed and reconstituted

Scheme I

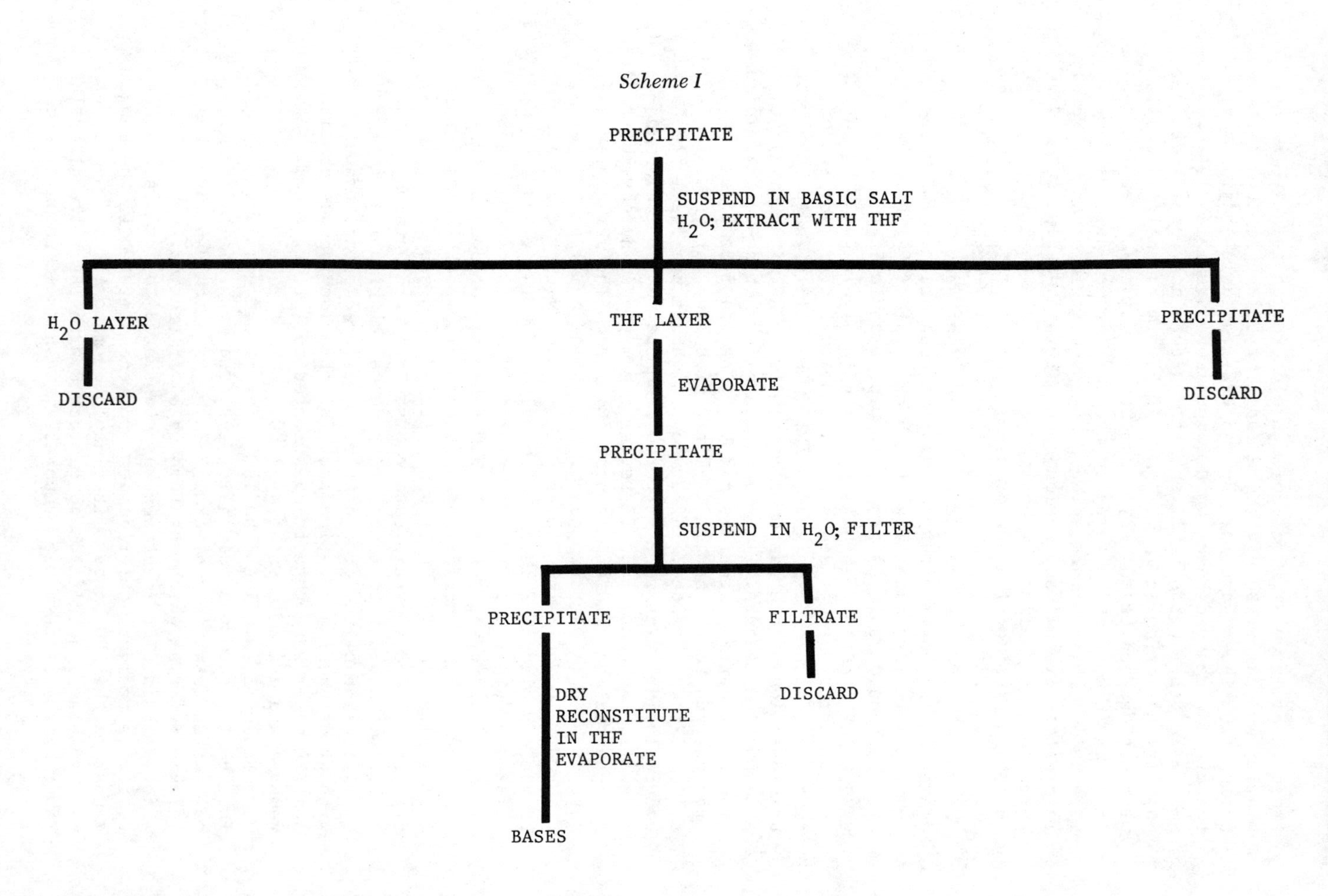
PRECIPITATE
SUSPEND IN BASIC SALT
H_2O; EXTRACT WITH THF
H_2O LAYER
DISCARD
THF LAYER
EVAPORATE
PRECIPITATE
SUSPEND IN H_2O; FILTER
PRECIPITATE
DRY
RECONSTITUTE
IN THF
EVAPORATE
BASES
FILTRATE
DISCARD
PRECIPITATE
DISCARD

Scheme II

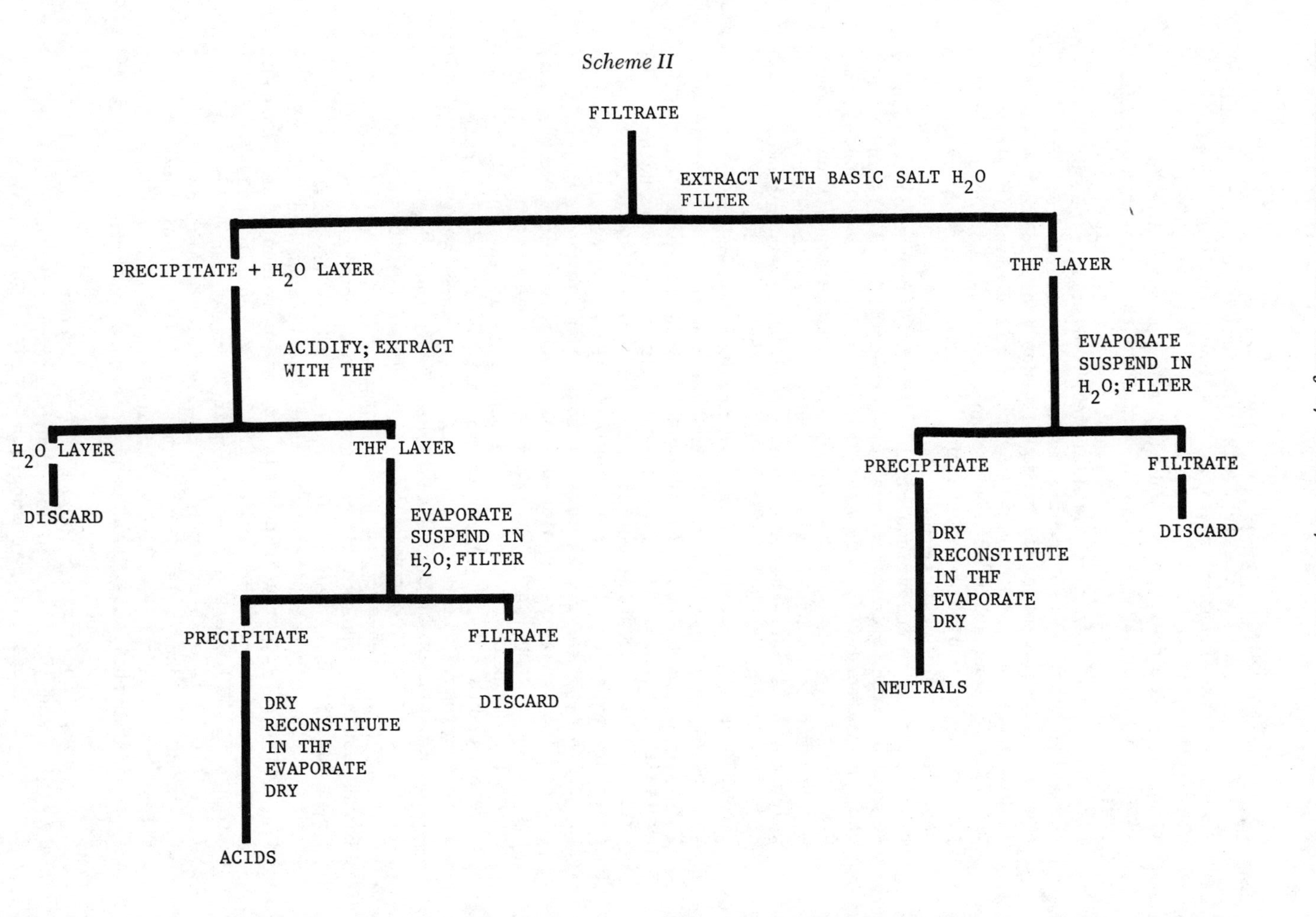
FILTRATE
EXTRACT WITH BASIC SALT H_2O
FILTER
PRECIPITATE + H_2O LAYER
THF LAYER
ACIDIFY; EXTRACT
WITH THF
EVAPORATE
SUSPEND IN
H_2O; FILTER
H_2O LAYER
THF LAYER
PRECIPITATE
FILTRATE
DISCARD
DRY
RECONSTITUTE
IN THF
EVAPORATE
DRY
DISCARD
EVAPORATE
SUSPEND IN
H_2O; FILTER
NEUTRALS
PRECIPITATE
FILTRATE
DRY
RECONSTITUTE
IN THF
EVAPORATE
DRY
DISCARD
ACIDS

as described above to yield SRC acids.

Table I lists the weight distributions, and ultimate analyses for some of these fractions. Chlorine analysis suggest very little, if any, incorporation of chloride into each fraction in spite of exposure to HCl(g) and aqueous chloride solutions. Chlorine analysis on SRC prior to the separation scheme revealed the presence of comparable amounts of chlorine. Both Pittsburgh and Amax give similiar weight distributions of acids-neutrals-bases. The percent acid/neutral compares very favorably with results obtained on asphaltenes isolated from various liquid products taken from the Synthoil pilot plant (6, 16). Our further separation into acids and neutrals indicates a preponderence of neutrals in both the Pittsburgh and Amax cases. Asphaltene acids/neutrals were not subjected to a finer separation. A cursory comparison of the weight distribution for bases suggests a significant discrepancy between SRC (~27%), and Synthoil asphaltene (~55%) bases. This observation may be explained when one notes that all HCl precipitate in the asphaltene case was assumed to be bases; whereas, only THF soluble material which could be extracted from basic salt water was presumed to be bases in our procedure. Certainly if one adds to the bases the THF insoluble residue the percentages are comparable. At this point we are unable to say whether the residue is additional bases or amphoterics or whether we have "salted-out" neutrals or mineral matter. Several points should be noted which support our acid-neutral-base designation. Acids and bases are considerably less soluble in THF and $CHCl_3$ than neutrals. Bases have 1 - 2% more nitrogen than acids or neutrals while acids probably contain more oxygen as reflected by the much lower percent carbon relative to the bases and neutrals. Molecular weights via vapor phase osmometry in THF and $CHCl_3$ have been obtained on selected components at three different concentrations. The molecular weights are surprisingly similar for Amax neutrals and bases in light of the fact that acid/neutrals were approximately 200 units higher than bases for Kentucky asphaltenes (16). These same workers recently (6) found just the opposite situation with acid/neutrals versus bases for Pittsburgh asphaltenes. In our case the higher molecular weight may constitute the insoluble residue and, therefore, not appear in either acid/neutral or base component. It is interesting to note, where solubility allows, that molecular weights in $CHCl_3$ were greater than in THF. As can be seen in Table I this is also observed for $CHCl_3$ soluble Amax and Pittsburgh SRC. Higher molecular weights in a solvent of lower donor strength suggests that these coal liquids, as some other recently studied asphaltanes (17), many associate via some intermolecular mechanism.

Neutral and basic components were further separated via GPC (Bio Beads SX-2) on the 36 inch column mentioned above. Figure 4 compares the chromatograms of THF soluble Amax SRC,

TABLE I

	ELEMENTAL ANALYSIS (%)				MOLECULAR WEIGHT	WEIGHT DISTRIBUTION (%)
	C	H	N	Cl		
Amax - Acids	78.86	7.70	0.52	1.27		
	78.60	7.63	0.48	1.31		5[c]
Amax - Neutrals	86.57	6.57	1.17	0.81	459[a]	
	86.73	6.44	1.22	0.65	(582)[b]	49
Amax - Bases	86.19	5.47	2.72		499[a]	
	85.51	5.58	2.67			16
Pittsburgh - Acids	76.53	5.53	0.87	0.75		
	76.57	5.35	0.86	0.88		2[d]
Pittsburgh - Neutrals	87.06	6.22	1.36	0.72	542[a]	
	36.51	6.06	1.23	0.65	(612)[b]	46
Pittsburgh - Bases	84.61	5.34	2.57	0.52		
	85.19	5.54	2.77	0.52		27
Amax - $CHCl_3$ Soluble	87.78	5.99	1.58	0.71	448[a]	
	88.14	6.02	1.60	0.41	(531)[b]	
Pittsburgh - $CHCl_3$ Soluble	88.44	5.69	1.68	0.42	450[a]	
	88.97	5.92	1.71	0.43	(524)[b]	

[a] Measured in tetrahydrofuran
[b] Measured in chloroform
[c] Initial weight of THF soluble SRC taken for separation = 10.02 gms
[d] Initial weight of THF soluble SRC taken for separation = 9.05 gms

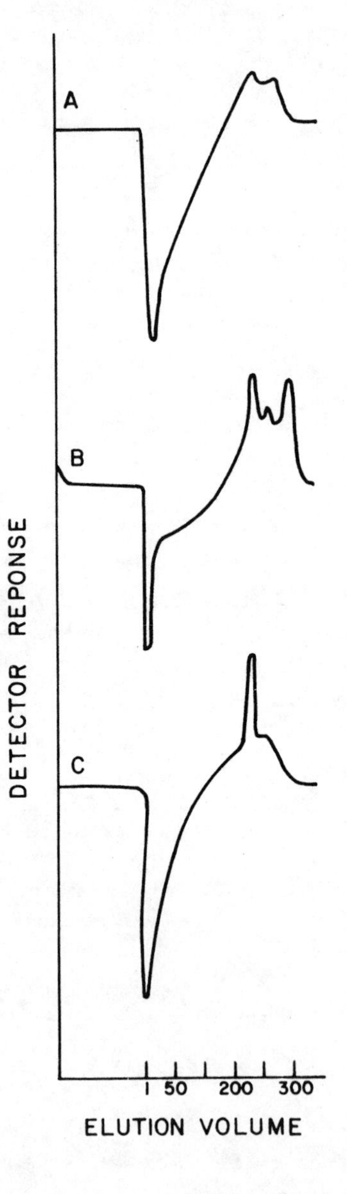

Figure 4. Chromatograms of (A) Amax SRC, (B) Amax neutrals and (C) Amax bases. See Figure 2 for experimental parameters.

Amax neutrals SRC and Amax bases SRC (refractive index change versus THF elution volume). Somewhat better chromatographic resolution is obtained on the neutrals and bases. This is especially apparent with the smaller "sized" molecules. Elution volumes are considerably larger for neutrals and bases than for THF and $CHCl_3$ soluble Amax SRC. This finding may be due to the fact that more intermolecular association is occurring in the latter products than in the neutrals or bases product. If this were the case one might conclude that separation of neutrals, acids and bases precludes an association possibility. Alternatively changes in column packing characteristics or coal liquid solution behavior between components may account for this. The similar number average molecular weights for each sample indicates that the latter may be the case.

While chromatograms may indicate the resolution efficiency in a separation they do not indicate the fractional abundance of each "sized" molecule when trying to separate what maybe very different type molecules. Figures 5 and 6 illustrate the weight profiles accompanying the neutrals and bases chromatograms. The amount of initially eluted material (e. g. larger size) appears to be about the same relative to the smaller sized material for neutrals and bases. In other words neutrals and bases consist of both large and small "sized" components. The enhance resolution suggested for the smaller neutral molecules from their chromatogram is conclusively supported by the weight profile data in the same region. Again, it should be emphasized that just because neutrals and bases in solution have similar number average molecular weights, chromatograms and chromatographic weight profiles, this does not imply that neutrals and bases are of approximately the same size. No doubt the insoluble residue referred to previously contains some THF insoluble larger "sized" molecules which maybe acids or bases.

Selected 1.0 ml fractions from the neutrals and bases separation were analyzed via ^1H FT n.m.r. analysis. Figures 7 - 11 illustrate some representative spectra. The solvent employed in obtaining the ^1H n.m.r. spectra was deutero-chloroform ($CDCl_3$) with the residual $CHCl_3$ peak apparent in these spectra at ~7.3 ppm. Figures 7 - 9 for the THF soluble Amax 1 ml sample cuts from the 3 foot column indicate the low aromatic to total hydrogen ratios (~0.1) for the higher molecular weight fractions (Figure 7) which increases to the nearly equal ratios (~0.5) found in the low molecular weight fractions (Figure 9). This general trend was also observed for larger cuts in a related study (18).

Cursory examination of these spectra suggest that the 1 ml fractions still represent a complex mixture(s) of compounds. However, a prominent feature apparent in most of these fractions is the broad, but nearly separate pattern between 3.5 to 4.5 ppm. A more informative comparison of this spectral region is evident for the 1 ml fractions obtained at approximately the same

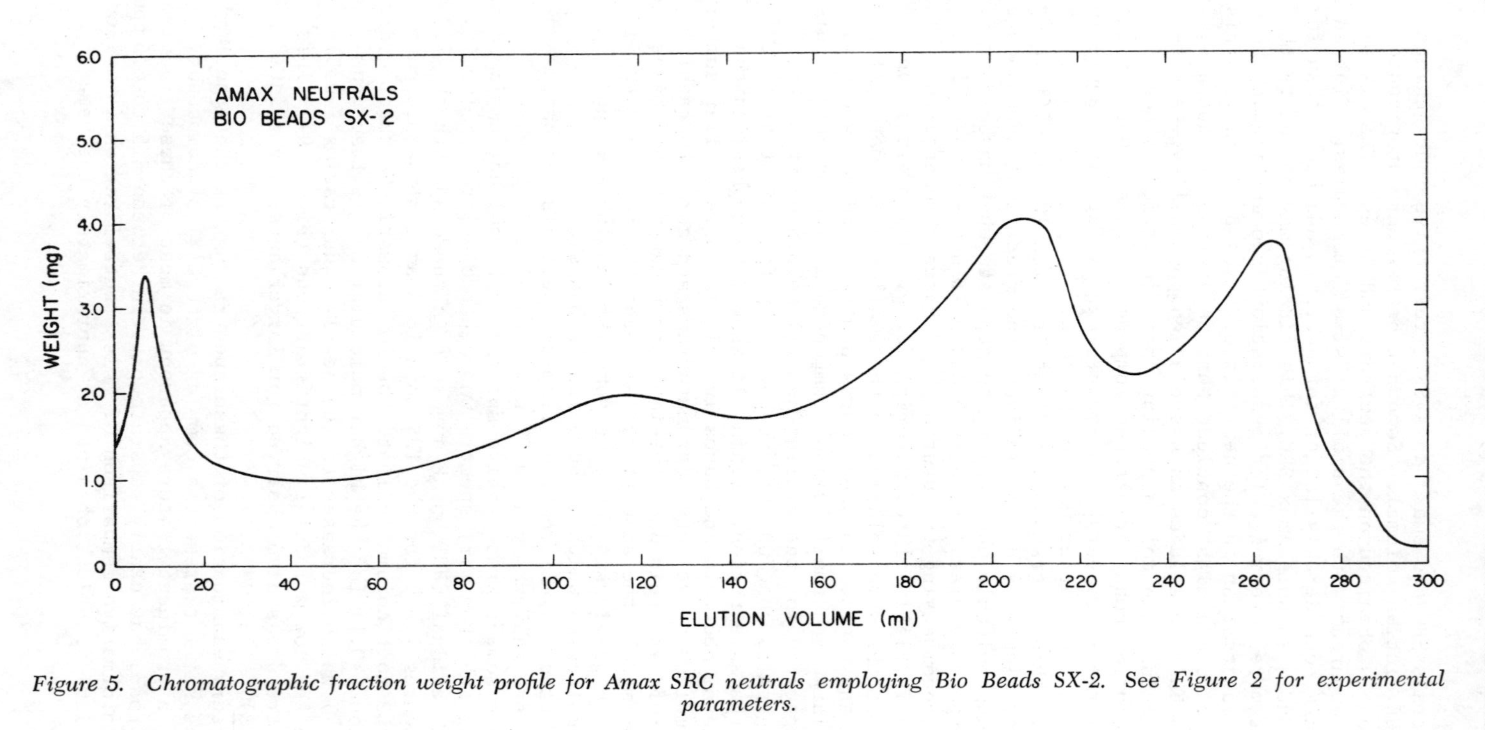

Figure 5. Chromatographic fraction weight profile for Amax SRC neutrals employing Bio Beads SX-2. See Figure 2 for experimental parameters.

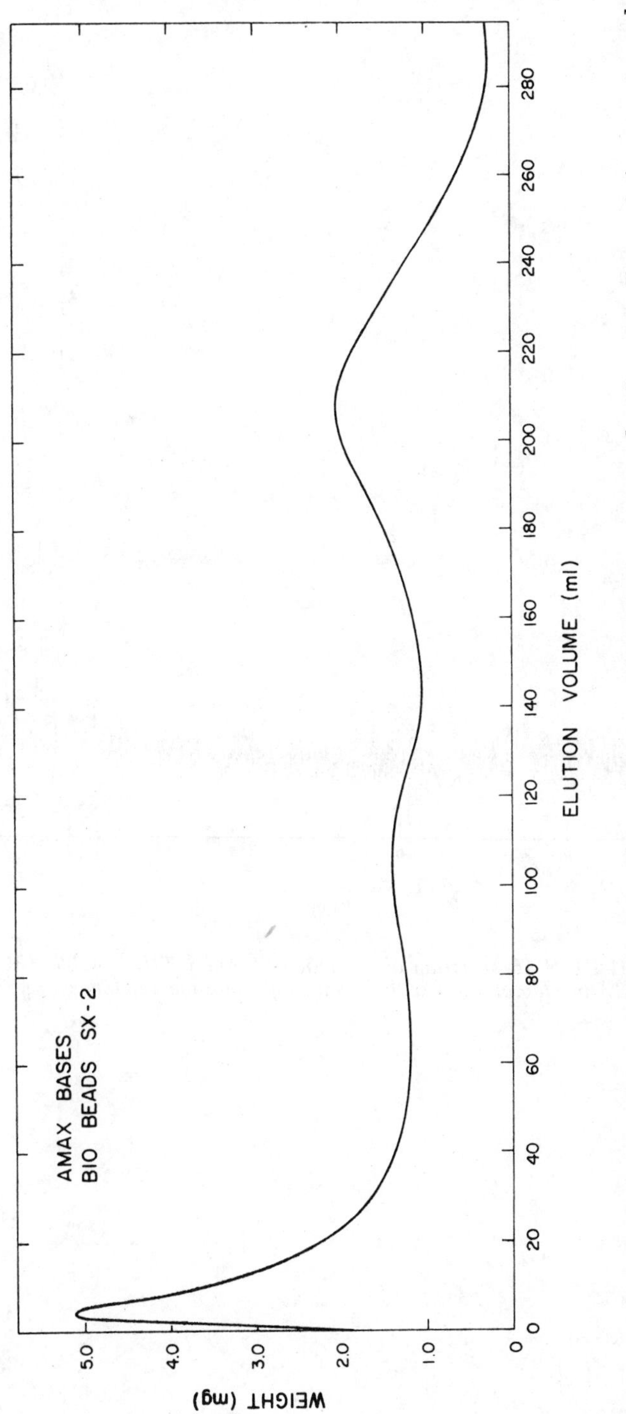

Figure 6. Chromatographic fraction weight profile for Amax SRC bases employing Bio Beads SX-2. See Figure 2 for experimental parameters.

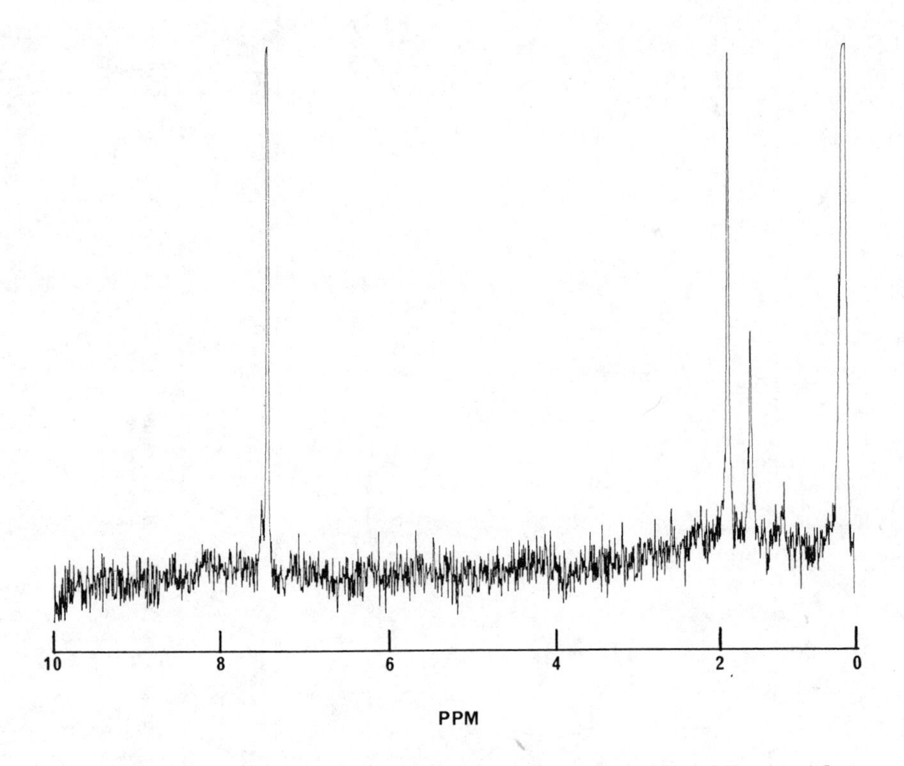

Figure 7. ^1H *FT NMR spectrum of a fraction of Amax THF soluble SRC (elution vol = 30 mL) in chloroform-*d *with hexamethyldisiloxane reference.* See *Figure 2.*

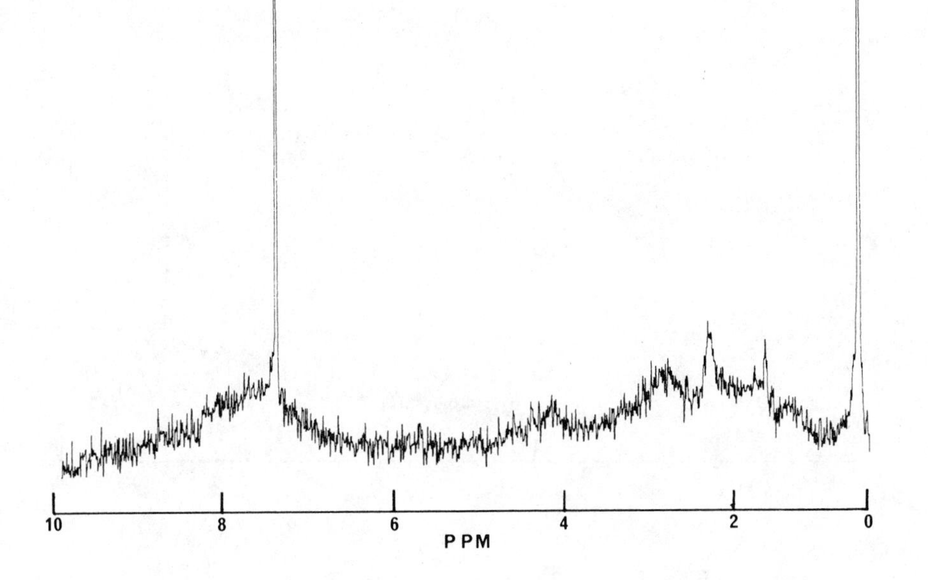

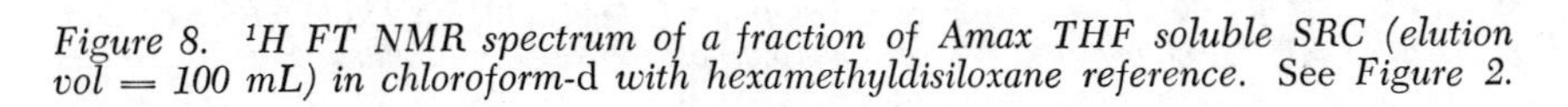

*Figure 8. ^{1}H FT NMR spectrum of a fraction of Amax THF soluble SRC (elution vol = 100 mL) in chloroform-*d *with hexamethyldisiloxane reference.* See *Figure 2.*

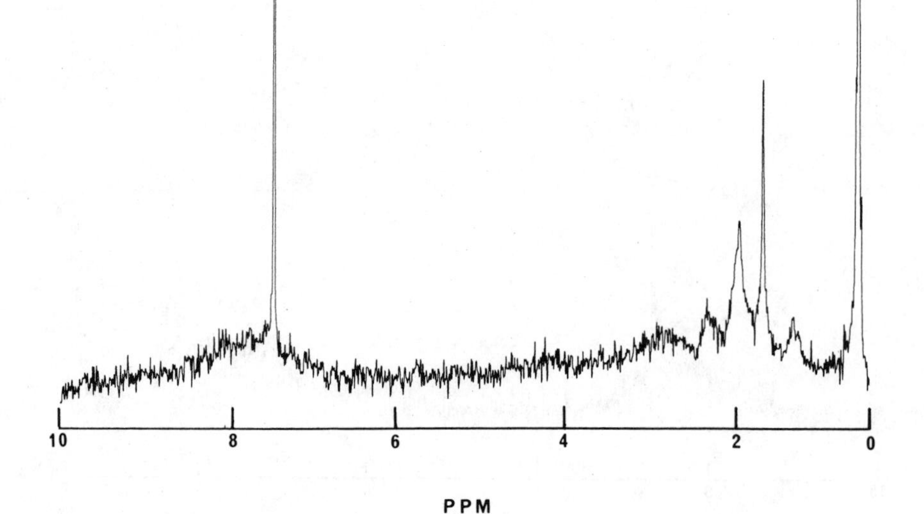

*Figure 9. ^1H FT NMR spectrum of a fraction of Amax THF soluble SRC (elution vol = 130 mL) in chloroform-*d *with hexamethyldisiloxane reference.* See Figure 2.

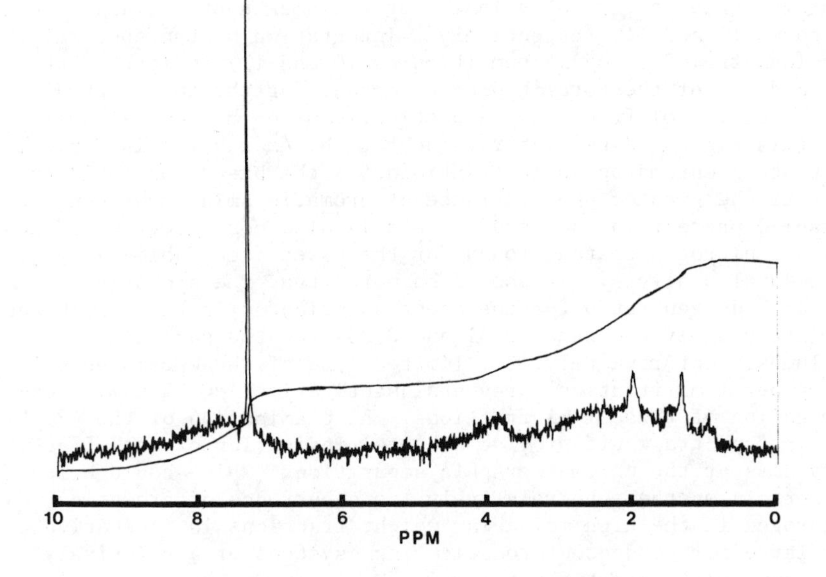

Figure 10. ^{1}H *FT NMR spectrum of a fraction of Amax neutrals (elution vol = 175 mL) in chloroform*-d. See *Figure 5.*

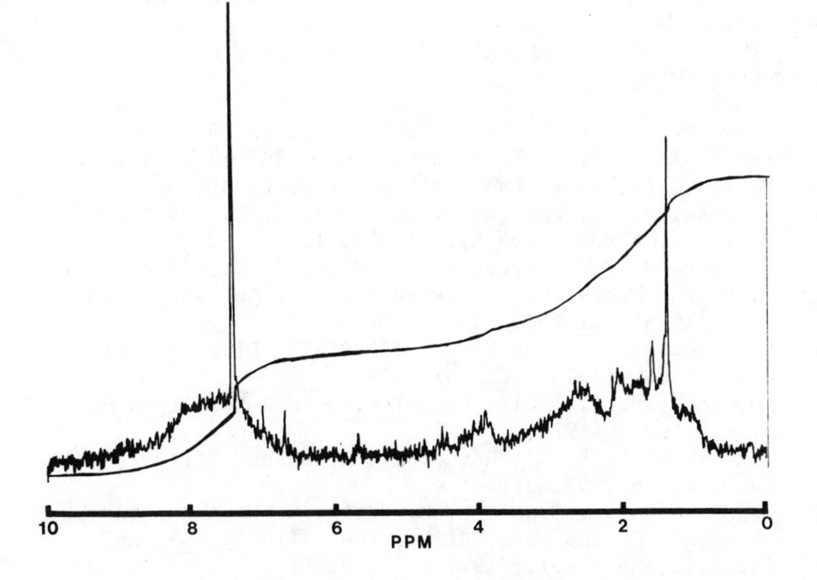

Figure 11. ^{1}H *FT NMR spectrum of a fraction of Amax bases (elution vol = 175 mL) in chloroform*-d. See *Figure 6.*

chromatographic elution volumes for the Amax neutrals and bases, Figures 10 and 11, respectively. Integration of the spectral regions from 3.5 to 4.5 ppm (Figures 10 and 11) indicate that 7% and 14% of the total hydrogen content for the Amax neutrals and bases, 1 ml fractions respectively, have ^{1}H chemical shifts in this region. The higher value for the Amax bases is typical for other chromatographic fractions for the bases and could indicate the greater preponderance of aromatic amine hydrogens assumed present in the bases. This is also consistent with the higher nitrogen content found for the bases (see Table I) by elemental analysis. It should be noted that the aromatic to total hydrogen ratio for the spectra indicated in Figures 10 and 11 are roughly the same (0.31 and 0.33) for the same elution volumes. Unfortunately, the limited size of these samples (~1 - 3 mg per 1 ml fraction) prevents quantitative ^{13}C FT n.m.r. examination of these 1 ml fractions. An examination of the ^{13}C FT n.m.r. spectra would provide a better indication of the effectiveness of the chromatographic separation. This should help ascertain whether the relatively low occurrence of aromatic hydrogen in the high molecular weight fractions is indicative of large polycondensed aromatic ring systems or a relatively low number of aromatic rings.

Acknowledgment

The financial assistance provided by Department of Energy Grant EF-77-G-01-2696 and the Commonwealth of Virginia is greatly appreciated.

Literature Cited

1. Schweighardt, F. K., Retcofsky, H. L. and Raymond, R., PREPRINTS, Div. of Fuel Chem., ACS, (1976), 21 (7), 27.
2. Dark, W. A., Amer. Lab., (1975), August, 50.
3. Coleman, W. M., Wooton, D. L., Dorn, H. C. and Taylor, L. T., J. Chromatography, (1976), 123, 419.
4. Prather, J. W., Tarrer, A. R., Guin, J. A., Johnson, D. R. and Neely, W. C., PREPRINTS, Div. of Fuel Chem., ACS, (1976), 21 (5), 144.
5. Schwager, I. and Yen T. F., PREPRINTS, Div. of Fuel Chem., ACS, (1976), 21 (5), 199.
6. Bockrath, B. C., Delle Donne, C. C. and Schweighardt, F. K., Fuel, (1978), 57, 4.
7. Schweighardt, F. K., Retcofsky, H. L. and Friedel, R. A., Fuel, (1976), 55, 313.
8. Pugmire, R. J., Grant, D. M., Zilm, K. W. Anderson, L. L., Oblad, A. G. and Wood, R. E., Fuel, (1977), 56, 295.
9. Farcasiu, M., Fuel, (1977), 56, 9.
10. Jewell, D. M., Albaugh, E. W., Davis, B. E. and Ruberto, R. G., Ind. Eng. Chem., Fundam., (1974), 13, 278.

11. Coleman, W. M., Wooton, D. L., Dorn, H. C. and Taylor L. T., Anal. Chem., (1977), 49, 533.
12. Wooton, D. L., Coleman, W. M., Taylor, L. T. and Dorn, H. C., Fuel, (1978), 57, 17.
13. Whitehurst, D. D. and Mitchell, T. O., PREPRINTS, Div. of Fuel Chem., ACS, (1976), 21 (5), 127.
14. Gerstein, B. C., Private Communication.
15. Vanderhart, D. L. and Retcofsky, H. L., Fuel, (1976), 55, 202.
16. Sternberg, H. W., Raymond, R. and Schweighardt, F. K., Science, (1975), 188, 49.
17. Schwager, I., Lee, W. C. and Yen, T. F., Anal. Chem., (1977), 49, 2365.
18 Wooton, D. L., Coleman, W. M., Glass, T. E., Dorn, H. C. and Taylor, L. T., "A ^{1}H and ^{13}C Nuclear Magnetic Resonance Study of the Organic Constituents in Different Solvent Refined Coals as a Function of the Feed Coal", ACS Advances in Chemistry Series, in press.

RECEIVED March 13, 1978

21

Comparison of Solvent-Refined Lignites with Solvent-Refined Bituminous Coals

R. J. BALTISBERGER, K. J. KLABUNDE, V. I. STENBERG, N. F. WOOLSEY, K. SAITO, and W. SUKALSKI

Department of Chemistry, University of North Dakota, Grand Forks, ND 58202

Comparison of SRL and SRC

Introduction. A considerable amount of work is currently being conducted on solvent refining of bituminous coals. The resulting solvent refined coal (SRC) can be used as a boiler fuel or as a feedstock for further refining. A parallel program on lignite coal is being carried out in our Chemical Engineering Department (Project Lignite). Considering the properties of the starting coals, it was initially assumed that SRC and solvent refined lignite (SRL) would be greatly different and, thus, second stage refining reactions and conditions would have to be developed and "fine tuned" for the different feedstocks. As these programs developed, however, it soon became apparent that SRL and SRC were more similar to each other than the starting coals were. In view of these similarities and because of a lack of definitive evidence to the contrary, it has been generally accepted that the second stage reactions and conditions can be simultaneously, rather than separately, developed.

In view of the importance of this tentative conclusion to our work, we have set about examining the similarities and differences between a wide variety of SRL and SRC samples. Our preliminary results (1) were consistent with the conclusion that SRC and SRL were nearly the same within the limits of the experiments and samples we were using. We have now obtained more representative samples of both SRL and SRC produced under more commercial conditions and report their preliminary comparison here.

Procedure. Samples refined from lignite, subbituminous and bituminous coals were obtained. (2,3) In addition, as controls, more deeply hydrogenated samples from a COSteam process and from second stage hydrogenation of SRL were included in the comparison. (3) The method of analysis was similar to that used previously (1) using the whole coal samples. Ultimate analyses, including neutron activation oxygen analysis, nonaqueous

0-8412-0427-6/78/47-071-294$05.00/0

titrations, uv, mw, nmr and esr spectroscopy were used to examine these samples. A difficulty rapidly developed. Several of the samples contained unreacted coal and ash. Initially we thought a comparison could be made by correcting for these insoluble materials on the basis of pyridine solubility of the sample. Neutron activation (naa) oxygen analysis showed, however, that oxygen by difference and by naa for maf samples were fairly close, but samples containing ash deviated considerably (Table 1). Thus, the variability of oxygen in the ash and unreacted coal led to large errors in the amount of oxygen in the maf material. Because the oxygen content could be critical to a lignite-bituminous coal comparison, we set out to develop an exact, reproducible laboratory deashing procedure. The deashing procedure was based on pyridine Soxhlet extraction, filtration of the eluate and removal of pyridine under standard conditions. The amount of pyridine remaining in each sample was checked by pmr. Several samples were crosschecked by mass spectroscopy. As little as 0.5% pyridine could have been detected. None was found. Table I shows the percentage of each sample found soluble in pyridine with and without 5μ filtration. The results show the 5 μ sizing procedure seems desirable, especially in view of the variable amounts of insoluble materials obtained from different samples by this technique. For laboratory studies we would like to propose this separatory technique as a standard procedure to define an SRC or SRL (cf. Experimental Section).

Discussion of Analytical - Spectroscopic Results

A. NMR. Proton nmr analysis of samples before and after deashing indicates no gross changes in the samples. Nevertheless, small systematic change did occur. Samples, both SRL and SRC, initially containing ash and unreacted coal had the same Har/(Hα + Ho) ratio within experimental error before and after deashing. (Table II). Maf samples showed changes of 0.07-0.26 in this ratio, which is outside the precision of the measurements. Another subtle, but consistent trend for nearly all samples was a decrease in the Hα/Ho ratio on deashing. This was true for initially maf, as well as samples with ash and unreacted coal. Furthermore, this change in ratio was 0.-0.48 for SRL's and larger, 0.36-1.27, for SRC's. Excluding the Amax sample, the range was 1.17-1.27. The Hα/Ho decreased by 0.83-1.17 for two samples when the pyridine insoluble fraction was simply filtered off (nmr's were run with all insoluble material present in non maf material).

Only experience will show if the change in Hα/Ho ratio, caused by laboratory deashing, can be used to identify the coal used for SRC or SRL preparation. Whether these small changes are caused by material in the ash or by chemical reaction during the deashing process also reamins to be investigated. Our

Table I

Comparison of Deashing Techniques

Sample	Percent soluble by: Simple Extraction	Filtration Extraction	Percent oxygen by: Difference	NAA**
M11-A*	100	96.8	5.80	2.35
Tacoma II*	99.7	95.4	3.81	3.22
Amax*	99.8	94.1	12.36	3.37
M5-C	80.1	84.2	3.88	7.31
M13-A	86.1	77.6	2.91	5.37
M21-A	75.8	68.0	7.04	5.57
Tacoma I	71.8	69.6	10.7	6.53
Wilsonville	76.4	59.3	8.49	7.71

*maf as received
**neutron activation analysis

Table II

Comparison of maf Samples Before Deashing

	M11-A	Amax	Tacoma II
%C	89.31	80.57	87.19
%H	5.80	5.50	5.45
%N	1.11	1.57	2.25
%S	0.86	< 0.02	1.3
%O(NA)	2.35	3.37	3.22
MW	450	597	495
Acid meq/g	1.58	2.22	1.63
Base meq/g	0.30	0.55	0.84
pyr. sol. (%)	96.80	94.10	95.38
Har/Hα+Ho	1.07	1.00	0.826
Hα/Ho	2.58	2.22	3.46
fa	0.813	0.797	0.897
σ	0.304	0.399	0.397
Haru/Car	0.707	0.849	0.701
Cl	1.39	1.45	1.29
Ra	5.28	3.41	5.82
Esr ΔH(pp)(G)	4.8		7.6
g value	2.0026		2.0028
Molecular Formula			
C	33.5	40.0	35.9
H	25.9	32.6	26.8
N	0.36	0.67	0.80
S	0.12	0	0.20
O	0.66	1.26	1.00
Acid eq/mole	0.71	1.33	0.81
Base eq/mole	0.14	0.33	0.42

experience with FT carbon-13 determined Car/C total (i.e. fa) ratios are consistent with a recent report (4) that the ratios determined from proton and carbon-13 nmr are very similar. (eg. for M11A Carbon-13 nmr = 0.815 ± 0.009 and from proton fa = 0.813).

B. Molecular Weight. The molecular weights (by VPO) of the samples before and after deashing were also measured. The precision on the single determinations before deashing is much less than for the three concentration extrapolated values determined on the laboratory deashed material. Even so, the two values for each sample were either the same within experimental error or very close to each other (with the exception of Tacoma II, which increased significantly). The ranges, in general, both before and after deashing are not grossly different, although the SRC's (before 420-597; after 460-620) are marginally higher in mw than the SRL's (before 400-598; after 428-481).

C. Ultraviolet Spectra. The ultraviolet spectra for these samples was run between 270 - 400 nm and was plotted vs $E^{1\%}$. These and other such samples have remarkably featureless spectra. All samples thus far examined, fall generally within the same range. These factors make it unlikely that SRL's and SRC's may be distinguished by such data. Comparison to the COSteam and J-1-11-87 samples, indicates that larger reductive changes, however, can be characterized by uv spectra. There also seems to be a direct relationship between the integrated uv absorption and Har from nmr data.

D. Acid Base Properties. Nonaqueous titration for acidity gave a range of 1.34-2.80 meq/g for SRL's and 1.45-2.95 for SRC's. The basic titer ranged for SRL's 0.30-0.63, and 0.52-0.84 meq/g for SRC's. Where comparisons can currently be made, the oxygen content is marginally lower for SRL's (despite the fact that initially it was much higher in lignite) than for SRC's. Thus, the acidity and basicity of both SRL's and SRC's seem to parallel roughly the percentage oxygen and nitrogen respectively in the sample.

The percentages of carbon and hydrogen are typical (Amax % C is low and is being checked) of other samples with carbon ranging 85-90% (maf) and hydrogen, 5-6% for SRL and SRC samples. The sulfur content ranges somewhat higher for SRC's, 1.30-3.68% (but only a trace for Amax) than for SRL's, 0.85-1.24; probably a reflection of the sulfur content of starting coal.

E. Electron Spin Resonance. The esr spectra of the samples was measured, with the results given in Table II. The range of g-values was very small for the samples ivnestigated, 2,0026-2.0028. The g values are in good agreement with those reported for coals having carbon contents over 80%. The linewidths for SRL's ranged 3.8-4.8 gauss; for SRC's 1.7 to 7.6. The linewidth appears marginally lower than that expected for a vitrain of comparable hydrogen content. (5)

Summary of Analytical Comparisons. While the comparisons are not yet complete, the gross makeup of the samples indicates that SRL's and SRC;s are quite similar. The variations noted in the uv, molecular weight, esr, and nmr analyses may be more a function of reaction conditions than of starting coal. The percentage composition and acid-base characteristics seem to indicate that the starting coal properties, particularly nitrogen, sulfur and oxygen percentages, may be carried over into the solvent refined products, although they are affected also, to a large extent, by reaction conditions.

Furthermore, in cases which we have investigated thus far, hydrotreating of SRL and SRC have shown similar trends, in that their reactivities and product distributions depend more on how they are made, stored and treated than on the starting coal.

Assessment of Analytical Procedure. Several difficulties have emerged in these analyses for whole solvent refined samples. A standardized laboratory deashing procedure which we have developed, appears necessary. Oxygen analysis on maf samples by difference may be generally acceptable, but should be checked with neutron activation analysis and is absolutely essential with samples containing unreacted coal and ash. A standardized procedure which we are developing needs to be uniformly applied to pulsed carbon-13 nmr analyses.

Future Work in Analysis of Gross Solvent Refined Coal Samples. Application of carbon-13 nmr techniques has been applied to coal derived samples, but generally only to that portion of the sample soluble in a "desirable" nmr solvent (CS_2, $CDCl_3$, etc.). Whole solvent refined samples containing large amounts of preasphaltenes (like SRL and SRC) are not soluble in these solvents. In order to properly characterize whole samples, either new solvents or new techniques have to be developed. We are working on both.

A second area of critical concern to us is the qualitative way comparisons currently must be made. A critical set of standardized measurements needs to be developed (which we have alluded to above). These measurements should then be reduced to a set of critical structural factors, probably through a computerized technique which will allow direct quantitative comparison of samples. While inroads are being made on this approach, (6) better methods need to be developed.

Experimental Section

Analyses were performed by Midwest Micro and Spang Laboratories. Nmr spectra were measured on an EM-390 and analyzed, as previously described. (1) Titrations were conducted, as previously indicated. (1)

Neutron activated analysis was carried out by Intelcom Rad Tech, San Diego, CA. Uv spectra were measured on a Cary 14 in dimethylacetamide. Esr were determined on a Bruker ER 420. Molecular weights were measured by Spang, and with a Corona Wescan 232 VPO in dimethylformamide, the latter at 74.8° at three different concentrations with extrapolation to infinite dilution. Very little association was noted in this solvent.

Laboratory Deashing Procedure. A Whatman no. 1 Soxhlet thimble was shrunk in acetone, dried at 110°C, cooled in a dessicator and weighed to constant weight. A 3 to 5 g sample of SRL or SRC was weighed into the thimble and extracted with pyridine for 24 hrs under nitrogen. The thimble was dried at 110° for 24 hrs, cooled in a dessicator and weighed. The pyridine solution was filtered through a preweighed 5 μ Teflon filter. The sum of the sample in the thimble and on the filter constituted the undissolved sample by definition. Most of the pyridine in the filtrate was removed at 50° (1mm), then at 50° (0.05mm) for 24 hrs. The sample was scraped into a drying boat and further dried at 56° (0.1mm) for 24 hrs. After grinding in a mortar, the sample was redried at 100° (0.1mm) for 24 hrs more. Less than 1% of pyridine could be observed by nmr in hexamethylphosphoramide. Mass spectrometry indicated less than 0.5% pyridine in several samples.

Literature Cited

1. Woolsey, N., Baltisberger, R., Klabunde, K., Stenberg, V. and Kaba, R., ACS Fuel Preprints, 2 (7), 33 (1976).
2. Samples were generously supplied by Project Lignite, University of North Dakota and the Grand Forks Energy Research Center.
3. Samples are designed as follows:
   M5-C-(undeashed) Conditions: 2500 psi of 1:1 $CO:H_2$ at 455°C (max).*
   M11-A-(deashed) Conditions: 2500 psi of 1:1 $CO:H_2$ at 479°C (max).*
   M13-A-(partly deashed) Conditions: 2500 psi of 1:1 $CO:H_2$ at 453°C (max).*
   M21-A-(undeashed) Conditions: 2000 psi of 1:3 $CO:H_2$ at 439°C (max).*
   Tacoma I(undeashed): Pittsburg and Midway Coal Co., Merriam, Kansas (a Gulf subsidiary). From a Kentucky no. 9 and no. 14 blend of bituminous coal from the

Colonial Mine. Conditions: PDU, continuous flow, 1500 psi of H_2 (85% min), 450°C (max), using recycle solvent; product not filtered and stored in the open.
Tacoma II(deashed): Same as Tacoma I except product filtered before solvent distilled. (maf)
Amax-Southern Services, Inc., Birmingham, Alabama (an ERPI contractor). From a sub-bituminous coal from the Bel Air Mine in Wyoming. Conditions: continuous flow, 2500 psi of H_2 and recycle gases, 460° (max) using recycle heavy in phenols with 1, 2 and 3 ring aromatics.
Wilsonville-Prepared essentially by the procedure for Amax.
J-1-11-87: Reduction fraction of KC-SRL. Conditions: batch autoclave, 4500 psi (max) of H_2 at 450° for 2 hrs. using presulfieded HT-100 (NiMo) catalyst b.p. 110-180°C (0.4mm) followed by removal of solid which precipitated on standing in the freezer for ca. month.
COSteam: Pittsburg Energy Research Laboratory (an ERDA Lab). N. D. Lignite treated in a tubular reactor with synthesis gas under COSteam conditions. Conditions: continuous flow, 4000 psi of $CO:H_2$. Whole product chromatographed on alumina. This fraction eluted with toluene (aromatics fraction). 126-2 (Separation done in Grand Forks Energy Research Laboratory)

4. Retcofsky, H. L., Schweighardt, F. K. and Hough, M., Analytical Chem., 49 (4), 585 (1977).
5. Retcofsky, H. L., Thompson, G.P., Raymond, R. and Friedel, R. A., Fuel 54, 126 (1975) and papers cited therein.
6. Oka, M., Chung, H.-C. and Gavalas, G. R., ibid, 56, 3 (1977).

*Solvent refined N.D. Lignite by Project Lignite of the University of North Dakota. PDU operation, FS-120 as make up solvent during recycle operation using ca. 1.8:1 solvent to coal.

RECEIVED February 10, 1978

22

Temperature Effects on Coal Liquefaction: Rates of Depolymerization and Product Quality as Determined by Gel Permeation Chromatography

CURTIS L. KNUDSON, JOSEPH E. SCHILLER, and ARTHUR L. RUUD

DOE, Grand Forks Energy Research Center, Grand Forks, ND 58202

In this paper the reactor temperature and residence time effects on the depolymerization of coal will be discussed. The ratio of the absorbances of the 950 molecular weight (MW) to 280 MW peaks observed in the gel permeation high pressure liquid chromatography (HPLC) separation of products has indicated use as a measure of depolymerization and product quality.

Research on the non-catalytic CO-Steam liquefaction of coal was initiated at the Grand Forks Energy Research Center (GFERC) in 1974 as a continuation of research started at the Pittsburgh Energy Research Center (PERC). The objective is to develop the process as a commercial, economic method to produce a stable coal liquid which will meet environmental requirements (ash, nitrogen, and sulfur contents) for use as a boiler fuel. The research program has followed three overlapping stages: 1) Batch autoclave research, 2) 3-lb/hr continuous process unit operation, and 3) process development unit scaleup based on findings from (1) and (2). Selected data from item (1) will be considered in this paper.

The batch autoclave consists of a batch reactor system (Fig. 1) that can be charged with room temperature slurry to a hot (up to 500° C) reactor which can be sampled (both liquid and gas phases) at various times to determine reactant and product changes. Figure 2 depicts pressure and temperature changes, as well as gas composition changes with time during a typical run where a hot reactor was charged with slurry and gas. Figure 3 depicts the major fractions that are obtained during analysis. This system is also being utilized to determine total gas content of slurry at up to 4500 psig and 480° C. The amount of the material charged that is found in the gas phase in the reactor at various temperatures and pressures has been determined, and the effects of slow (1 hour) or rapid (3 minute) heatup of slurries to reaction temperature on molecular weight distributions and coking is being studied.

0-8412-0427-6/78/47-071-301$05.00/0

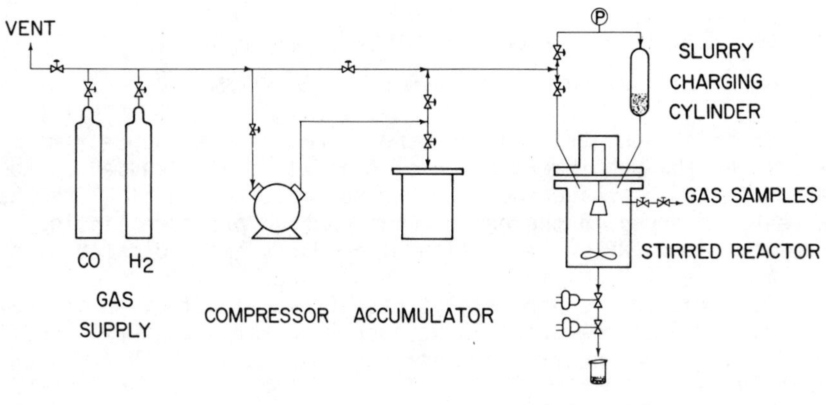

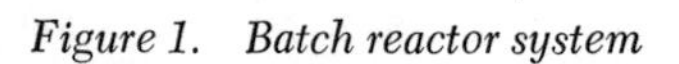

Figure 1. Batch reactor system

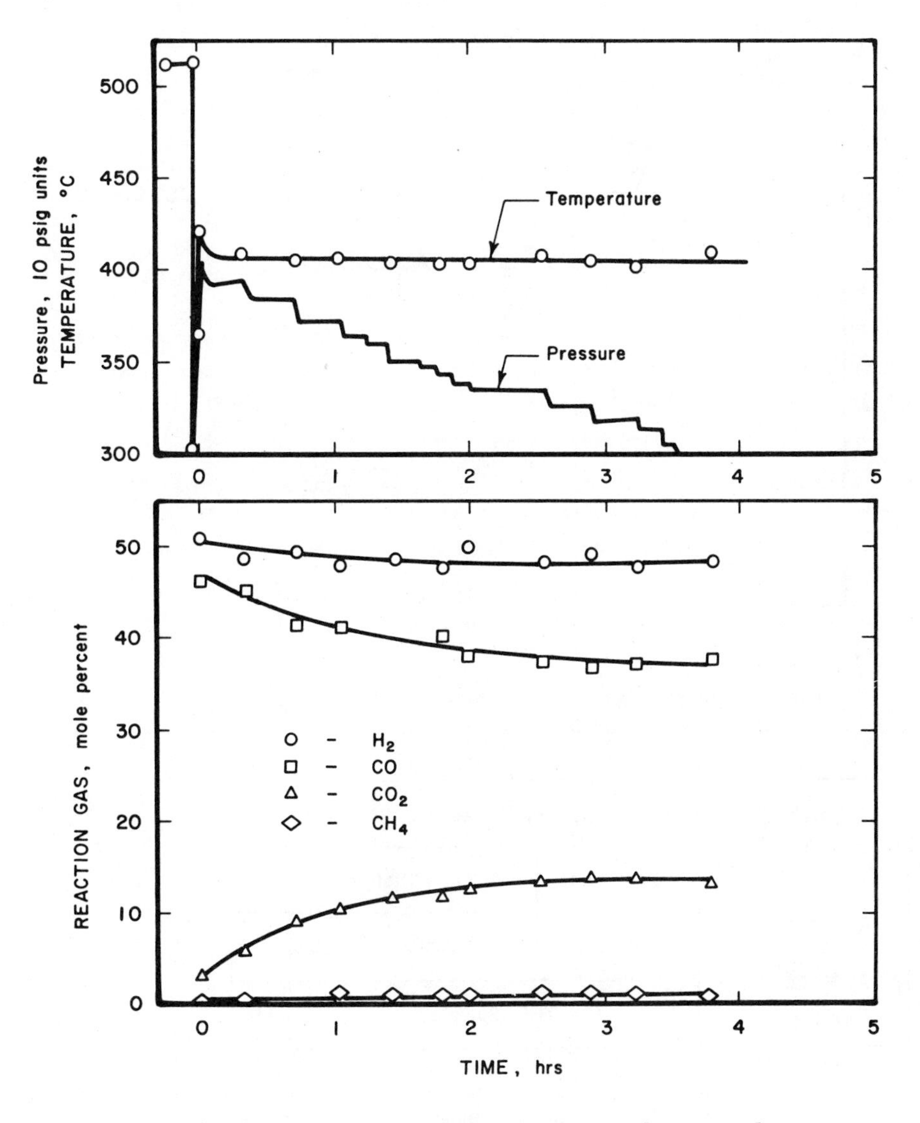

Figure 2. Pressure, temperature, and gas composition changes with time for a rapid heat-up experiment

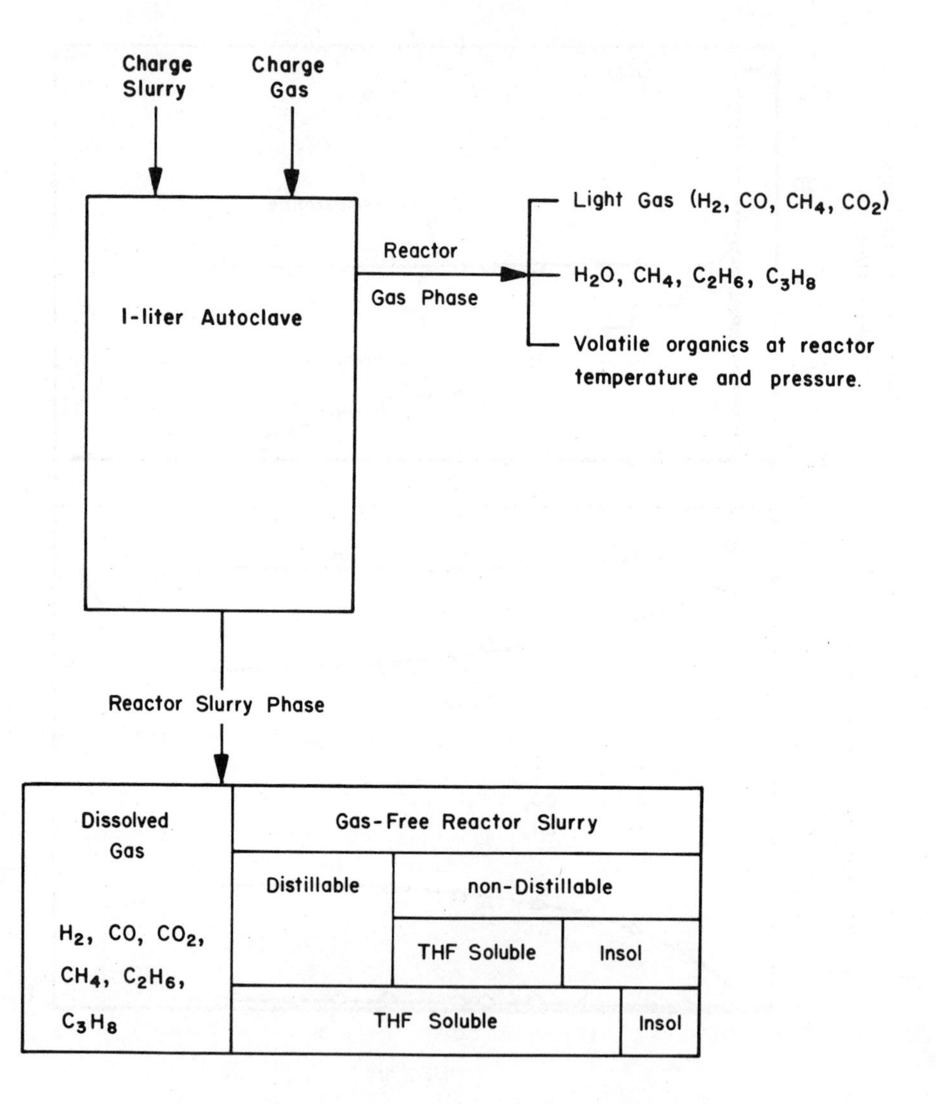

Figure 3. Major product fractions obtained during analysis

Undistilled coal products (solvent refined coals and coal liquids) analyzed at GFERC have contained 10 to 80 pct volatiles at 1 torr and 250° C. This volatile material is readily analyzed by GC, GC-MS, or low voltage MS. Approximately 50 to 80 pct of the non-volatile fraction was readily soluble in tetrahydrofuran (THF). Analysis of a 1 w/w pct solution of this non-volatile fraction (after passing through a 0.5 μ filter) has yielded molecular weight distributions which are not masked by solvent or light organic material. Figure 4 depicts the molecular weight distribution of an undistilled sample. The solvent peak at total retention volume (Vt) is essentially removed by distillation, resulting in enhancement of the region between 280 MW and the exclusion volume (Vo). Observing this enhanced region has enabled the determination of the rate of disappearance of high molecular weight materials as a function of reactor temperature and residence time.

Gas-Slurry Charge Composition

The one-liter reactor was typically charged with 250 gms of slurry (from a cylinder to the reactor using gas pressure). The amount of gas charged was approximately 3.5 moles. Molecular weight distributions have been found to be essentially independent of whether the charge gas was CO or synthesis gas (50:50 hydrogen to carbon monoxide). For the data reported, 250 gms of slurry was charged. The slurry composition was as follows:

27.3% as-received coal
8.0% of additional water added
58.2% anthracene oil
6.5% tetralin

Proximate and ultimate analysis for the lignite used in these studies is presented in Table I.

Molecular Weight Distribution Determinations

The equipment used to determine molecular weight distributions was a Water's Model ALC-GPC-201 system fitted with one 500 and three 100 Å microstyragel columns. CO-Steam product samples were first distilled at 250° C and 1 torr to remove volatile material. A 1 w/w pct tetrahydrofuran (THF) solution of the non-distillable material was passed through a 0.5 μ filter prior to injection. The percent THF insoluble residue was compared to THF solubility data obtained for a sample that was not subjected to distillation. A quotient of stability to distillation can be calculated which reflects the fraction of the original THF soluble material that was still soluble after distillation at 250° C and 1 torr. This stability quotient has been found to increase with reactor residence time and temperature.

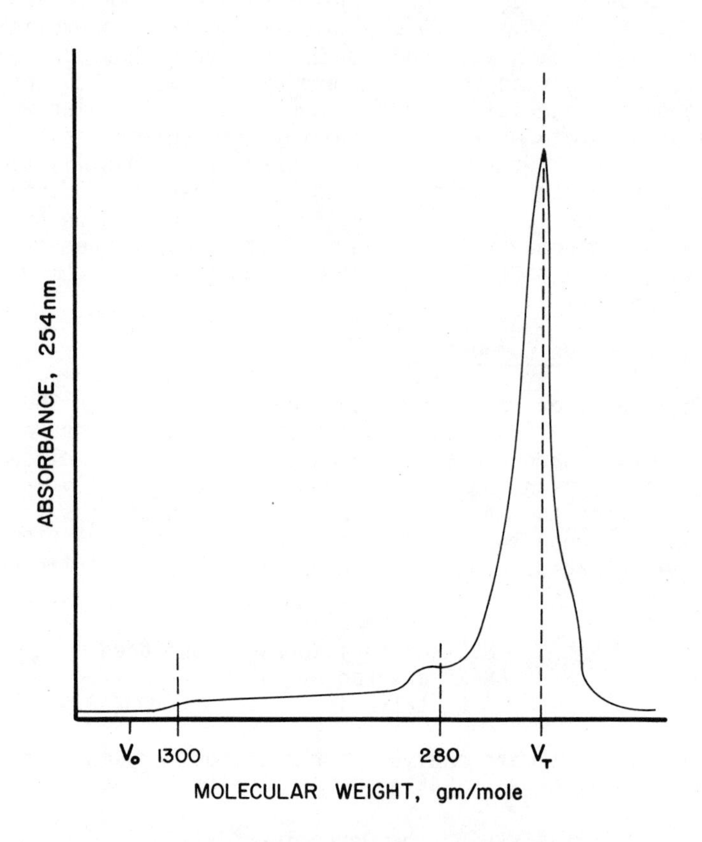

Figure 4. Molecular weight distribution of an undistilled sample showing solvent masking (peak at V_t) of the higher MW. material

Table I. Beulah Standard II; GF-77-712

	Coal (As-received)	Coal (Moisture-free)	Coal (Moisture- and ash-free)
Proximate analysis, pct:			
Moisture................	27.80	--	--
Volatile matter.........	30.45	42.18	46.93
Fixed carbon............	34.45	47.71	53.07
Ash.....................	7.30	10.11	--
TOTAL	100.00	100.00	100.00
Ultimate analysis, pct:			
Hydrogen................	6.09	4.16	4.63
Carbon..................	46.30	64.14	71.35
Nitrogen................	0.72	1.00	1.11
Oxygen..................	39.03	19.82	22.05
Sulfur..................	0.56	0.77	0.86
Ash.....................	7.30	10.11	--
TOTAL	100.00	100.00	100.00

In run 30, the reactor was charged cold and heated to 400 ± 2° C in 92 minutes and held at this temperature for 1/2 hour. The reactor was subsequently heated to and held in steps at 435 ± 4° C and 475 ± 5° C.

In Table II, data from run 30 are presented. The stability quotient increased from 22 to 56 pct in 19 minutes at 400° C. In going to 437° C the quotient goes to 71 pct and for the next 104 minutes remained essentially constant. The stability quotient also reflects the quantity of sample that is analyzed by HPLC after distillation.

Table II. Product Stability to Distillation Versus Residence Time and Temperature in Run 30

Sample No.	Temp., °C	Time	Product Fractions (1) THF Insoluble, pct[a]	Product Fractions (2) Non-distillable, pct[a]	Stability quotient, pct[b]
	400° ± 2°	92			
1		94	11	22	22
3		113	6	18	56
	435° ± 4°	141			
4		142	7	22	71
6		182	9	20	78
	475° ± 5°	200			
7		212	8	23	74
9		246	11	29	73

[a] The values for (1) and (2) are percents of total reactor slurry product obtained from the reactor at temperature and pressure.

[b] The stability quotient is the percent of (2) minus (1) that was soluble after distillation at 250° C and 1 mm Hg.

The molecular weight distributions were measured relative to commercially available polyglycol standards (purchased from Waters) to enable comparison by other researchers. Detection was by UV at 254 nm.

Temperature Effects

The temperature at which the reactor was operated has had dramatic effects on the molecular weight distribution obtained for the non-distillable, THF soluble fraction of coal liquefaction products.

Figure 5 depicts the molecular weight distributions obtained at 400, 435 to 450, and 460 to 480° C for residence times under 1/2 hour. These distributions have been observed in experiments where the reactor was heated to temperature in 3 minutes or in 1 hour. Figure 5 illustrates the rapid decrease in the absorbance at molecular weights greater than 750 and the shift of the average molecular weight of this fraction from over 600 at 400° C to about 500 at 440-450° C and finally to about 300 at 460 to 480° C. Since the absorbance is proportional to concentration, ratios of the absorbance maximum between 1200 and 950 to the value at 280 MW (A_{950}/A_{280}) have been used to correlate data from various experiments. This ratio has also been found to be independent of experimental errors in preparing 1 w/w pct solutions. In Figure 6, the ratios obtained for samples removed during the initial 1/2 hour of reaction are depicted versus reactor temperature for a number of experiments.

High ratios indicate large concentrations of over 750 MW molecular weight material. The rapid decrease of the ratio with temperature indicates that reactors operated at higher temperatures produce product distillation residues with lower average molecular weights.

Rates of Depolymerization

The change in molecular weight distribution of distillation residue with reactor residence time can be used as a qualitative measure of the rates of depolymerization at different temperatures. At 400° C the molecular weight distribution changes little with time (Fig. 5). However, the stability to distillation does increase, (Table II) which indicates that the quality of the product increases with residence time. Product changes at 400° C have not been studied to any degree, since the molecular weight distribution observed is similar to that of solvent refined coals, which are not liquids at room temperature.

At temperatures of 435 to 450° C, notable residence time effects have been observed, especially in the greater than 750 MW material. Figure 7 presents two ratios (A_{950}/A_{280} and A_{450}/A_{280}) versus residence time for an extended experiment in which the reactants were initially heated to 440° C in three minutes and held there for 1/2 hour, followed by stepwise heating during 5-10 minutes and holding for 1/2 hour at 450, 460, and 470° C. In Figure 6 a number of asphaltenes and preasphaltenes are tabulated versus their 950/280 MW absorbance ratio. Asphaltenes yield ratios lower than those obtained for preashphaltenes obtained from the same sample. However, the viscosity of distilled samples (25 wt pct in anthracene oil) has not correlated with either the asphaltene or preasphaltene ratios, indicating that both contribute to the final viscosity. After 30 minutes at 440° C, the A_{950}/A_{280} ratio tends to level off. The changes observed were on similar fractions of total product. The mole-

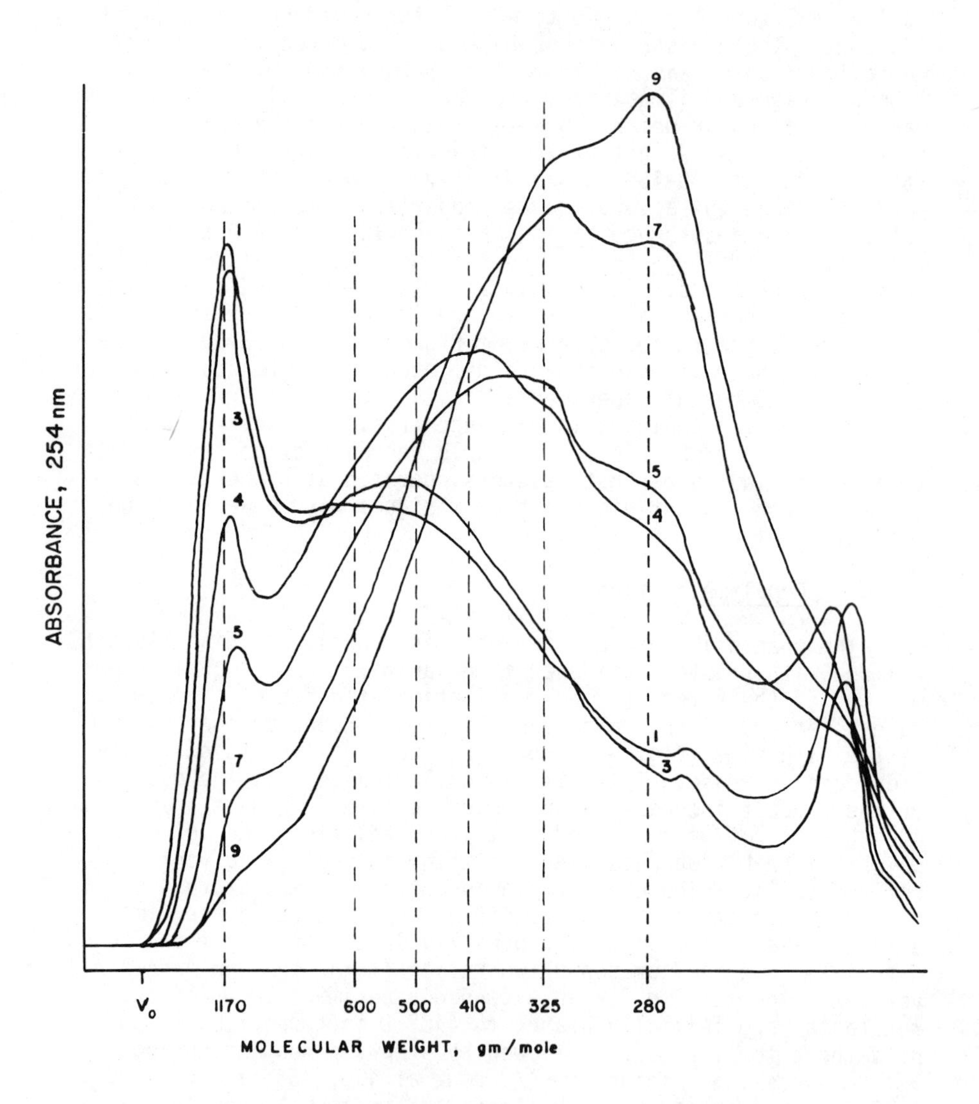

Figure 5. Molecular weight distribution changes with temperature. (1) 2 min at 400°C, (3) 21 min at 400°C, (4) 1 min at 435°C, (5) 13 min at 435°C, (7) 12 min at 475°C, (9) 46 min at 475°C.

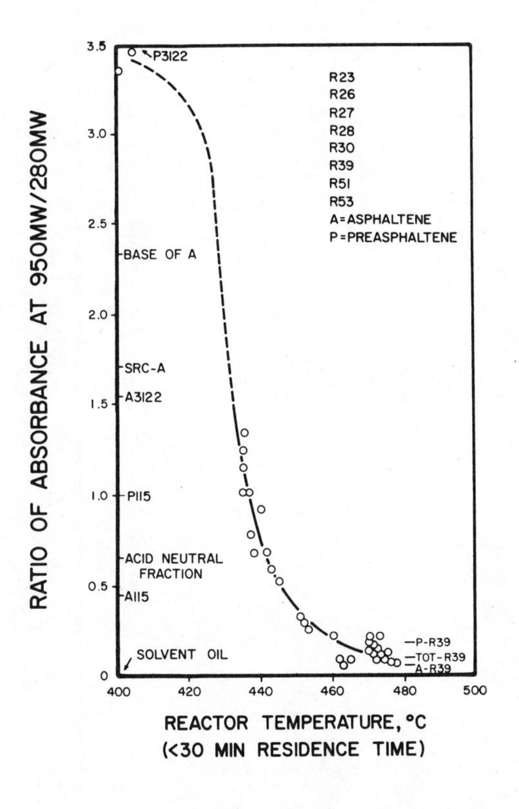

Figure 6. Change in absorbance ratio with reactor temperature as well as comparison with various asphaltene and preasphaltene fractions

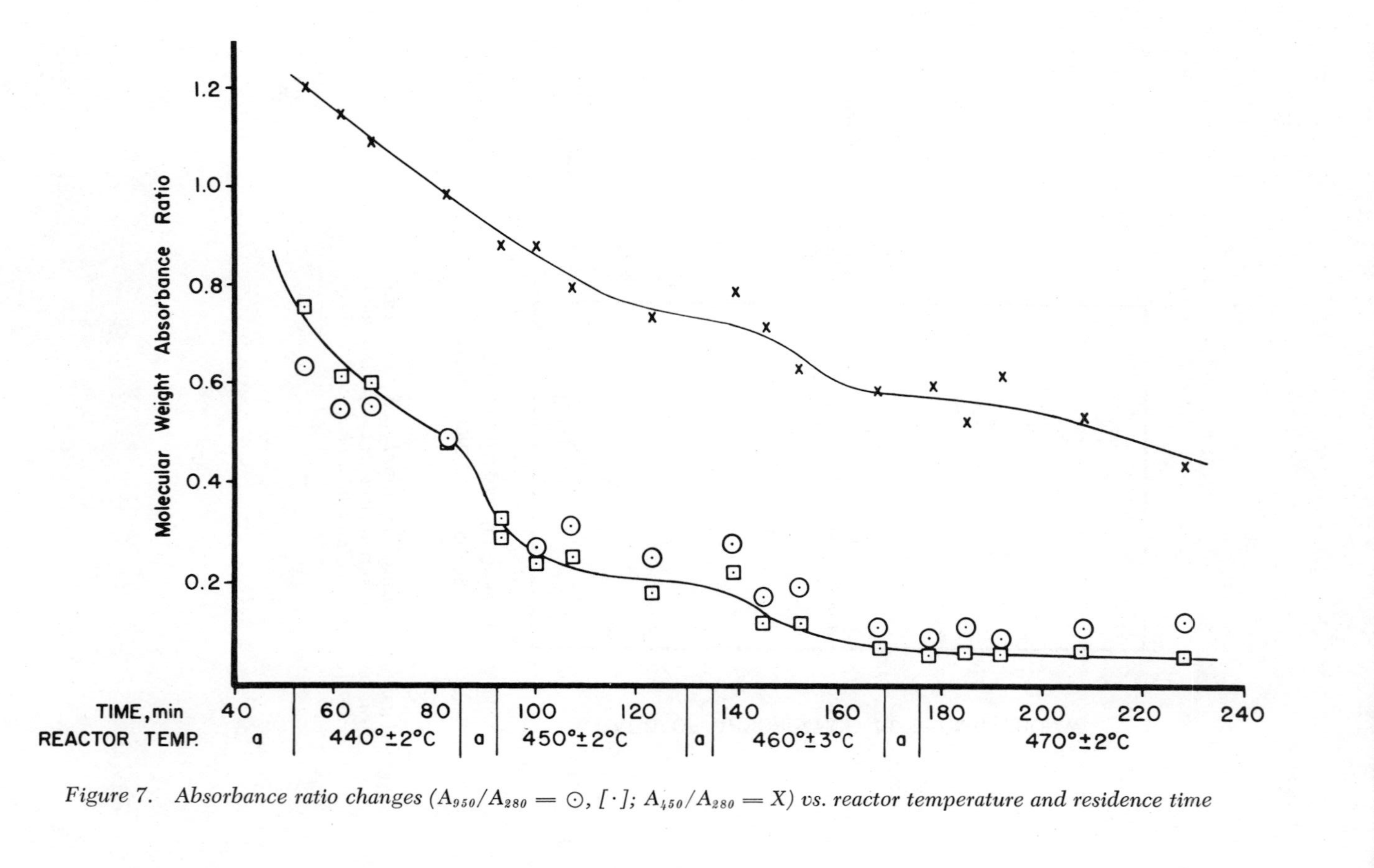

Figure 7. Absorbance ratio changes (A_{950}/A_{280} = ⊙, [·]; A_{450}/A_{280} = X) *vs. reactor temperature and residence time*

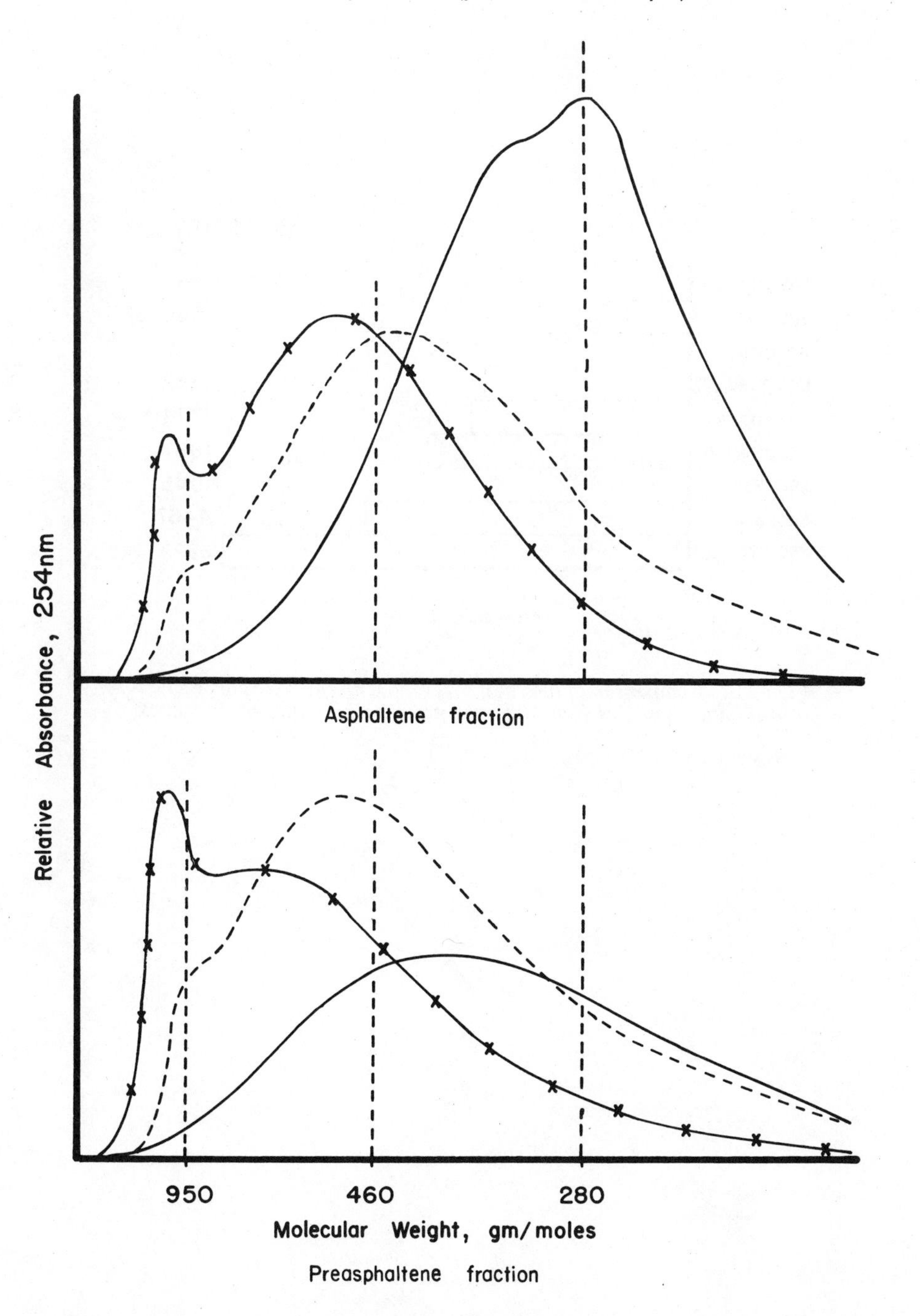

Figure 8. Molecular weight distributions of asphaltene and preasphaltene fractions obtained from PERC CO–Steam (–×–×–), SRC II (– – –), and GFERC (——) products

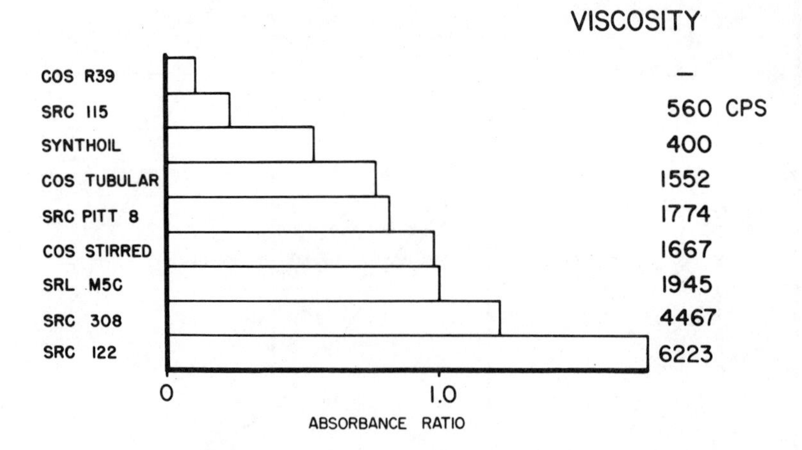

Figure 9. Comparison of the A_{950}/A_{280} ratio with the viscosity of 25 wt % of distillation residues from different products in anthracene oil at 25°C

cular weight distribution and the ratios observed at 440° C were similar to those observed for CO-Steam product prepared at PERC in a stirred pot reactor at 450° C, 4000 psi.

A rapid decrease in the A_{950}/A_{280} ratio occurred during 7 minutes of heating from 440° to 450° C. Additional time at 450° C yielded little change in the ratio. The A_{450}/A_{280} ratio also leveled off. Molecular weight distributions for this material indicated that no recoverable material greater than 1000 MW was present; however, the distribution did have a shoulder at 950 MW. These results are similar to those obtained for coal liquids produced by the Synthoil Process.

Subsequent heating to 460° C caused an additional decrease in both the A_{950}/A_{280} and A_{460}/A_{280} ratios, with little change continuing after 15 minutes. Little further change has been observed in these molecular weight distributions in other experiments at 470° C for 1 hour or at 480° C for 1/2 hour. Similar distributions have been obtained for samples made by the SRC II Process.

The most dramatic decrease in molecular weight occurred during from 440 to 450° C. However a further beneficial decrease occurs between 450 to 470° C. Depolymerization occurs very rapidly at 460° C, and the 300 MW material produced does not depolymerize further with time beyond 15 minutes.

Product Quality

Other research being conducted at GFERC has indicated direct dependence of viscosity on the average molecular weight and the concentration of pre-asphaltene material (1). Asphaltene content in higher concentrations has also effected increased viscosities of coal products (2,3,4). Figure 8 depicts the molecular weight distributions of preasphaltene and asphaltene fractions obtained from PERC CO-Steam, SRC II, and the GFERC product produced at 470° C. A complete description of production conditions as well as analyses of the volatile fractions can be found elsewhere (5). For each product, the asphaltene fraction has a lower average molecular weight than the preasphaltene fraction. Absorbance at 950 MW is most indicative of the presence of the higher MW preasphaltene and asphaltene material, and at 280 MW of lighter non-distillable material.

The A_{950}/A_{280} ratio for distillation residues obtained from a number of products made by different processes has correlated with viscosity (Fig. 9). The comparison was made on 25 wt pct of residue in anthracene oil. Comparing only the distillation residue removes dependence on the amount of non-distillable material in a particular product.

The A_{950}/A_{280} ratio is currently being used to determine product quality changes during Continuous Process Unit operation with different processing conditions. Lower values mean the final product will have a lower viscosity.

ACKNOWLEDGEMENTS

The authors wish to thank Richard Payfer for operation of the batch autoclave system and Randy Molmen for initial sample analysis.

This paper is based in part on HPLC techniques developed by Arthur Ruud as partial fulfillment of the requirements for a Master of Science degree at the University of North Dakota.

LITERATURE CITED

1. J.E. Schiller, B.W. Farnum, and E.A. Sondreal. To be published.
2. H.W. Sternberg, R. Raymond, and S. Akhtar, Hydrocracking and Hydrotreating (Ed. J.W. Ward and S.A. Qader), Am. Chem. Soc. Symposium Series No. 20, Washington, DC, pp. 111-122, 1975.
3. C. Mack, J. Phys. Chem., 36, 2901, (1932).
4. G.W. Echert and B. Weetman, Ind. and Eng. Chem., 39 (11), 1512 (1947).
5. J.E. Schiller, Hydrocarbon Processing, 1, 147, (1977).

RECEIVED April 6, 1978

INDEX

INDEX

D

E

F